D1432104

Josef Bigun

Vision with Direction

Josef Bigun

Vision with Direction

A Systematic Introduction
to Image Processing and Computer Vision

With 146 Figures, including 130 in Color

 Springer

Josef Bigun
IDE-Sektionen
Box 823
SE-30118, Halmstad
Sweden
josef.bigun@ide.hh.se
www.hh.se/staff/josef

Library of Congress Control Number: 2005934891

ACM Computing Classification (1998): I.4, I.5, I.3, I.2.10

ISBN-10 3-540-27322-0 Springer Berlin Heidelberg New York
ISBN-13 978-3-540-27322-6 Springer Berlin Heidelberg New York

Springer is a part of Springer Science+Business Media

springer.com

© Springer-Verlag Berlin Heidelberg 2006
Printed in Germany

Typeset by the author using a Springer TeX macro package
Production: LE-TeX Jelonek, Schmidt & Vöckler GbR, Leipzig
Cover design: KünkelLopka Werbeagentur, Heidelberg

Printed on acid-free paper   45/3142/YL - 5 4 3 2 1 0

To my parents, H. and S. Bigun

Preface

Image analysis is a computational feat which humans show excellence in, in comparison with computers. Yet the list of applications that rely on automatic processing of images has been growing at a fast pace. Biometric authentication by face, fingerprint, and iris, online character recognition in cell phones as well as drug design tools are but a few of its benefactors appearing on the headlines.

This is, of course, facilitated by the valuable output of the resarch community in the past 30 years. The pattern recognition and computer vision communities that study image analysis have large conferences, which regularly draw 1000 participants. In a way this is not surprising, because much of the human-specific activities critically rely on intelligent use of vision. If routine parts of these activities can be automated, much is to be gained in comfort and sustainable development. The research field could equally be called *visual intelligence* because it concerns nearly all activities of awake humans. Humans use or rely on pictures or pictorial languages to represent, analyze, and develop abstract metaphors related to nearly every aspect of thinking and behaving, be it science, mathematics, philosophy, religion, music, or emotions.

The present volume is an introductory textbook on signal analysis of visual computation for senior-level undergraduates or for graduate students in science and engineering. My modest goal has been to present the frequently used techniques to analyze images in a common framework–directional image processing. In that, I am certainly influenced by the massive evidence of intricate directional signal processing being accumulated on human vision. My hope is that the contents of the present text will be useful to a broad category of knowledge workers, not only those who are technically oriented. To understand and reveal the secrets of, in my view, the most advanced signal analysis "system" of the known universe, primate vision, is a great challenge. It will predictably require cross-field fertilizations of many sorts in science, not the least among computer vision, neurobiology, and psychology.

The book has five parts, which can be studied fairly independently. These studies are most comfortable if the reader has the equivalent mathematical knowledge acquired during the first years of engineering studies. Otherwise, the lemmas and theorems can be read to acquire a quick overview, even with a weaker theoretical background. Part I presents briefly a current account of the human vision system with short notes to its parallels in computer vision. Part II treats the theory of linear systems, including the various versions of Fourier transform, with illustrations from image signals. Part III treats single direction in images, including the tensor theory for direction representation and estimation. Generalized beyond Cartesian coordinates, an abstraction of the direction concept to other coordinates is offered. Here, the reader meets an important tool of computer vision, the Hough transform and its generalized version, in a novel presentation. Part IV presents the concept of group direction, which models increased shape complexities. Finally, Part V presents the grouping tools that can be used in conjunction with directional processing. These include clustering, feature dimension reduction, boundary estimation, and elementary morphological operations. Information on downloadable laboratory exercises (in Matlab) based on this book is available at the homepage of the author (**http://www.hh.se/staff/josef**).

I am indebted to several people for their wisdom and the help that they gave me while I was writing this book, and before. I came in contact with image analysis by reading the publications of Prof. *Gösta H. Granlund* as his PhD student and during the beautiful discussions in his research group at Linköping University, not the least with Prof. *Hans Knutsson*, in the mid-1980s. This heritage is unmistakenly recognizable in my text. In the 1990s, during my employment at the Swiss Federal Institute of Technology in Lausanne, I greatly enjoyed working with Prof. *Hans du Buf* on textures. The traces of this collaboration are distinctly visible in the volume, too.

I have abundantly learned from my former and present PhD students, some of their work and devotion is not only alive in my memory and daily work, but also in the graphics and contents of this volume. I wish to mention, alphabetically, *Yaregal Assabie, Serge Ayer, Benoit Duc, Maycel Faraj, Stefan Fischer, Hartwig Fronthaler, Ole Hansen, Klaus Kollreider, Kenneth Nilsson, Martin Persson, Lalith Premaratne, Philippe Schroeter,* and *Fabrizio Smeraldi.* As teachers in two image analysis courses using drafts of this volume, Kenneth, Martin, and Fabrizio provided, additionally, important feedback from students.

I was privileged to have other coworkers and students who have helped me out along the "voyage" that writing a book is. I wish to name those whose contributions have been most apparent, alphabetically, *Markus Bĕckman, Kwok-wai Choy, Stefan Karlsson, Nadeem Khan, Iivari Kunttu, Robert Lamprecht, Leena Lepistö, Madis Listak, Henrik Olsson, Werner Pomwenger, Bernd Resch, Peter Romirer-Maierhofer, Radakrishnan Poomari, Rene Schirninger, Derk Wesemann, Heike Walter,* and *Niklas Zeiner.*

At the final port of this voyage, I wish to mention not the least my family, who not only put up with me writing a book, often invading the private sphere, but who also filled the breach and encouraged me with appreciated "kicks" that have taken me out of local minima.

I *thank you* all for having enjoyed the writing of this book and I hope that the reader will enjoy it too.

August 2005   *J. Bigun*

Contents

Part III Vision of Single Direction

Part IV   Vision of Multiple Directions

Part V   Grouping, Segmentation, and Region Description

Abbreviations and Symbols

$\daleth(f)$	$(D_x f + i D_y f)^2$	infinitesimal linear symmetry tensor (ILST[1])
$\delta(x)$		Dirac delta distribution, if x is continuous
$\delta(m)$		Kronecker delta function, if m is an integer
C_N		N-dimensional complex vector space
BCC		brightness constancy constraint
E_N		real vectors of dimension N; Euclidean space
∇f	$(D_x f, D_y f, \cdots)^T$	gradient operator[1]
-	$(D_x + i D_y)^n f$	symmetry derivative operator of order n
CT		coordinate transformation
DFD		displaced frame difference
DFT		discrete Fourier transform
FC		Fourier coefficients
FD		Fourier descriptors
FE		finite extension functions
FF		finite frequency functions; band–limited functions
FIR		finite impulse response
FT	$\mathcal{F}$	Fourier transform
GHT		generalized Hough transform
GST	$\mathbf{S}$, or $\mathbf{Z}$	generalized structure tensor
HFP	$\{\xi, \eta\}$	harmonic function pair
ILST		see $\daleth(f)$
KLT		KLT Karhunen–Loève transform, see PCA
LGN		lateral geniculate nucleus
MS		mean squares
OCR		optical character recognition
ON		orthonormal
PCA		principal component analysis, see KLT
SC		superior colliculus
ST	$\mathbf{S}$, or $\mathbf{Z}$	structure tensor
SNR		signal-to-noise ratio
SVD		singular value decomposition
TLS		total least squares
V1		primary visual cortex, or striate cortex
WGSS		within-group sum of squared error

---

[1] The symbols $\daleth$ and ∇ are pronounced as "doleth" and "nabla", respectively.

Human and Computer Vision

Enlighten the eyes of my mind
that I may understand my place
in Thine eternal design!

St. Ephrem (A.D. 303–373)

Neuronal Pathways of Vision

Humans and numerous animal species rely on their visual systems to plan or to take actions in the world. Light photons reflected from objects form images that are sensed and translated to multidimensional signals. These travel along the visual pathways forward and backward, in parallel and serially, thanks to a fascinating chain of chemical and electrical processes in the brain, in particular to, from, and within the visual cortex. The visual signals do not just pass from one neuron or compartment to the next, but they also undergo an incredible amount of signal processing to finally support among others, planning, and decision–action mechanisms. So important is the visual sensory system, in humans, approximately 50% of the cerebral cortex takes part in this intricate metamorphosis of the visual signals. Here we will present the pathways of these signals along with a summary of the functional properties of the cells encountered on these. Although they are supported by the research of reknowned scientists that include Nobel laurates, e.g., Santiago Ramon y Cajal (1906), and David Hubel and Thorsten Wiesel (1983), much of the current neurobiological conclusions on human vision, including what follows, are extrapolations based on lesions in human brains due to damages or surgical therapy, psychological experiments, and experimental studies on animals, chiefly macaque monkeys, and cats.

1.1 Optics and Visual Fields of the Eye

The eye is the outpost of the visual anatomy where the light is sensed and the 3D *spatio–temporal signal*, which is called image, is formed. The "spatial" part of the name refers to the 2D part of the signal that, at a "frozen" time instant, falls as a picture on light-sensitive retinal cells, *photoreceptors*. This picture is a spatial signal because its coordinates are in length units, e.g., millimeters, representing the distance between the sensing cells. As time passes, however, the amount of light that falls on a point in the picture may change for a variety of reasons, e.g., the eye moves, the object in sight moves, or simply the light changes. Consequently the sensed amount of photons at every point of the picture results in a 3D signal.

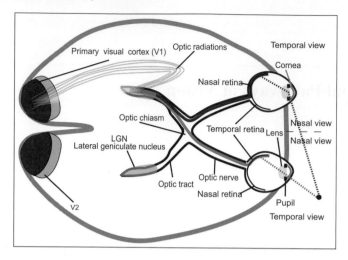

Fig. 1.1. The anatomic pathways of the visual signals

The spatial 2D image formed on the retina represents the light pattern reflected from a thin[1] plane in the 3D spatial world which the eye observes. This is so, thanks to the deformable *lens* sitting behind the *cornea*, a transparent layer of cells that first receives the light. The thickness of the cornea does not change and can be likened to a lens with fixed focal length in a human-made optical system, such as a camera. Because the lens in the eye can be contracted or decontracted by the muscles to which it is attached, its focal length is variable. Its function can be likened to the zooming of a telephoto objective. Just as the latter can change the distance of the plane to be imaged, so can the eye *focus on objects* at varying distances. Functionally, even the cornea is thus a lens, in the vocabulary of technically minded. Approximately 75% of the refraction that the cornea and the eye together do is achieved by the cornea (Fig. 1.1). The *pupil*, which can change the amount of light passing into the eye, can be likened to a diaphram in a camera objective.

The light traverses the liquid filling the eye before it reaches the retinal surface attached to the inner wall of the eyeball. The light rays are absorbed, but the sensitivity to light amount, that is the *light intensity*,[2] of the retinal cells is adapted in various ways to the intensity of the light they usually receive so as to remain operational despite an overall decrease or increase of the light intensity, e.g., on a cloudy or a sunny day. A ubiquous tool in this adaptation is the pupil, which can contract or decontract, regulating the amount of light reaching the retina. There is also the

---

[1] The thickness of the imaged 3D plane can be appreciated as thin in comparison with its distance to the eye.

[2] The light consists of photons, each having its own wavelength. The number of photons determines the light intensity. Normally, light contains different amounts of photons from each wavelength for *chromatic light*. If however there is only a narrow range of wavelengths among its photons, the light is called *monochromatic*, e.g., laser light.

night vision mechanism in which the light intensity demanding retinal cells (to be discussed soon) are shut off in favor of others that can function at lower amounts of light. Although two-dimensional, the retinal surface is not a flat plane; rather, it is a spherical surface. This is a difference in comparison to a human-made camera box, where the sensing surface is usually a flat plane. One can argue that the biological image formed on the retina will in the average be better focused since the surfaces of the natural objects the eye observes are mostly bent, like the trunks of trees, although this may not be the main advantage. Presumably, the great advantage is that an eye can be compactly rotated in a spherical socket, leaving only a small surface outside of the socket. Protecting rotation-enabled rectangular cameras compactly is not an easy mechanical feat.

1.2 Photoreceptors of the Retina

In psychophysical studies, it is customary that the closeness of a retinal point to the center O' is measured in degrees from the optical axis; this is called the *eccentricity* (Fig. 1.2). Eccentricity is also known as the *elevation*. The eccentricity angle is represented by ϵ in the shown graphs and every degree of eccentricity corresponds to ≈ 0.35 mm in human eyes. The locus of the retinal points having the same eccentricity is a circle. Then there is the *azimuth*, which is the polar angle of a retinal point, i.e., the angle relative the positive part of the horizon. This is shown as α in the figure on the right, where the azimuth radii and the eccentricity circles are given in dotted black and pink, respectively. Because the diameter $O'O$ is a constant, the two angles ϵ, α can then function as retinal coordinates. Separated by the vertical *meridian*, which corresponds to $\alpha = \pm\frac{\pi}{2}$, the left eye retina can roughly be divided into two halves, the *nasal* retina, which is the one farthest away from the nose, and the *temporal* retina, which is the one closest to the nose. The names are given after their respective views. The nasal retina "sees" the *nasal hemifield*, which is the view closest to the nose, and the temporal retina sees the *temporal hemifield*, which is the view on the side farthest away from the nose. The analogous names exist for the right eye.

In computer vision, the closest kinn of a photoreceptor is a *pixel*, a picture element, because the geometry of the retina is not continuous as it is in a photographic film, but discrete. Furthermore, the grid of photoreceptors sampling the retinal surface is not equidistant. Close to the *optic axis* of the eye, which is at 0° eccentricity, the retinal surface is sampled at the highest density. In *macula lutea*, the retinal region inside the eccentricity of approximately 5° on the retina, the highest concentration of photoreceptors are found. The view corresponding to this area is also called *central vision* or *macular vision*. The area corresponding to 1° eccentricity is the *fovea*.

The photoreceptors come in two "flavors", the color-sensitive *cones* and light intensity-sensitive *rods*. The cones are shut off in night vision because the intensity at which they can operate exceeds those levels that are available at night. By contrast, the rods can operate in the poorer light conditions of the night, albeit with little or no sensitivity for color differences. In the fovea there are cones but no rods. This is one

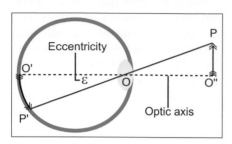

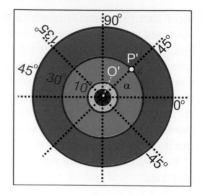

Fig. 1.2. Given the diameter $O'O$, the eccentricty ϵ (*left*), and the azimuth α, one can determine the position of a point P' on the retina (*right*)

of the reasons why the *spatial resolution*, also called *acuity*, which determines the picture quality for details that can be represented, is not very high in night vision. The peak resolution is reserved for day vision, during which there is more light available to those photoreceptors that can sense such data. The density of cones decreases with high eccentricity, whereas that of rods increases rapidly. Accordingly, in many night-active species, the decrease in rod concentration towards the fovea is not as dramatic as day-active animals, e.g. in owl monkey [171]. In fovea there are approximately 150,000 cones per mm^2 [176]. The concentration decreases sharply with increased eccentricity. To switch to night vision requires time, which is called *adaptation*, and takes a few minutes in humans. In human retinae there are three types of cones, sensitive to long, medium, and short wavelengths of the received photons. These are also known as "red", "green", and "blue" cones. We will come back to the discussion of color sensitivity of cones in Chap. 2.

The retina consists of six layers, of which the photoreceptor layer containing cones and rods is the first, counted from the eye wall towards the lens. This is another remarkable difference between natural and human-made imaging systems. In a camera, the light-sensitive surface is turned towards the lens to be exposed to the light directly, whereas the light-sensitive rods and cones of the retina are turned away from the lens, towards the wall of the eye. The light rays pass first the other five layers of the retina before they excite the photoreceptors! This is presumably because the photoreceptors bleach under the light stimuli, but they can quickly regain their light-sensitive operational state by intaking organic and chemical substances. By being turned towards the eye walls, their supply of such materials is facilitated while their direct exposure to the light is reduced (Fig. 1.3). The light stimulus is translated to electrical pulses by a photoreceptor, rod, or cone, thanks to an impressive chain of electrochemical process that involves hyperpolarization [109]. The signal intensity of the photoreceptors increases with increased light intensity, provided that the light is within the operational range of the photoreceptor in terms of its photon amount (intensity) as well as photon wavelength range (color).

1.3 Ganglion Cells of the Retina and Receptive Fields

The ganglion cells constitute the last layer of neurons in the retina. In between the ganglion cells and photoreceptor layer, there are four other layers of neuronal circuitry that implement electro-chemical signal processing. The processing includes photon amplification and local neighborhood operation implementations. The net result is that ganglion cells outputs do not represent the intensity of light falling upon photoreceptors, but they represent a signal that can be comparable to a bandpass-filtered version of the image captured by all photoreceptors. To be precise, the output signal of a ganglion cell responds vigorously during the entire duration of the stimulus only if the light distribution on and around its closest photoreceptor corresponds to a certain light intensity pattern.

There are several types of ganglion cells, each having its own activation pattern. Ganglion cells are *center–surround* cells, so called because they respond only if there is a difference between the light intensity falling on the corresponding central and the surround photoreceptors [143]. An example pattern called $(+/-)$ is shown in Fig. 1.3, where the central light intensity must exceed that in the annulus around it. The opposite ganglion cell type is $(-/+)$, for the surround intensity must be larger than the central intensity. The opposing patterns exist presumably because the neuronal operations cannot implement differences that become negative.

There are ganglion cells that take inputs from different cone types in a specific fashion that make them color sensitive. They include $(r+g-)$-type, reacting when the intensities coming from the central L-cones are larger than the intensities provided by the M-cones in the surround, and its opposite type $(r-g+)$, reacting when the intensities coming from the central L-cones are smaller than the intensities provided by the M-cones in the surround. There are approximately 125 million rods and cones, which should be contrasted to about 1 million ganglion cells, in each eye. After a bandpass filtering the sampling rate of a signal can be decreased (Sect. 6.2), which in turn offers a signal theoretic justification for the decrease of the sampling rate at the ganglion cell layer. This local comparison scheme plays a significant role in color constancy perception, which allows humans to attach the same color label of a certain surface seen under different light sources, e.g., daylight or indoor light. Likewise, this helps humans to be contrast-sensitive rather than gray-sensitive at first place, e.g., we are able to recognize the same object in different black and white photographs despite the fact that the object surface does not have the same grayness.

The output of a ganglion cell represents the result of computations on many photoreceptor cells, which can be activated by a part of the visual field. To be precise, only a pattern within a specific region in the visual field is projected to a circular region on the retina, which in turn steers the output of a ganglion cell. This retinal region is called the *receptive field* of a ganglion cell. The same terminology is used for other neurons in the brain as well, if the output of a neuron is steered by a local region of the retina. The closest concept in computer vision is the *local image* or the *neighborhood* on which certain computations are applied in parallel. Consequently, the information on absolute values of light intensity, available at the rod and cone level, never leaves the eye, i.e., gray or color intensity information is not available

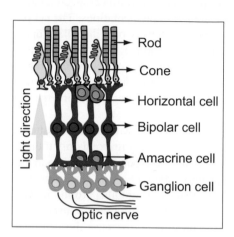

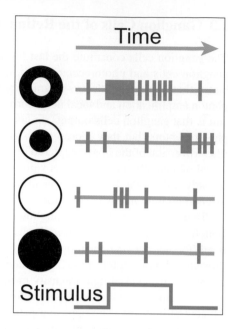

Fig. 1.3. The graph on left illustrates the retinal cells involved in imaging and visual signal processing. On the right the response pattern of a $(+/-)$-type ganglion cell is shown

to the brain. All further processing in the brain takes place on "differential signals", representing local comparisons within and between the photoreceptor responses, not on the intensity signals themselves.

The outputs of the ganglion cells converge to eventually form the *optic nerve* that goes away from the eye. Because the ganglion layer is deep inside the eye and farthest away from the eye wall, the outputs come out of the eye through a "hole" in the retina that is well outside of the fovea. There are no photoreceptors there. The visual field region that projects on this hole is commonly known as the *blind spot*. The hole itself is called the *optic disc* and is about 2 mm in diameter. Humans actually do not see anyting at the blind spot, which is in the temporal hemifield, at approximately 20° elevation close to the horizontal meridian.

Exercise 1.1. *Close your left eye, and with your right eye look at a spot far away, preferably at a bright spot on a dark background. Hold your finger between the spot and the eye with your arm stretched. Move your finger out slowly in a half circle without changing your gaze fixation on the spot. Do you experience that your finger disappears and reappears? If so, explain why, and note at approximately what elevation angle this happens. If not, retry when you are relaxed, because chances are high that you will experience this phenomenon.*

The ganglion cells are the only output cells of the eye reaching the rest of the brain. There is a sizable number of retinal ganglion cell types [164], presumably to equip the brain with a rich set of signal processing tools, for, among others, color, texture, motion, depth, and shape analysis, when the rest of the brain has no access to the original signal. The exact qualities that establish each type and the role of these are still debated. The most commonly discussed types are the small *midget cells*, and the large *parasol cells*. There is a less-studied third type, frequently referred to when discussing the lateral geniculate nucleus connections, the *koniocelullar cells*.

The midget cells are presumed to process high spatial frequency and color. They have, accordingly, small receptive fields and total about 80% of all retinal ganglion cells. The large majority of midget cells are color-opponent, being excited by red in the center and inhibited by green in the surround, or vice versa. Parasol cells, on the other hand, are mainly responsible for motion analysis. Being color indifferent, they total about 10% of ganglion cells, and have larger receptive fields than the midget cells. There are few parasol cells in the fovea. The ratio of parasol to midget cells increases with eccentricity. Parasol cells are insensitive to colour, i.e., they are luminance-opponent. This is a general tendency; the receptive fields of ganglion cells increase with eccentricity. This means that bandpass filtering is achieved at the level of retina. Accordingly, the number of ganglion cells decreases with eccentricity. Since ganglion cells are the only providers of signals to the brain, the cerebral visual areas also follow such a spatial organization.

The koniocelullar cells are much fewer and more poorly understood than midget and parasol cells. They are not as heterogenous as these either, although a few common properties have been identified. Their receptive fields lack surround and they are color sensitive! In the center, they are excited by blue, whereas they are inhibited (in the center) by red or green [104]. Presumably, they are involved in object/background segregation.

1.4 The Optic Chiasm

The optic nerve is logically organized in two bundles of nerves, carrying visual signals responsible for the nasal and temporal views, respectively. The two optic nerves coming from both eyes meet at the *optic chiasm*, where one bundle of each sort travels farther towards the left and the right brain halves. The temporal retina bundle crosses the midline, whereas the nasal retina bundle remains on the same side for both eyes. The bundle pair leaving the chiasm is called the *optic tract*. Because of the midline crossing arrangement of only the temporal retina outputs, the optical tract that leaves the chiasm to travel to the left brain contains only visual signal carriers that encode the patterns appearing on the right hemifield. Similarly, the one reaching the right brain carries visual signals of the left hemifield. The optic tract travels chiefly to reach the lateral geniculate nucleus, LGN to be discussed below. However, some 10% of the connections in the bundle feed an area called *superior colliculus,*[3] (SC). From the SC there are outputs feeding the primary visual cortex at the back of the brain, which we will discuss further below. By contrast, SC will not be discussed

---

[3] This area is involved in visual signal processing controlling the eye movements.

further here; see [41, 223]. We do this to limit the scope but also because this path to the visual cortex is much less studied than the one passing through the LGN.

1.5 Lateral Geniculate Nucleus (LGN)

The lateral geniculate[4] nucleus (LGN) is a laminated structure in the thalamus. Its inputs are received from the ganglion cells coming from each eye (Fig. 1.4). The input to the layers of LGN is organized in an orderly fashion, but the different eyes remain segregated. That is there are no LGN cells that react to both eyes, and each layer contains cells that respond to stimuli from a single eye. The left eye (L) and the right (R) eye inputs interlace when passing from one layer to the next, as the figure illustrates. Being R,L,L,R,L,R for the left LGN, the left–right alternation reverses between layers 2 and 3 for reasons that are not well understood. Layer 1 starts with the inputs coming from the eye on the other side of the LGN, the so called *contralateral*[5] eye, so that for the right eye the sequence is L,R,R,L,R,L. Each LGN receives signals representing a visual field corresponding to the side opposite their own, that is a *contralateral view*. Accordingly, the left and right LGNs cope only with, the right and left visual fields, respectively.

Like nearly all of the neural visual signal processing structures, LGN also has a *topographic organization*. This implies a continuity (in the mathematical sense) of the mapping between the retina and the LGN, i.e., the responses of ganglion cells that are close to each other feed into LGN cells that are located close to each other.[6]

The small ganglion cells (midget cells) project to the cells found in the *parvocellular layers* of LGN. In Fig. 1.4 the parvocellular cells occupy the layers 3–6. The larger cells (parasol cells) project onto the *magnocellular layers* of the LGN, layers 1–2 of the figure. The koniocellular outputs project onto the layers K1–K6. The koniocellular cells, which are a type of cells found among the retinal ganglion cells, have also been found scattered in the entire LGN. Besides the bottom–up feeding from ganglion cells, the LGN receives significant direct and indirect feedback from the V1 area, to be discussed in Sect. 1.6. The feedback signals can radically influence the visual signal processing in LGN as well as in the rest of the brain. Yet the functional details of these connections are not well understood. Experiments on LGN cells have shown that they are functionally similar to those of the retinal ganglion cells that feed into them. Accordingly, the LGN is frequently qualified as a relay station between the retina and visual cortex, and its cells are also called *relay cells*. The outputs from LGN cells form a wide band called *optic radiations* and travel to the primary visual cortex (Fig. 1.1).

---

[4] Geniculate means kneelike, describing its appearance.

[5] The terms *contralateral* and *ipsilateral* are frequently used in neurobiology. They mean, respectively, the "other" and the "same" in relation to the current side.

[6] Retrospectively, even the ganglion cells are topographically organized in the retina because these are placed "behind" the photoreceptors from which they receive their inputs.

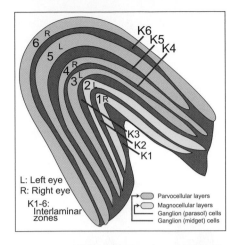
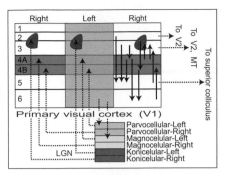

Fig. 1.4. The *left* graph illustrates the left LGN of the macaque monkey with its six layers. The *right* graph shows the left V1 and some of its connections, following Hassler labelling of the layers [47, 109].

1.6 The Primary Visual Cortex

Outputs from each of the three LGN neuron types feed via optic radiations into different layers of the *primary visual cortex*, also known as *V1*, or *striate cortex*. The V1 area has six layers totalling ≈ 2 mm on a few cm^2. It contains the impressive ≈ 200 million cells. To compare its enormous packing density, we recall that the ganglion cells total ≈ 1 million in an eye. The V1 area is by far the most complex area of the brain, as regards layering of the cells and the richness of cell types.

A schematic illustration of its input–output connections is shown in Fig. 1.4 using Hassler notation [47]. Most of the outputs from magnocellular and parvocellular layers of the LGN arrive at layer 4, but to different sublayers, 4A and 4B, respectively. The cells in layer 4A and 4B have primarily receptive field properties that are similar to magnocellular and parvocellular neurons, which feed into the former. The receptive field properties of other cells will be discussed in Sect. 1.7. The koniocellular cell outputs feed narrow volumes of cells spanning layers 1–3, called *blobs* [155]. The blobs contain cells having the so-called *double-opponent color property*. These are embedded in a center–surround receptive field that is presumably responsible for color perception, which operates fairly autonomously in relation to V1. We will present this property in further detail in Sect. 2.3. Within V1, cells in layer 4 provide inputs to layers 2 and 3, whereas cells in layers 2 and 3 project to layers 5 and 6. Layers 2 and 3 also provide inputs to adjacent cortical areas. Cells in layer 5 provide inputs to adjacent cortical areas as well as nonadjacent areas, e.g., the superior colliculus. Cells in layer 6 provide feedback to the LGN.

As to be expected from the compelling evidence coming from photoreceptor, ganglion, and LGN cell topographic organizations, the visual system devotes the largest amount of cortical cells to fovea even cortically. This is brilliant in the face

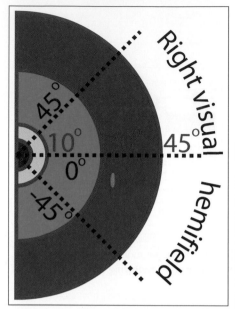
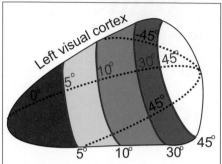

Fig. 1.5. On the *left*, a model of the retinal topography is depicted. On the *right*, using the same color code, a model of the topography of V1, on which the retinal cells are mapped, is shown. Adapted after [217]

of the limited resources that the system has at its disposal, because there is a limited amount of energy available to drive a limited number of cells that have to fit a small physical space. Because the visual field, and hence the central vision, can be changed mechanically and effectively, the resource-demanding analysis of images is mainly performed in the fovea. For example, when reading these lines, the regions of interest are shuffled in and out of the fovea through eye motions and, when necessary, by a seamless combination of eye–head–body motions.

Half the ganglion cells in both eyes, are mapped to the V1 region. Geometrically, the ganglion cells are on a quarter sphere, whereas V1 is more like the surface of a pear [217], as illustrated by Fig. 1.5. This is essentially equivalent to a mathematical deformation, modeled as a coordinate mapping. An approximation of this mapping is discussed in Chap. 9. The net effect of this mapping is that more of the total available resources (the cells) are devoted to the region of the central retina than the size of the latter should command. The over-representation of the central retina is known as *cortical magnification*. Furthermore, isoeccentricity half circles and isoazimuth half-lines of the retina are mapped to half-lines that are approximately orthogonal.

Cortical magnification has also inspired computer vision studies to use log–polar spatial-grids [196] to track and/or to recognize objects by robots with artificial vision systems [20, 187, 205, 216]. The log–polar mapping is justified because it effectively models the mapping between the retina and V1, where circles and radial half-lines

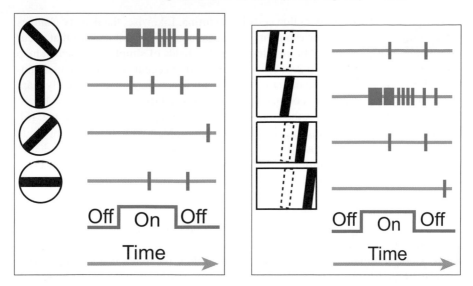

Fig. 1.6. On the *left*, the direction sensitivity of a cell in V1 is illustrated. On the *right*, the sensitivity of simple cells to position, which comes on top of their spatial direction sensitivity, is shown

are mapped to orthogonal lines in addition to the fact that the central retina is mapped to a relatively large area in V1.

1.7 Spatial Direction, Velocity, and Frequency Preference

Neurons in V1 have radically different receptive field properties compared to the center–surround response pattern of the LGN and the ganglion cells of the retina. Apart from the input layer 4 and the blobs, the V1 neurons respond vigorously only to edges or bars at a particular *spatial direction*, [114], as illustrated by Fig. 1.6. Each cell has its own spatial direction that it prefers, and there are cells for (approximately) each spatial direction. The receptive field patterns that excite the V1 cells consist in lines and edges as has been illustrated in Fig. 1.8. Area V1 contains two types of direction-sensitive cells, *simple cells* and *complex cells*. These cells are insensitive to the color of light falling in their receptive fields.

Simple cells respond to bars or edges having a specific direction at a specific position in their receptive fields, Fig. 1.6. If the receptive field contains a bar or an edge that has a different direction than the preferred direction, or the bar is not properly positioned, the firing rate of a simple cell decreases down to the biological zero firing rate, spontaneous and sporadic firing. Also, the response is maintained for the entire duration of the stimulus. The density of simple cells decreases with increased eccentricity of the retinal positions they are mapped to. Their receptive fields increase in size with increased eccentricity. This behavior is in good agreement with that of the receptive field sizes of ganglion cells in the retina. Likewise, the density changes of the simple cells reflect corresponding changes in ganglion cell density that occur with increased eccentricity. The smallest receptive fields of simple cells, which map to fovea, are approximately $0.25° \times 0.25°$, measured in eccentricity and azimuth angles. This is the same as those of ganglion cells, on which they topographically map. The farthest retinal periphery commands the largest receptive field sizes of $\approx 1° \times 1°$ for simple cells. Furthermore, the simple cell responses appear to be linear, e.g. [6]. That is, if the stimulus is sinusoidal so is the output (albeit with different amplitude and phase, but with the same spatial frequency). This is a further evidence that at least a sampled local spectrum for all visual fields is routinely available for the brain when it analyzes images. In Sect. 9.6, we will study the signal processing that is afforded by local spectra in further detail.

Complex cells, which total about 75% of the cells in V1, respond to a critically oriented bar, moving *anywhere* within their receptive fields (Fig. 1.7). They share with simple cells the property of being sensitive to the spatial directions of lines, but unlike them, stationary bars placed anywhere in their receptive fields will generate vigorous responses. In simple cells, excitation is conditioned to the bar or edge with the critical direction be precisely placed in the center of the receptive field of the cell. Complex cells have a tendency to have larger receptive fields than the comparable simple cells, $0.5° \times 0.5°$ in the fovea. The bar widths that excite the complex cells, however, are as thin as those of simple cells, $\approx 0.03°$. Some complex cells (as well as some simple cells) have a sensitivity to the *motion-direction* of the bar, in addition to the spatial direction of it. Also, the complex cell responses are nonlinear [6].

In neurobiology the term *orientation* is frequently used to mean what we here called the spatial direction, whereas the term direction in these studies usually represents the motion-direction of a moving bar in a plane. Our use of the same term for both is justified because, as will be detailed in Chap. 12, these concepts are technically the same. Spatial direction is a direction in 2D space, whereas velocity (direction + absolute speed information) is a direction in the 3D spatio–temporal signal space (see Fig. 12.2). Accordingly, the part of the human vision system that determines the spatial direction and the one that estimates the velocity mathematically solve the same problem but in different dimensions, i.e., in 2D, and 3D, respectively.

The cells that are motion-direction sensitive in V1 are of lowpass type, i.e., they respond as long as the amplitude of the motion (the speed) is low [174]. This is in contrast to some motion-direction sensitive cells found in area V2, which are of bandpass-type w.r.t. the speed of the bar, i.e., they respond as long as the bar speed is within a narrow range. There is considerable specialization in the way the the cortical cells are sensitive to motion parameters. Those serving the fovea appear to be of lowpass character, hence they are maximally active during the eye fixation, in all visual areas of the cortex, although those in V2 have a clear superiority for coding both the absolute speed and the motion-direction. Those cells serving peripherial vision appear to have large receptive fields and are of high-pass type, i.e., they are active when the moving bar is faster than a certain speed. Area V1 motion-direction cells are presumably engaged in still image analysis (or smooth pursuit of objects in motion), whereas those beyond V1, especially V2, are engaged in analysis and

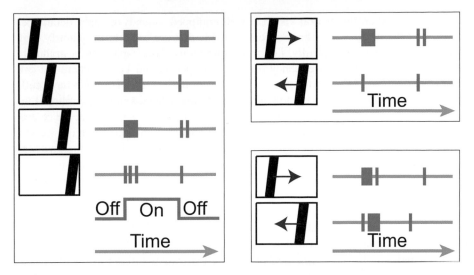

Fig. 1.7. The graph on the *left* illustrates that the complex cells of V1 are insensitive to bar position. On the *right, top* and *right, bottom* the responses of motion-direction sensitive and a motion-direction insensitive complex cell responses are shown

tracking of moving objects. Except for those which are of high-pass type, the optimal velocity of velocity-tuned cells increases with visual eccentricity and appears to range from $2°$ to $90°$ per second. To limit the scope of this book and also because they are less studied, we will not discuss cells beyond area V1 further, and refer to further readings, e.g., [173].

Complex cells are encountered later in the computational processing chain of visual signals than are simple cells. Accordingly, to construct their outputs, the complex cells presumably receive the outputs of many simple cells as inputs. As in the case of simple cells, the exact architecture of input–output wiring of complex cells has not been established experimentally, but there exist suggested schemes that are being debated.

There is repeatedly convincing evidence, e.g., [4, 6, 45, 46, 159, 165], suggesting the existence of well-organized cells in V1 that exhibit a spatial frequency selectivity to moving and/or still sinusoidal gratings, e.g., the top left of Fig. 10.2. The cells serving fovea in V1, have optima in the range of 0.25–4 cycles/degree and have bandwiths of approximately 1.5 octaves [165]. Although these limits vary somewhat between the different studies that have reported on frequency selectivity, even their very existence is important. It supports the view that the brain analyzes the visual stimuli by exploding the original data via frequency, spatial direction, and spatio temporal direction (velocity) channels in parallel before it actually reduces and simplifies them, e.g., to yield a recognition of an object or to generate motor responses such as those of catching a fast ball.

Taken together, the central vision is well equipped to analyze sharp details because its cells in the cortex have receptive fields that are capable to quantify high spatial frequencies isotropically, i.e., in all directions. This capability is gradually replaced with spatial low-frequency sensitivity at peripherial vision where the cell receptive fields are larger. In a parallel fashion, in the central vision we have cells that are more suited to analyze slow moving patterns, whereas in the peripherial vision the fast moving patterns can be analyzed most efficiently. Combined, the central vision has most of its resources to analyze high spatial frequencies moving slowly, whereas the peripheral vision devotes its resources to analyze low spatial frequencies moving fast. This is because any static image pattern is equivalent to sinusoidal gratings, from a mathematical viewpoint, since it can be synthesized by means of these.[7]

The spatial directional selectivity mechanism is a result of interaction of cells in the visual pathway, presumably as a combination of the LGN outputs which, from the signal processing point of view, are equivalent to time-delayed outputs of the retinal ganglion cells. The exact mechanism of this wiring is still not well understood, although the scheme suggested by Hubel and Wiesel, see [113], is a simple scheme that can explain the simple cell recordings. It consists in an additive combination of the LGN outputs that have overlapping receptive fields. In Fig. 1.8, this is illustrated for a bar-type simple cell, which is synthesized by pooling outputs of LGN cells having receptive fields along a line.

A detailed organization of the cells is not yet available, but it is fairly conclusive that depthwise, i.e., a penetration perpendicular to the visual cortex, the cells are organized to prefer the same spatial direction, the same range of spatial frequencies, and the same receptive field. Such a group of cells is called an *orientation column* in the neuroscience of vision. As one moves along the surface of the cortex, there is *locally* a very regular change of the spatial direction preference in one direction and ocular dominance (left or right eye) in the other (orthogonal to the first). However, this orthogonality does not hold for long cortical distances. Accordingly, to account for the spatial direction and ocular dominance changes as one moves along the surface, a rectangular organization of the orientation columns in alternating stripes of ocular dominance is not observed along the surface of the cortex. Instead, a structure of stripes, reminiscent of the ridges and valleys of fingerprints, is observed. Across the stripes, ocular dominance and along the stripes, spatial direction preference changes occur [222].

The direction, whether it represents the spatial direction or the motion, is an important feature for the visual system because it can define the boundaries of objects as well as encode texture properties and corners.[8] Also, not only patterns of static images but also motion patterns are important visual attributes of a scene because object background segregation is tremendously simplified by motion information by motion information compared to attempting to resolve this in static images. Likewise,

---

[7] We will introduce this idea in further detail in Chap. 5.
[8] The concept of direction can be used to represent texture and corners in addition to edges. In Chaps. 10 and 14 a detailed account of this is given

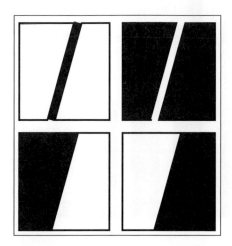 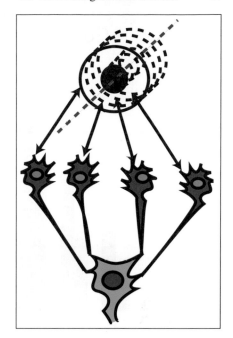

Fig. 1.8. The patterns that excite the cells of V1 are shown on the *left*. On the *right*, a plausible additive wiring of LGN cell responses to obtain a directional sensitivity of a simple cell is shown

the spatial frequency information is important because, on one hand, it encodes the sizes of objects, while on the other hand, it encodes the granularity or the scale of repetitive patterns (textures).

1.8 Face Recognition in Humans

Patients suffering from *prosopagnosia* have a particular difficulty in recognizing face identities. They can recognize the identities of familiar persons, if they have access to other modalities, e.g., voice, walking pattern, length, or hairstyle. Without nonfacial cues, the sufferers may not even recognize family members, or even their own faces may be foreign to them. They often have good ability to recognize other objects that are nonfaces. In many cases, they become prosopagnosic after a stroke or a surgical intervention.

There is significant evidence, both from studies of prosopagnosia and from studies of brain damage, that face analysis engages special signal processing in visual cortex that is different from processing of other objects [13, 68, 79]. There is a general agreement that approximately at the age of 12, the performance of children in face recognition reaches adult levels, that there is already an impressive face recognition ability by the age of 5, and that measurable preferences for face stimuli exist

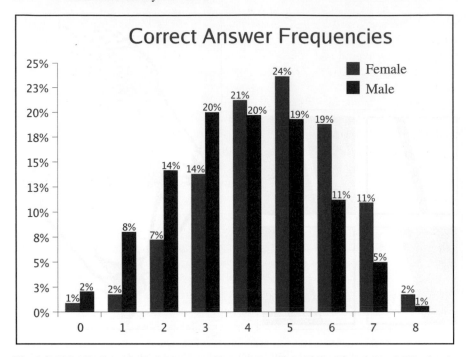

Fig. 1.9. Distribution of correct answers. Example reading: 11% of females and 5% of males had 7 correct answers out of 8 they provided

in babies even younger than 10 minutes [66]. For example, human infants a few minutes of age show a preference to track a human face farther than other moving nonface objects [130]. While there is a reasonable rotation-invariance to recognize objects though it takes longer times, turning a face upside down results usually in a dramatic reduction of face identification [40]. These and other findings indicate that face recognition develops earlier than other object recognition skills, and that it is much more direction sensitive than recognition of other objects.

Perhaps recognition of face identities is so complex that encoding the diversity of faces demands much more from our general-purpose, local direction, and frequency-based feature extraction system. If so, that would explain our extreme directional sensitivity in face recognition. One could even speculate further that the problem is not even possible to solve in real time with our general object recognition system, that it has an additional area that is either specialized on faces or helps to speed-up and/or to robustify face recognition performance. This is more than an experiment of thought because there is mounting evidence that faces [13, 99, 181, 232], just like color (Chap. 2), disposes its own "brain center". *Face sensitive cells* have been found in several parts of the visual cortex of monkeys, although they are found in most significant numbers in a subdivision of *inferotemporal cortex* in the vicinity of the superior temporal sulcus. Whether these cells are actually necessary and sufficient to establish the identity of a face, or if they are only needed for gaze-invariant general human face recognition (without person identity) is not known to sufficient accuracy. In humans, by using magneto resonance studies, the face identity establishing system engages a brain region called *fusiform gyrus*. However, it may not exclusively be devoted to face identification, as other sub-categorization of object tasks activate this region too.

As pointed out above, humans are experts in face recognition, at an astonishing maturity level, even in infancy. Our expertise is so far-reaching that we remember hundreds of faces, often many years later, without intermediate contact. This is to be contrasted to the difficulty in remembering their names many years later, and to the hopeless task of remembering their telephone numbers. Yet this specialization appears to have gone far in some respects and less so in others. We have difficulty recognizing faces of another ethnic group versus own group [36, 38, 65, 158]. For an african-american and a caucasian, it is easier to recognize people of their own ethnicity as compared to cross-ethnic person identification, in spite of the fact that both groups are exposed to each other's faces. Besides this cross-ethnic bias, hair style/line is another distraction when humans decide on face similarities, [24, 39, 201]. Recently [24], another factor that biases recognition has been evidenced (Fig. 1.9). Women had systematically higher correct answer frequencies than men in a series of face recognition tests (Fig. 1.10), taken by an excess of 4000 subjects. A possible explanation is that face identification skill is more crucial to women than men in their social functioning.

1.9 Further Reading

Much of the retinal cell connections and cell types were revealed by Santiago Ramon y Cajal [44] at the begining of the 1900s using a cell staining technique known as Golgi staining. However, the current understanding of human vision is still debated at various levels of details (and liveliness) as new experimental evidence accumulates. An introductory, yet rich description of this view is offered by [113, 121] whereas [173] offers discussions of the neuronal processing and organization w.r.t. the experimental support. The study [109] offers a more recent view of mammalian vision with extrapolations to human vision, whereas [235] presents the current view on human color vision. The study in [189] provides support for Hubel and Wiesel's wiring suggestion to model simple cell responses from LGN responses, whereas that of [206] offers an alternative view that also plausibly models the various types of simple cells that have different decreases in sensitivity when stimulated with nonoptimal directions. The reports in [40, 69], provide a broad overview of the human facerecognition results. The study of [196] suggested a nonuniform resource allocation to analyze static images in computer vision. In analogy with the cortical cell responses to moving patterns, one could differentiate resource allocations in motion image processing too. This can be done by designing filter banks containing elements that can analyze high spatial frequencies moving slowly, as well as low spatial frequencies moving fast at the cost of other combinations. A discussion of this is given in [21].

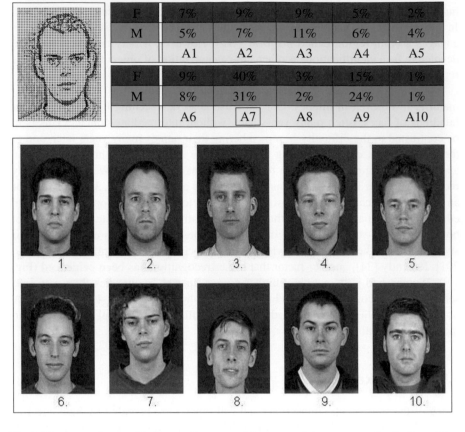

F	7%	9%	9%	5%	2%
M	5%	7%	11%	6%	4%
	A1	A2	A3	A4	A5
F	9%	40%	3%	15%	1%
M	8%	31%	2%	24%	1%
	A6	A7	A8	A9	A10

Fig. 1.10. A question used in human face recognition test and the response distribution of the subjects. The *rows F and M* represent the female and the male responses, respectively. The rectangle indicates the alternative that actually matches the stimulus

2

Color

Color does not exist as a label inherent to a surface, but rather it is a result of our cerebral activity, which constructs it from further processing of the photoreceptor signals. However perceptional, the total system also relies on a sensing mechanism, which must follow the strict laws of physics regulating the behavior of light in its interaction with matter. These laws apply from the moment it is reflected from the surface of an object, until the light photons excite the human photoreceptors after having passed through the eye's lens system. The light stimulates the photoreceptors, and after some signal processing both in the retina and in other parts of the brain, the signals result in a code representing the color of the object surface. At the intersection of physics, biology, psychology, and even philosophy, color has attracted many brilliant minds of humanity: Newton, Young, Maxwell, and Goethe to name but a few. Here we discuss the color sensation and generation along with the involved physics, [166,168], and give a brief account of the signal processing involved in color pathways, evidenced by studies in physiology and psychology [146,235].

2.1 Lens and Color

The role of the lens is to focus the light coming from a plane (a surface of an object) at a fixed distance on the retina which contains light-sensitive sensors. Without a lens the retina would obtain reflected rays coming from different planes, at all depths, thereby blurring the retinal image. For a given lens curvature, however, the focal length varies slightly, depending on the wavelength of the light. The longer wavelengths have longer focal lengths. A light ray having a wavelength interpreted by humans as red has the longest focal length, whereas bluelight has the shortest focal length. Humans and numerous other species have dynamically controlled lens curvatures. If human eyes are exposed to a mixture of light having both red and blue wavelengths, e.g., in a graph, the eyes are exposed to fatigue due to the frequent lens shape changes. The lens absorbs light differently as a function of the wavelength. It absorbs roughly twice as much the blue light as it does the red light. With aging this absorption discrepancy with wavelengths is even more accentuated. As a result,

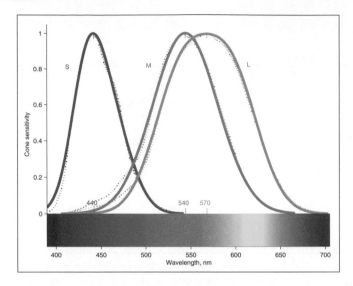

Fig. 2.1. The graph illustrates the sensitivity curves of S-, M-, and L-cones of the retina to the wavelength of the stimulating light

humans become more sensitive to longer wavelengths (yellow and orange) than short wavelengths (cyan and blue). In addition to decreased sensitivity to blue, with aging our general light sensitivity also decreases.

2.2 Retina and Color

The retina has sensor cells, called *cones* and *rods*, that react to photons. Rods require very small amounts of photons to respond compared to cones. Rods also respond to a wider range of wavelengths of photons. Within their range of received photon amounts, that is the *light intensity*, both cone and rod cells respond more intensely upon arrival of more light photons. Humans rely on cones for day vision, whereas they use rod sensors, which are wavelength-insensitive in practice, for night vision. This is the reason that we have difficulty perceiving the color of an object at dark, even though we may be perfectly able to recognize the object. By contrast, the cones, which greatly outnumber the rods in the retina, are not only sensitive to the amount of light, but are also sensitive to the wavelength of the light. However, they also require many more photons to operate, meaning that the cones are switched "off" for nightvision, and they are "on" for dayvision.

Cones belong to either L-, M-, or S- types representing long, middle and short wavelengths. These categories have also been called red, green and blue types making allusion to the perceived colors of the respective wavelengths of the cells. However, studies in neurobiology and psychology have shown that the actual colors the top sensitivity of the cones represent do not correspond to the perceptions of red, green, and blue but rather to perceptions of colors that could be called yellowish-green, green and blue-violet, respectively. Figure 2.1 illustrates the average sensitivity of the cones to photon wavelengths along with perceptions of colors upon reception of photons with such wavelengths. Note that hues associated with pink/rose are absent at the bottom of the diagram. This is because there are no photons with such wavelengths in the nature. Pink is a sensation response of the brain to a mixture of light composed of photons predominantly from short (blue) and long (red) wavelengths.

The dotted sensitivity curves in the graph are published experimental data [80], whereas the solid curves are Gaussians fitted by the author.

Long (570 nm): $(\exp(-\frac{(\omega-\omega_1)^2}{2\sigma^2}) + \exp(-\frac{(\omega-\omega_2)^2}{2\sigma^2}))/C$,
where $C = 1.32$, $\omega_1 = 540$, $\omega_2 = 595$, and $\sigma = 30$.

Middle (540 nm): $\exp(-\frac{(\omega-\omega_1)^2}{2\sigma^2})$,
where $\omega_1 = 544$, and $\sigma = 36$.

Short (440 nm): $(\exp(-\frac{(\omega-\omega_1)^2}{2\sigma_1^2}) + C_E \exp(-\frac{(\omega-\omega_2)^2}{2\sigma_2^2}))/C$,
where $C_E = 0.75$, $C = 1.48$, $\omega_1 = 435$,
$\omega_2 = 460$, $\sigma_1 = 18$ and $\sigma_1 = 23$.

More than half of the cones are L-cones (64%). The remaining cones are predominantly M-cones (32%), whereas only a tiny fraction are S-cones (4%). It is in fovea, within appproximately $1°$ of eccentricity, where humans have the densest concentration of cones. The fovea has no rods, and with increased density towards higher eccentricities, the cone density decreases while the rod density increases. Even the cones are unevenly distributed in the central part, with M-cones being the most frequent at the very center, surrounded by a region dominated by L-cones. The S-cones are mainly found at the periphery, where the rods are also found. The center of the retina is impoverished in S-cones (and rods). The minimum amounts of photons required to activate rods, S-cones, M-cones, and L-cones are different, with the rods demanding the least. Among the cones, our M-type need the least amount of photons for activation, meaning that more intense blues and reds, compared to green-yellows, are needed in order to be noticed by humans.

The coarseness of a viewed pattern matters to the photoreceptors too. A retinal image with a very coarse pattern has small variations of light intensities in a given area of the retina than does a fine pattern that varies more. A repeating pattern is also called *texture*. Coarse textures contain more low spatial-variations than fine textures. Silhouettes of people viewed through bathroom glass belong to the coarse category. A retinal image with "fine" texture is characterized by rapid spatial changes of the luminosity such as edge and line patterns. This type of patterns is responsible for the rich details and high resolution of the viewed images. We will discuss the coarseness and fineness with further precision, when discussing the Fourier transform and the spectrum (Chap. 9). Generally, the photoreceptors at high eccentricities, i.e., basically S-cones and rods, respond to low spatial-variations (spatial frequencies), whereas those in the central area, i.e., basically M- and L-cones, respond best to high spatial-variations. The fineness (spatial frequency) at which a photoreceptor has its peak sensitivity decreases with increased eccentricity of the receptors. At the periphery, where we find rods and S-cones, the photoreceptors respond to low spatial variations (silhouettes) whereas the central vision dominated by M- and L-cones responds better to high spatial variations.

2.3 Neuronal Operations and Color

Color perception is the result of comparisons, not direct sensor measurements. The amount of photons with a narrow range of wavelengths reflected from a physical surface changes greatly as a function of the time of the day, the viewing angle, the age of the viewer,..., etc., and yet humans have developed a code that they attach to surfaces, *color*. Human color encoding is formidable because, despite severe illumination variations (including photon wavelength composition), it is capable of responding with constant color sensation for the viewed surface. This is known as *color constancy*. It has been demonstrated by Land's experiments [145] that the color of a viewed patch is the result of a comparison between the dominant wavelength of the reflected photons from the patch and those coming from its surrounding surface patches.

The signals coming from the L-, M-, and S-cones of the retina, represented by L, M, and S here, arrive at the two *lateral geniculate nucleus* (LGN) areas of the brain. At the LGN the signals stemming from the 3 cone types in the same retinal proximity are presumably added and subtracted from each other as follows:

$$\tilde{L} + \tilde{M} \qquad \text{Lightness sensation}$$
$$\tilde{L} - \tilde{M} \qquad \text{Red–green sensation}$$
$$\tilde{L} + \tilde{M} - \tilde{S} \qquad \text{Blue–yellow sensation}$$

The $\tilde{}$ represents a local, weighted spatial summation of the respective cone type responses [113]. The local window weighting is qualitatively comparable to a 2D probability distribution (summing to 1), e.g., a "bell"-like function (*Gaussian*). The positive terms in the three expressions have weight distributions that are much larger at the center than those of the negative terms. Accordingly, the net effect of $\tilde{L} - \tilde{M}$ is a *center–surround* antagonism between red and green, where red excites the center as long as there is no green in the surround. If there is green in the surround the response attenuates increasingly. This signal processing functionality is found among parvocellular cells in layers 4–6 of the *LGN*, called $(r + g-)$-cells. However, the mathematical expression $\tilde{L} - \tilde{M}$ above can result in negative values if $\tilde{L} < \tilde{M}$. In that case another group of cells, the $(g + r-)$-cells which are also found among the parvocellular cells, will deliver the negative part of the signal $\tilde{L} - \tilde{M}$, while the $(r + g-)$-cells will be inactive. The $(g + r-)$-cells function in the same way as $(r + g-)$-cells except that they are excited by green in the center and inhibited by r in the surround. Accordingly, $(r + g-)$- and $(g-r+)$-cells together implement $\tilde{L} - \tilde{M}$. Likewise, $\tilde{L} + \tilde{M} - \tilde{S}$ results in an antagonism between blue and yellow. This scheme is presumably implemented by two groups of parvocellular LGN cells,

$(y + b-)$ and $(b + y-)$, where y is a shorthand way of saying "red plus green". The latter is perceived as yellow light if the amount of light in the red wavelength range is approximately the same as that of green. Together, the $(r+g-)$-, $(g+r-)$-, $(y+b-)$-, and $(b+y-)$-cells populate the vast majority of the cells in layers 3–6 of LGN. Albeit in minority, there is another significant cell type in these layers that is of the *center–surround* type. Cells of this type differ from the other cells in that they are color-insensitive, and presumably implement the $\tilde{L} + \tilde{M}$-scheme. Additionally, the entire layers 1 layers 2 are populated by this type of cells, the *magnocellular cells*, albeit these are larger than the *parvocellular cells* populating layers 3–6.

It is worth noting that in the perception of lightness, or luminosity, the blue color does not play a significant role. Details only differing in the amount of blue do not show up very well because such changes do not contribute to the perception of edges and lines.

In LGN, most of neurons are wavelength-selective while being *center–surround*. They are excited by one wavelength pattern of the stimulus light falling in one region of their receptive field and inhibited by another in the other. However, they do not measure wavelength differences between the light falling into their center, and surround. Merely, they express the difference in the amount of light quanta with specific wavelengths captured by the center and surround regions. In a way, it is a matter of spatial subtraction that these cells perform, not wavelength subtraction. The blobs encountered in layers 1–3 of the V1 area contain the so-called *double-opponent color cells*, which are sensitive to wavelength differences in the center and surround regions [155]. They respond vigorously to one wavelength in the center of their receptive field, while they are inhibited by another (still in the center). The same cells are excited by this second wavelength in the surround and depressed by the first. A double-opponent color cell can thus be excited by the wavelength of red and inhibited by that of green in its center, while it will be excited by the wavelength of green and inhibited by the wavelength of red in the surround. This behavior has been denoted as $(r + g - /g + r-)$ in neurobioligical studies. Consequently, a large patch reflecting red will generate zero response from these cells because the wavelength pattern in the center is "subtracted" from that of the surround. In fact, not only red colored light, but any colored light, including white, that shines up a large patch observed by an $(r + g - /g + r-)$-cell will generate zero response. An $(r + g-)$-cell of LGN, by contrast, will be excited, if the wavelength pattern matches either the one it prefers in the center or the one in the surround. The following types of double-opponent color cells have been experimentally observed in blobs: $(r + g - /r - g+)$, $(r - g + /r + g-)$, $(b + y - /y + b-)$, $(b - y + /y - b+)$, where b corresponds to the light with wavelength patterns of S-cones (blue) and y is light with an additive combination of wavelength patterns represented by L-cones (red) and M-cones (green). It is presumably the double-opponent color cells that are largely responsible for color constancy observed in many fish species, macaque monkey and humans, although these cells appear in the retina in fish.

Simplified, there are three color axes (lightness, red–green and yellow–blue) along which color processing takes place in humans. However, there are only three independent measurements, represented by the signals L, M, and S, that drive our color perception system. The comparisons are carried out on spatially filtered L-, M-, and S-signals combined linearly (additions or subtractions preceded by spatial summation) rather than the original cone signals. One can therefore expect that a color perception model can be built by a spatial summation filtering of L, M, and S signals combined with pointwise operations that probably include addition, subtraction, and normalization to achieve color constancy. Next, we outline such a plausible theory.

In Land's retinex theory [146], which is in part found in that of Ewald Hering (1834–1918) [105], the color sensation algorithm is suggested as

$$R_i(x, y) = \log \frac{f_i(x, y)}{g(x, y) * f_i(x, y)} \tag{2.1}$$

where g is a spatial lowpass filter that is used to average large areas of the retina, $*$ is the operation that performs local averaging (we will discuss such operations further in Section 7.3), and f_i is one of the cone signal response combinations, $\{\tilde{L} + \tilde{M}, \tilde{L} + \tilde{M} - \tilde{S}, \tilde{L} - \tilde{M}\}$, above. There exist simulation studies of this model, including on how the order of convolution and log functions affects the result, and how a Gaussian and other functions perform [127], confirming a fairly accurate prediction of the color constancy.

2.4 The 1931 CIE Chromaticity Diagram and Colorimetry

The *chromaticity diagram*, constructed in 1931 by the Committe International de l'Eclairage,[1] *CIE*, links the wavelength of light to perceived colors as an international standard, (Fig. 2.2). It is used for a variety of purposes, including to compare colors produced by color-producing devices, e.g., PC monitors, printers, and cameras. The science of quantifying color is called *colorimetry*.

The CIE diagram is a projection of a 3D color space, called *XYZ color space*, to 2D. The X, Y, Z coordinates are found as follows. The light emitted by a device, or light reflected from a surface consists of photons with different wavelengths. The amount of photons with a certain wavelength, λ, in a given light composition is represented by the function $C(\lambda)$. The CIE diagram comprises three functions $\mu_X(\lambda)$, $\mu_Y(\lambda)$, $\mu_Z(\lambda)$ (Fig. 2.3). With these functions one can calculate three scalars, called X, Y, Z,

$$\begin{aligned} X &= \int C(\lambda)\mu_X(\lambda)d\lambda \\ Y &= \int C(\lambda)\mu_Y(\lambda)d\lambda \\ Z &= \int C(\lambda)\mu_Z(\lambda)d\lambda \end{aligned} \tag{2.2}$$

There are devices that can measure X, Y, and Z by use of filters and photosensors. Because the functions C and μ are positive, the scalars X, Y, Z are real and nonnegative. These measurements represent the color coordinates of the observed light in the CIE–XYZ color system. The projection to the CIE diagram is obtained via

---

[1] French for illumination

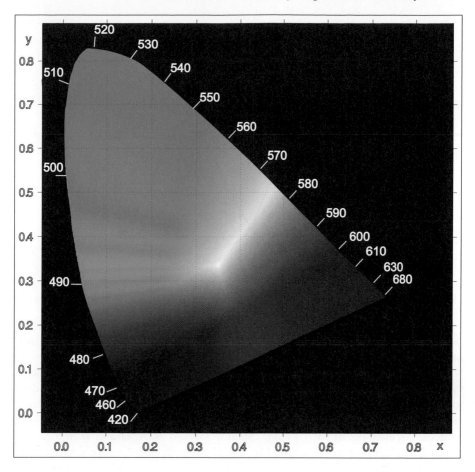

Fig. 2.2. The perceivable colors and their wavelengths, according to CIE (1931) standard on color

$$x = \frac{X}{X + Y + Z} \tag{2.3}$$

$$y = \frac{Y}{X + Y + Z} \tag{2.4}$$

$$z = \frac{Z}{X + Y + Z} \tag{2.5}$$

where $x + y + z = 1$, its so that only two of x, y, z are independent, making the projection a planar surface. In Fig. 2.2 x and y are the coordinate axes. After having projected its X, Y, Z values via Eqs. (2.3)–(2.5), each perceived color is a point on the CIE diagram. The projection amounts to a normalization of the 3D XYZ space with respect to luminosity, $X + Y + Z$. The 2D *xy color space* represents the colors appearing in the CIE diagram.

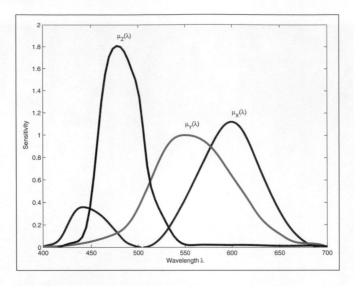

Fig. 2.3. The functions used in projecting the wavelength distribution to CIE XYZ space

The aim of the CIE diagram is to model color, e.g., those generated by a TV set, as if generated by mixing three types of light sources, each composed of photons with different ranges of wavelengths. This is called *additive[2] color model*. Each of the threee light sources alone will produce a different color sensation, corresponding to the three "primary" colors, i.e., three points in the CIE diagram. A new color is produced by changing the relative amount of light emitted by the primary light sources. An example of such a color triplet is marked as R, G and B in the copy of the CIE diagram represented by Fig. 2.4. If these three colors are appropriately placed by the manufacturer of the device, then most colors will be reproducable. The points R, G, B on the diagram will define a triangle, so that any new color made by mixing these three (primary) colors will be within the triangle. It should, however, be emphasized that it is impossible to find three such points so that all perceivable colors of the CIE diagram can fit into the corresponding triangle, since the form of the diagram is not strictly triangular. In consequence, there will always be a fraction not included in the triangle, if the CIE diagram is to be approximated by three points. The colors included in the triangle are called the gamut of the three primaries. As a special case, one can produce a limited range of "color" by mixing only two primaries. In this case the produced colors will be limited to those to be found on the line joining the two primaries, the gamut of them.

The point marked as W in Fig. 2.4 is the color white. Note that we have three color components, XYZ, but these are normalized to yield xy coordinates.[3] As a result, the colors in the CIE diagram are normalized so that colors differing by only

---

[2] The subtractive color model, e.g., used by painters, also exists and achieves the same result.

[3] Because $x + y + z = 1$, computing z is not useful in practice.

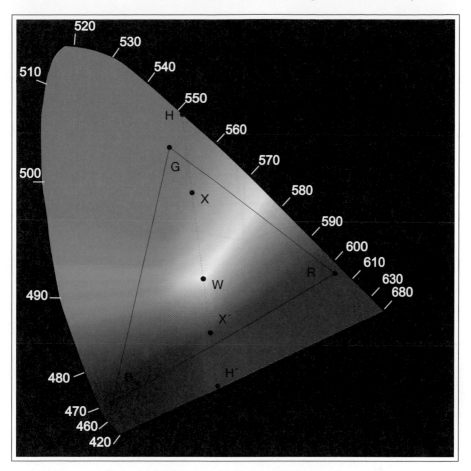

Fig. 2.4. The definition of hue and purity

luminosity are represented by the same point. The pure hue is defined via the normalized color components x, y and is to be found at the boundaries, where the dominant wavelength of photons producing this color is also read out. The point marked as H in the figure is a pure hue, whereas less pure colors of the same hue are to be found on the line segment between W and H. The point H has the purity 1, while the point marked as X has the purity $\frac{\|WX\|}{\|WH\|}$ with $\|WX\|$ representing the distance from W to X. The point X' is the *complementary color* of X, and is found in the opposite direction, but with the same purity, $\frac{\|WX'\|}{\|WH'\|} = \frac{\|WX\|}{\|WH\|}$. The point H' is the boundary point obtained by mirroring H through W.

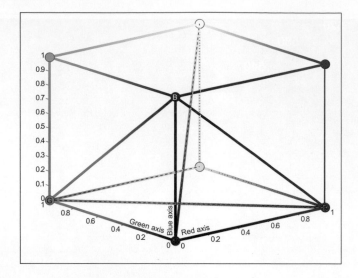

Fig. 2.5. The edges of the RGB color space and some important lines with the corresponding colors. The edges behind the visible faces are marked as *dotted*

2.5 RGB: Red, Green, Blue Color Space

There exist other color spaces, each offering a desired convenience in applications it is intended for, but all relate to CIE–XYZ space. The RGB color space is one such space that is widely used by millions of devices, including nearly all TV sets, computer displays, projectors. From experimentally determined XYZ values of the three primaries, the RGB space is obtained via the linear (coordinate) transformation

$$\begin{pmatrix} R \\ G \\ B \end{pmatrix} = \begin{pmatrix} 2.36461 & -0.89654 & -0.46807 \\ -0.51517 & 1.42641 & 0.08876 \\ 0.00520 & -0.01441 & 1.00920 \end{pmatrix} \cdot \begin{pmatrix} X \\ Y \\ Z \end{pmatrix} \tag{2.6}$$

In the additive model, a mixture of colors using positive amounts from the primaries R,G,B results in a new color. Assuming that the amount mixed from each primary is in the interval $[0, 1]$, the resulting color will be represented by a point in a cube, having the red, green, blue basis as the orthogonal edges of a cube (Fig. 2.5). The vertices marked as R, G and B define a triangle in the CIE color diagram above. The triangle represents the gamut of the printer having its primary colors as those marked with R, G and B. The gray line in one of the main diagonals represents the luminosity variation available to the RGB space. It consists of colors having equal components of red, green and blue. Two colors that are complementary have red, green and blue components that sum to 1, 1, 1, respectively.

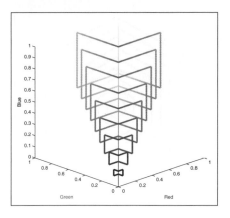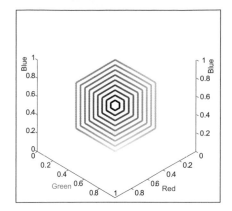

Fig. 2.6. (*Left*) The isocurves of the HSB coordinates when the saturation is 0.5. The nine twisted hexagons represent colors for different values of brightness. (*Right*) The same isocurves viewed through the brightness axis, $(1, 1, 1)^T$

2.6 HSB: Hue, Saturation, Brightness Color Space

Whereas the colors of the RGB space are usually produced by varying three colored light emitters, it is not easy for humans to interpret the thus-obtained colors. Derived from the RGB space, the *hue*, *saturation* and *brightness* space,[4] *HSB color space* is, by contrast, intuitive. It interprets the color in a way that resembles artists' way of describing or perceiving it. Assuming to have a color with the components $(r, g, b)^T$, in the RGB space, the HSB representation of this color yields

$$H = \begin{cases} (\frac{g-b}{\max - \min})\frac{\pi}{3}, & \text{if } r = \max, \\ (2 + \frac{b-r}{\max - \min})\frac{\pi}{3}, & \text{if } g = \max, \\ (4 + \frac{r-g}{\max - \min}\frac{\pi}{3}, & \text{if } b = \max, \end{cases} \qquad (2.7)$$

$$S = \max - \min \qquad (2.8)$$

$$B = \max \qquad (2.9)$$

where $\max$ and $\min$ refer to the largest and smallest values of r, g, b, respectively. The HSB coordinates resemble[5] the cylindrical coordinates of the RGB color point around the *brightness axis*, the main diagonal represented by the direction $(1, 1, 1)^T$ of the *RGB cube*, although there are significant differences. First, the coordinate curves and sufaces of the HSB colors are contained within the RGB cube, but they are neither cylindrical nor circular. By that we mean the curves and surfaces generated when one or two of the variables H,S,B of the color is kept constant, i.e., the isocurves and surfaces of the coordinate transformation above. They are curves that generate

---

[4] Another name for this is hue, saturation, value *(HSV) color space*.

[5] There are also studies in which HSB coordinates are presented as cylindrical coordinates, although the "hexagon" version here appears to be the most common, because of its conversion equations using only arithmetic operations, which result in simple graphics hardware.

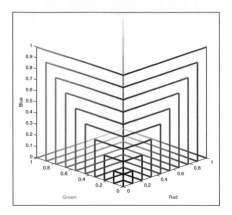 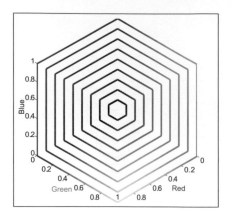

Fig. 2.7. The same as in Fig. 2.6 except the saturation is now full, $S = 1$

cones around the brightness axis instead of cylinders. In Fig. 2.6 and Fig. 2.7 two such cones, corresponding to $S = 0.5$ and $S = 1$, respectively, are sampled. Second, on such hexagonal cones, the color curves corresponding to iso-brightness, i.e., when both saturation and brightness are constant, consist of twisted hexagons i.e., each such hexagon is a true 3D curve, does not lie in a single plane. The hexagons are twisted in such a way that the resulting curve is continuous, but the curve zigzags in 3D, while always being a straight line parallel to the edges of the RGB cube, as depicted by the figures.

The hue component tells where on the twisted hexagon a color is in analogy with the angle in cylindrical coordinates. The "zero" value of hue points at red, as does the value 1, to be contrasted to 2π an angle of cylindrical coordinates. The saturation component S is similar to the radial coordinate, i.e., it represents the "radius" of the hexagon around the brightness axis. By changing it, we effectively change the size of the twisted hexagon. The brightness component resembles the height component of the cylindrical coordinates. Increasing it at the same saturation moves the twisted hexagon towards the white color, $(1, 1, 1)^T$ in RGB, with the value 1 representing the intersection of the cone with the three walls of the RGB cube neighboring the white point.

All human labels for color are reachable by changing the hue variable, whereas changing the other two variables for a fixed hue does not change the human label for the sensed color. Decreasing the saturation component amounts to adding more black color in the minds of artists. A 100% saturated color of a given hue and brightness is the most vivid color of that hue/brightness combination. Fully saturated colors, $S = 1$, appear on any of the three boundary surfaces of the RGB cube (Fig. 2.7). A 0% saturated color is perceived as colorless, i.e., gray, regardless the hue. It is the result of collapsing the hexagons to the brightness axis. Increasing the brightness component amounts to adding white in artists' terminology. This has the effect of making the hue shine.

Part II

Linear Tools of Vision

Let books be your dining table
And you shall be full of delights
Let them be your mattress
And you shall sleep restful nights

St. Ephrem (A.D. 303–373)

3

Discrete Images and Hilbert Spaces

A *Hilbert space* is a mathematical space that is equipped with a scalar product. A Hilbert space is also referred to as an *inner product space*. The elements of Hilbert spaces are typically sequences of scalars or functions, also called *vectors*. Provided that certain rules are respected, matrices, which are studied in linear algebra, can be viewed as vectors, and their space as a vector space. A scalar product for matrices, which can easily represent discrete images, can also be defined. Most important, however, is that the space of all images can be viewed as a Hilbert space. By using the scalar product the angles between vector pairs of the Hilbert space can be determined. This allows us to quantify how similar two images are. A Hilbert space contains numerous orthonormal bases. Once a basis is given, every element of the Hilbert space can be written uniquely as a sum of multiples of the basis elements. In this way, it is feasible to represent an image as a sum of the basis vectors weighted with appropriate coefficients.

3.1 Vector Spaces

In linear algebra courses one studies vectors that represent points in spaces. A vector is often visualized as an arrow from the center of the coordinate system, *the origin*, to the point. The point itself can be seen as equivalent to the vector, or vice versa. Although we will work with other vector spaces mostly in this book, we start with a vector space with which we have the most familiarity, the environment in which we move about, i.e., ordinary space. Without the time aspect, it is referred to as a 3D *Euclidean space,* or simply as E_3. The points in this space are numerically represented as an ordered set of real numbers $\mathbf{x}(1), \mathbf{x}(2), \mathbf{x}(3)$ standing one on the top of the other:

$$\mathbf{x} = \begin{pmatrix} \mathbf{x}(1) \\ \mathbf{x}(2) \\ \mathbf{x}(3) \end{pmatrix} \tag{3.1}$$

The elements, $\mathbf{x}(i)$ are to be interpreted relative to a set of abstract quantities, $\mathbf{e}_1$, $\mathbf{e}_2$, and $\mathbf{e}_3$, called the *basis*, such that they are coordinates, i.e., $\mathbf{x}(i)$ act as coefficients "resizing" the basis, to "add" up to the vector $\mathbf{x}$. Assuming that we know how to resize, and add vectors, this idea is expressed as follows.

$$\mathbf{x} = \mathbf{x}(1)\mathbf{e}_1 + \mathbf{x}(2)\mathbf{e}_2 + \mathbf{x}(3)\mathbf{e}_3 \tag{3.2}$$

In related literature as well as later in this book, the basis set attached to an origin is also called a *coordinate frame,* or just *frame* when there is no risk of interference with other concepts. Since a basis set is the same for all points in the space, $\mathbf{e}_i$ do not have to be written each time one needs to represent a vector. This is why they are omitted in (3.1). Because any point P in E_3 can be represented by a triplet of real numbers relative to an origin and a basis set, as in Eq. (3.2), all points of E_3 are "vector" quantities. In that, the point P's being a vector depends on another point, which is often not mentioned, the origin, O. Without O the point P would not have been a vector. Indeed vectors are quantities with directions so that any two points A and B in E_3 define a vector, also written as $\overrightarrow{AB}$ or $\overline{AB}$ (or even as AB when there is no risk of confusion) in an alternative notation to $\mathbf{x}$.

The above concept does not need to be limited to our 3D space. The vectors $\mathbf{e}_i$, and thereby $\mathbf{x}$, do not need to be basis vectors in our ordinary 3D space. They can be any abstract quantities or "points" as long as the set they are part of is a set that has the vector space properties defined below:

Definition 3.1. *A set of quantities* $\{\mathbf{x}\}$ *is called a* vector space *if the members obey two rules*

1. **Scaling with a scalar** *is defined and the space is* closed *under this operation, also called* vector scaling*;*
2. **Addition** *is defined and the space is closed under this operation, also called* vector addition.

The term *closed* represents the fact that the result of an operation (vector scaling or vector addition) never leaves the original space (the set $\{\mathbf{x}\}$), no matter which quantities are involved in the operation.

Scaling and addition in our 3D Euclidean space are visually meaningful. Scaling represents making the vectors "longer" or "shorter" while adding two vectors means that we concatenate them at their ends by translation (moving one of the vectors without rotation). In a given basis of the E_3, the mechanism of scaling is to multiply all three vector components with the scalar α, while the mechanism performing addition is to add the three vector components. In other words, given a representation of E_3, there is a method of performing scaling and addition that obeys the scaling and addition properties without leaving the original space. Accordingly, E_3 is a vector space.

Using the same idea, we can construct similar scaling and addition mechanisms for N-tuple arrays with $N > 3$. The N-tuple of real numbers,

$$\begin{pmatrix} \mathbf{x}(1) \\ \mathbf{x}(2) \\ \vdots \\ \mathbf{x}(N) \end{pmatrix} \tag{3.3}$$

also constitutes a vector space, E_N. This is called the N-dimensional Euclidean space (or room) and behaves very much like E_3 as far as addition and scaling is concerned, albeit it is not easy to visualize its members.

This concept can be easily extended to comprise arrays with more than one index, e.g., with four indices:

$$\mathbf{A} = \{\mathbf{A}(k, l, m, n)\} \tag{3.4}$$

where $A(k, l, m, n)$ is a scalar (i.e., a real or a complex number). To fix the ideas, the indicies k, l, m, n would be typically positive integers that can at most take the values K, L, M, N, respectively. An addition and scaling rule that can make the set of all $\mathbf{A}$s having the same K, L, M, N, a vector space is as follows¡:

$$\mathbf{A} + \mathbf{B} = \{\mathbf{A}(k, l, m, n) + \mathbf{B}(k, l, m, n)\} \tag{3.5}$$

$$\alpha\mathbf{A} = \{\alpha\mathbf{A}(k, l, m, n)\} \tag{3.6}$$

Matrices, the double-indiced arrays studied in linear algebra courses, constitute a special case of the above. Evidently, those having the same size are a vector space too!

3.2 Discrete Image Types, Examples

There are many types of discrete images, and new types are being constructed as the science studying their production from light, by sensors and other hardware advances. This science is also called *imaging*. The simplest image type is a 2D array, where the array elements represent gray values

$$\mathbf{A} = \{\mathbf{A}(k, l)\} \tag{3.7}$$

where k, l take K, L different values. The size of the image is $K \times L$, and typically k, l corresponds to row (top–down) and column (left–right) counts. However, it is also common that k, l represent the horizontal (left–right) and vertical (down–up) counts. The pair k, l represents the coordinates of a point, also called *pixel*, on the discrete 2D grid of size $K \times L$. The scalar $\mathbf{A}(k, l)$ is typically an integer ranging 0–255 and represents the *gray value* or the *pixel value* of the pixel. For convenience we can assume that $\mathbf{A}(k, l)$ is a scalar (real or complex-valued), although it is not possible to interpret negative values as gray values. Other names for gray value, are *intensity*, *brightness*, and *luminance*. Negative and complex pixel values are common and can be obtained from computations in image processing, even from sensors. For example, we can have an array with two indices representing a map of altitudes w.r.t. the sea level. The points under the sea level will be negative, and those above will be positive. There are examples of complex-valued images that can be obtained from sensors as well as from computations. To limit the scope, we do not discuss the taxonomy of images at that level here. The space of discrete images of the same size taking scalar values is thus a vector space. The necessary rules of vector addition and scaling can be defined via the general elementwise addition and scaling rules discussed in the previous section. Accordingly, image addition is achieved by adding the image values at corresponding points, whereas image scaling is achieved by multiplying the image values with a scalar, e.g., as given by Eqs. (3.8) and (3.9) for gray images:

$$\mathbf{A} + \mathbf{B} = \{\mathbf{A}(i,j) + \mathbf{B}(i,j)\} \tag{3.8}$$

$$\alpha \mathbf{A} = \{\alpha \mathbf{A}(i,j)\} \tag{3.9}$$

Exercise 3.1. *Is the space of images having the same size but taking the gray values* $0, 1, \cdots, 255$ *a vector space?*

Extending the number of indices to three in a discrete array

$$\mathbf{A} = \{\mathbf{A}(k,l,m)\} \tag{3.10}$$

has also counterparts in the world of images. The most commonly known example is discrete color images, where the first two indices of $\mathbf{A}$ represent the coordinate point and the third index defines the color components. In other words, at every *spatial position*, which is another common name for the coordinate pair k, l, we have an M-dimensional array of scalars (instead of a single scalar). In the case of ordinary color images, M is 3, representing the color components, typically in the RGB color space. There are large banks of discrete image data of the earth where M is > 3; these are *multispectral images*. Here each "color" represents a specific photon wavelength range of the light, some visible, many invisible to humans. Such images with scalar values of $\mathbf{A}(k,l,m)$ constitute a vector space with the addition and scaling rules:

$$\mathbf{A} + \mathbf{B} = \{\mathbf{A}(k,l,m) + \mathbf{B}(k,l,m)\} \tag{3.11}$$

$$\alpha \mathbf{A} = \{\alpha \mathbf{A}(k,l,m)\} \tag{3.12}$$

For certain image types, there is also a different interpretation of the third index. In computer tomography images one also obtains an array of discrete data with three indices, as discussed above. In this case the indices k, l, m represent horizontal, vertical, and depth coordinates of a spatial point, commonly called a *voxel*. The scalar $\mathbf{A}(k,l,m)$ typically represents absorption.

There is yet another interpretation of the third index for *black and white motion images*. The first two indices k, l correspond to the spatial position, whereas the third index m corresponds to the temporal instant. Accordingly, k, l, m are called the *spatio–temporal coordinates*.[1] With the same size and scalar image values these constitute vector spaces too.

Increasing the number of indices to four brings into focus images that are commonly produced by consumer electronics, *color motion images*. These are also called *image sequences*. The first three indices are the spatio–temporal coordinates, whereas the fourth index encodes the color. Evidently, even these images constitute a vector space, with the addition and scaling rules analogous to Eqs. (3.5)–(3.6).

The list of discrete image types can be made longer, but we stop here to proceed with an example illustrating what one can do with vector spaces.

---

[1] A suitable name for these coordinates would be *stixels*, making allusion to the spatio–temporal nature of the data.

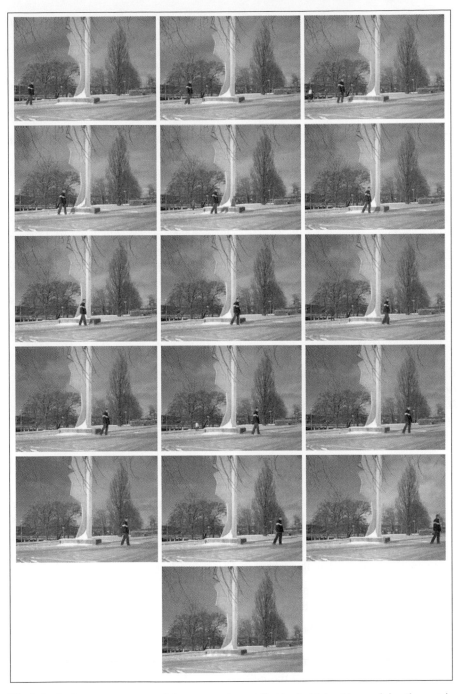

Fig. 3.1. An image sequence and the average image (*bottom*) obtained by applying the matrix vector space rules of addition and multiplication by a scalar according to Eq. (3.13)

Example 3.1. *The 2D discrete color images (of the same size) are vector spaces if we treat them as matrices having elements consisting of arrays representing the three color components, R, G, B. In the right–down direction, the digital pictures in Fig. 3.1 illustrate the frames of an image sequence except the last frame (bottom). The last frame is the average of all frames observed by the camera (shown in the figure).*

$$\bar{\mathbf{A}} = (\mathbf{A}_1 + \mathbf{A}_2 + \cdots \mathbf{A}_{15})/15 \tag{3.13}$$

where $\mathbf{A}_i$ is the ith frame. The frames are summed and finally multiplied with a scalar $(1/15)$, according to the addition and scaling rules of Eqs. (3.11)–(3.12)

In the observed sequence the camera is static (immobile) and the passing human is present in every frame, and yet nearly no human is present in the averaged frame. This is because at any given point, the color is nearly unchanged over the time because the camera is fixed. Though seldom, a change of the color does occur at that point, due to the passage of the human. This happens in, roughly 1 out of 15 time instants at the same pixel coordinate. Consequently, the time averages of color are not sufficiently influenced by one or two outliers, to the effect that the means closely approximate the background colors. Another effect of averaging is that the mean represents the color more accurately than any of the constituent images. Similar uses of matrix averaging, sometimes combined with robust statistics (e.g., clipped average, median, etc.) are at the heart of many image processing applications, e.g noise reduction [85, 192], super resolution, and the removal of moving objects, [9, 119].

Exercise 3.2. *Can we use the HSB space to represent the color arrays under the rules of (3.11)–(3.12)?*
HINT: Do the points on a circle represent a vector space? Is the numerical average of two numbers representing hue always a hue (e.g., two complementary colors)?

3.3 Norms of Vectors and Distances Between Points

Adding or scaling vectors are tools that are not powerful enough for our needs. We must be able to measure the "length" of the vectors as well. The length of a vector is also known as the norm or the magnitude of a vector. The symbol for the norm of a vector $\mathbf{u}$ is $\|\mathbf{u}\|$.

In analogy with the length concept in E_3,

- The norm should not be negative.
- Only the null vector has the norm zero.

Despite its simplicity, the second requirement constitutes the backbone of numerous impressive proofs in science and mathematics. It expresses when *only* the knowledge of the norm is sufficient to identify the vector itself fully. The vector space properties guarantee that the null vector is the only vector which enjoys the privilege that the knowledge of its norm automatically determines the vector itself. Paraphrasing this, we have the right to "throw away" the norm symbol only when we are sure that the norm is zero.

Exercise 3.3. *In two dimensions, how many points are there such that their coordinate vectors have the norm (i) 5, (ii) 0?*
HINT: Can we count them in both cases?

Once it is defined on a *vector space*, an important usage of the norm is in the computation of the distance between two points of that space. The difference

$$\mathbf{u} - \mathbf{v} \tag{3.14}$$

is a vector if the vectors $\mathbf{u}$ and $\mathbf{v}$ belong to the same vector space, because in such a space addition and scaling are well defined and the space is closed. The distance between two points is defined as

$$\|\mathbf{u} - \mathbf{v}\| \tag{3.15}$$

This is natural because when $\|\mathbf{u} - \mathbf{v}\| = 0$, then we know that the two vectors are one and the same. This follows from the properties of the norm, namely that if the norm is the null then the vector is null vector, $\mathbf{u} - \mathbf{v} = \mathbf{0}$, which is the same as $\mathbf{u} = \mathbf{v}$.

There are two other properties that a norm must have before it can properly be called a norm:

- When a vector is scaled by a scalar, its norm must scale with the magnitude of the scalar.
- The vector space must yield the *triangle inequality* under the norm, i.e., the shortest distance between two points is the norm of the vector joining them.

We summarize the *norm properties* that every norm, regardless of the vector space on which it is defined, must have as follows

$$0 \leq \|\mathbf{u}\|, \qquad\qquad\qquad \text{Nonnegativity} \quad (3.16)$$
$$\|\mathbf{u}\| = 0 \Leftrightarrow \mathbf{u} = 0, \qquad\qquad \text{Nullness} \quad (3.17)$$
$$\|\alpha\mathbf{u}\| = |\alpha|\|\mathbf{u}\|, \qquad\qquad\qquad \text{Scaling} \quad (3.18)$$
$$\|\mathbf{u} + \mathbf{v}\| \leq \|\mathbf{u}\| + \|\mathbf{v}\|. \qquad \text{Triangle inequality} \quad (3.19)$$

We already know that discrete color images constitute a vector space. By using the rules above, we can add a norm to a vector space that can be used to represent discrete images having vector-valued pixels.

Exercise 3.4. *Let* $\mathbf{A}$ *be an image with vector-valued pixels. Show that the expression*

$$\|\mathbf{A}\| = [\sum_{ijk} \mathbf{A}^*(i, j, k)\mathbf{A}(i, j, k)]^{1/2} \tag{3.20}$$

where * *is the complex conjugate, is a norm.*
HINT: This rule obeys the triangle inequality under the addition and scaling rules of Eq. (3.11)–(3.12).

Fig. 3.2. Norms under scaling of 2D vector-valued images are illustrated by a digital color image. On the *left* the original image **A**, having three color components at each image point, with the norm $\|\mathbf{A}\| = 874$, is shown. On the *right*, $0.5A$ with the norm $\|0.5\mathbf{A}\| = 437$ is displayed

Example 3.2. *The left image in Fig. 3.2 has three real valued color components at each pixel, i.e., RGB color values. Each color component varies between 0 and 1, representing the lowest and highest color component value, respectively. The image is thus a vector-valued image that we can call* **A**. *The darker right image in Fig. 3.2 is* $0.5\mathbf{A}$*, i.e., it is obtained by multiplying all color components by the scalar 0.5. The norm of* **A** *is computed to be 874, by using the expression in Eq. (3.20), whereas the norm of* $0.5\mathbf{A}$ *using the same expression is computed to be 437. As expected from a true norm, scaling all vector pixels by the scalar 0.5 results in a scaling of the image norm with the same scalar.*

Exercise 3.5. *If the image* **A** *shown in Fig. 3.2 had been white, i.e., all three color components were equal to 1.0, then one obtains* $\|\mathbf{A}\| = 1536$*, where the norm is in the sense of Eq. (3.20). How many pixels are there in* **A***? Can you find how many rows and columns there are in* **A***?*
HINT: Use a ruler.

The images in Fig. 3.3 illustrate the triangle inequality of the norm. They represent, clockwise from top left, $\mathbf{A}_1$, $\mathbf{A}_2$, $\mathbf{A}_3$, and $\mathbf{A}_1 + \mathbf{A}_2 + \mathbf{A}_3$, respectively. At any image point the pixel is a vector consisting of three color components. The norms of these images are $\|\mathbf{A}_1\| = 517$, $\|\mathbf{A}_2\| = 527$, $\|\mathbf{A}_3\| = 468$, and $\|\mathbf{A}_1 + \mathbf{A}_2 + \mathbf{A}_3\| = 874$, respectively. As expected from a norm, we obtain $\|\mathbf{A}_1 + \mathbf{A}_2 + \mathbf{A}_3\| \leq \|\mathbf{A}_1\| + \|\mathbf{A}_2\| + \|\mathbf{A}_3\|$.

To illustrate both the triangle inequality and the scaling property of the norms, we study the quotient Q

$$Q(\mathbf{A}_1, \mathbf{A}_2) = \frac{\|\mathbf{A}_1 + \mathbf{A}_2\|}{\|\mathbf{A}_1\| + \|\mathbf{A}_2\|} \leq 1 \qquad (3.21)$$

which equals 1 if both $\mathbf{A}_1$ and $\mathbf{A}_2$ are vectors that share the same direction. *In other words, if*

Fig. 3.3. Norms under addition of 2D vector-valued images are illustrated by adding three digital color images. The *bottom right* is the addition of the other images. The norms, counter clockwise from *top left*, are $\|\mathbf{A}_1\| = 517$, $\|\mathbf{A}_2\| = 527$, $\|\mathbf{A}_3\| = 468$, $\|\mathbf{A}_1 + \mathbf{A}_2 + \mathbf{A}_3\| = 874$, respectively

$$\mathbf{A}_2 = \alpha \mathbf{A}_1 \tag{3.22}$$

where α is a positive scalar, then the quotient Q will equal 1, which is a consequence of the triangle inequality and the scaling property of the studied vector space. When $\mathbf{A}_1$ and $\mathbf{A}_2$ are those displayed in Fig. 3.2, the quotient Q equals to 1, because by computation we obtain $\|\mathbf{A}_1 + \mathbf{A}_2\| = 1311$, $\|\mathbf{A}_1\| = 874$, and $\|\mathbf{A}_2\| = 437$. In view of the norm represented by Eq. (3.20) and the triangle inequality, relation (3.19), this is expected.

However, because of the continuity of Q w.r.t. $\mathbf{A}_2$, even when $\mathbf{A}_2$ approximately equals to $\alpha \mathbf{A}_1$, we may expect that Q will be close to 1. To see whether this is true, we study the corresponding quotients for the four images in Fig. 3.4, which are $Q(\mathbf{A}_1, \mathbf{A}_2) = 0.960$, $Q(\mathbf{A}_1, \mathbf{A}_3) = 0.943$, $Q(\mathbf{A}_1, \mathbf{A}_4) = 0.955$. These suggest that image $\mathbf{A}_1$ was likely obtained by multiplying $\mathbf{A}_2$ with a positive constant, although, in fact, this is not true. The images A_1 and A_2 are instead different views of the same scene, and are not obtainable from each other by scaling the color components. As suggested already by the example, i.e., that all three quotients are close to each other and not far away from 1, the quotient Q as a means to establish general similarities of images may not be reliable, although it can be used to test certain types of similarities

Fig. 3.4. Four images $\mathbf{A}_1$, $\mathbf{A}_2$, $\mathbf{A}_3$, $\mathbf{A}_4$, clockwise from the *top left*, are to be compared with each other by using the triangle inequality and the scaling rule of the norms. The pairwise quotients Q suggest that A_1 is most likely obtained from A_2 by a positive multiplicative constant

i.e., those that are obtained by positive scaling. Furthermore, in case α is negative Q will not be reliable at all because it will not only be less than 1, but it might even vanish, i.e., when $\alpha = -1$. We stress therefore that Q illustrates the triangle inequality and the scaling, but that it cannot be used as a means to test all types of image similarities.

3.4 Scalar Products

The norms are useful to measure distances. Next, we present another tool that is useful for "navigation" in abstract vector spaces. This is the *scalar product* which will be used to measure the "angles" between vectors in vector spaces. A scalar product is a particular operation between two vectors that results in a scalar (real or complex) that "somehow" represents the angle φ between the two vectors involved in the product. The symbol of this operation is written as:

$$\langle \mathbf{u}, \mathbf{v} \rangle \qquad (3.23)$$

Any rule that operates on two vectors and produces a number out of them is not good enough to be qualified to be a scalar product. To be called a scalar product, such an operator must obey the *scalar product rules*:

1. $\langle \mathbf{u}, \mathbf{v} \rangle \;=\; \langle \mathbf{v}, \mathbf{u} \rangle^*$
2. $\langle \alpha\mathbf{u}, \mathbf{v} \rangle \;=\; \alpha^* \langle \mathbf{u}, \mathbf{v} \rangle$
3. $\langle \mathbf{u} + \mathbf{v}, \mathbf{z} \rangle \;=\; \langle \mathbf{u}, \mathbf{z} \rangle + \langle \mathbf{v}, \mathbf{z} \rangle$
4. $\langle \mathbf{u}, \mathbf{u} \rangle \;> 0$ if $\mathbf{u} \neq 0,$ and $\langle \mathbf{u}, \mathbf{u} \rangle \;= 0$ iff $\mathbf{u} = 0$

Here the star is the complex conjugate, and "iff" is a short way of writing "if and only if". Remembering the first requirement, we note that the second is equivalent to

$$\langle \mathbf{u}, \alpha\mathbf{v} \rangle = \alpha \langle \mathbf{u}, \mathbf{v} \rangle \tag{3.24}$$

As a byproduct, the last relationship offers a natural way of producing a norm for a vector space having a scalar product:

$$\|\mathbf{u}\| = \sqrt{\langle \mathbf{u}, \mathbf{u} \rangle} \tag{3.25}$$

Since the scalar product can be used to express the norm, the distance between two vectors is easily expressed by scalar products as well:

$$\|\mathbf{u} - \mathbf{v}\|^2 = \langle \mathbf{u} - \mathbf{v}, \mathbf{u} - \mathbf{v} \rangle \tag{3.26}$$

Definition 3.2. *A vector space which has a scalar product defined in itself is called a Hilbert Space.*

We use the term "scalar product" and not the term "real" since this allows us to use the same concept of Hilbert Space for vector spaces having complex-valued elements. The scalar products can thus be complex-valued in general, but never the norm associated with it. Such vector spaces are represented by the symbol C_N. Hence a scalar product for C_N must be defined in such a way that the auto–scalar product of a vector represents the square of the length. The norm must be strictly positive or be zero if and only if the vector is null. As a scalar product for E_N would be a special case of the scalar product for C_N, it should be inherited from that of C_N. For this reason the definition of a scalar product on C_N must be done with some finesse.

Lemma 3.1. *A scalar product for vectors* $\mathbf{u}, \mathbf{v} \in C_N$, *i.e., vectors with complex elements having the same finite dimension, yields*

$$\langle \mathbf{u}, \mathbf{v} \rangle = \sum_i \mathbf{u}(i)^* \mathbf{v}(i) \tag{3.27}$$

where * *is the complex conjugate and* $\mathbf{u}(i)$, $\mathbf{v}(i)$ *are the elements of* $\mathbf{u}$, $\mathbf{v}$.

♦

We show the lemma by first observing that conditions 1–3 on scalar products are fulfilled because these properties follow from the definition of $\mathbf{u}$ and $\mathbf{v}$, and because $\mathbf{u}$ and $\mathbf{v}$ belong to the vector space. For any complex number $\mathbf{u}(i)$, the product

$\mathbf{u}(i)^*\mathbf{u}(i)$ is a real number which is strictly positive unless $\mathbf{u}(i) = 0$, in which case $\mathbf{u}(i)^*\mathbf{u}(i) = 0$. Accordingly,

$$\|\mathbf{u}\|^2 = \langle \mathbf{u}, \mathbf{u} \rangle = \sum_i |\mathbf{u}(i)|^2 \qquad (3.28)$$

is strictly positive except when $\mathbf{u} = \mathbf{0}$. Conversely, when $\|\mathbf{u}\| = 0$, the vector $\mathbf{u}$ equals the null vector so that even condition 4 on scalar products is fulfilled.

Evidently, this scalar product definition can be used also for E_N because E_N is a subset of C_N. Using this scalar product yields a natural generalization of the customary length in E_2 and E_3. Because of this, the norm associated with the scalar product is also called the Euclidean norm even if $\mathbf{u}$ is complex and has a dimension higher than 3. Another common name for this norm is the $\mathcal{L}^2$ norm.[2]

In applied mathematics Hilbert spaces are used for approximation purposes. A frequently used technique is to leave out some basis elements the space, that a priori are known to have little impact on the problem solution. To do similar approximations in image analysis, we need to introduce the concept of orthogonality.

3.5 Orthogonal Expansion

In our E_3 space, orthogonality is easy to imagine: Two vectors have the right angle ($\frac{\pi}{2}$) between them. Orthogonal vectors occur most frequently in human-made objects or environments. In a ceiling we normally have four corners, each being an intersection of three orthogonal lines. The books we read normally rectangular, and at each corner there are orthogonal vectors. Fish-sticks do not swim around in the ocean (perhaps because they have orthogonal vectors), but instead they are encountered in the human-made reality, etc.

In fact, orthogonality is even encountered in nonvisible spaces constructed by man, the Hilbert spaces. In these spaces, which include nonfinite dimensional spaces such as function spaces, one can express the concepts of distance and length. Additionally, one can express the concepts of orthogonality and angle, as we discuss below. First, let us introduce the definition of *orthogonality*.

Definition 3.3. *Two vectors are orthogonal if their scalar product vanishes*

$$\langle \mathbf{u}, \mathbf{v} \rangle = 0 \qquad (3.29)$$

In E_3, if we have three *orthogonal vectors* we can express all points in the space easily by means of these vectors, the basis vectors. This is one of the reasons why orthogonality appears in the human-made world, although it does not explain why the fish are, post mortem, forced to appear as rectangular (frozen) blocks. The other important reason, which does explain the fish-sticks, is that orthogonality reduces or eliminates redundancy. Humans can store and transport more fish if fish have corners with orthogonal vectors. For the same reason, pixels in images are quadratic and

---

[2] It is pronounced as "L two norm".

orthogonal functions that eliminate redundancies in small rectangular image patches have been constructed as one way to compress image data so that we can store and transport images efficiently (e.g., JPEG images).

Orthogonal sets are often used as bases; that is, they are used to synthesize as well as to analyze vector spaces. To illustrate this, we go back to our ordinary space of E_3 and ask ourselves the following: How do we find the coordinates of a point $\mathbf{x}$ if we have three orthogonal vectors $\mathbf{e}_1,\mathbf{e}_2,\mathbf{e}_3$ that we use as a basis? We do this by the so-called *projection* operation, which measures the "orthogonal shadow" of the vector representing the point (which starts at the origin and ends, at the point itself) on the three basis vectors. More formally, a projection is an operator that equals to identity if it is repeated more than once. Paraphrased, the resulting vector of a projection does not change by further projections. The point is projected orthogonally to define the tips of the three vectors that lie on the basis vectors, which when added vectorially end up with a result that is identical to the coordinate vector of the point, the one that goes from the origin to the point.

To teach projection to a computer we should have a more precise mathematical procedure of making projections than the term "orthogonal shadows". To do that we work backwards. We write the vector $\mathbf{x}$ as if we knew its coordinates or coefficients, c_1, c_2 and c_2 in the basis $\{\mathbf{e}_i\}$, that is

$$\mathbf{x} = \sum_i c_i \mathbf{e}_i \tag{3.30}$$

Now we take the scalar product of the left- as well as the right-hand side of this equation with one of the bases, e.g., $\mathbf{e}_1$.

$$\langle \mathbf{x}, \mathbf{e}_1 \rangle = \langle [\sum_i c_i \mathbf{e}_i], \mathbf{e}_1 \rangle = c_1 \langle \mathbf{e}_1, \mathbf{e}_1 \rangle + c_2 \langle \mathbf{e}_2, \mathbf{e}_1 \rangle + c_3 \langle \mathbf{e}_3, \mathbf{e}_1 \rangle \tag{3.31}$$

Since our basis vectors are orthogonal, the scalar products between different bases will vanish $\langle \mathbf{e}_1, \mathbf{e}_2 \rangle = \langle \mathbf{e}_1, \mathbf{e}_3 \rangle = \langle \mathbf{e}_2, \mathbf{e}_3 \rangle = 0$. So we have

$$\langle \mathbf{x}, \mathbf{e}_1 \rangle = c_1 \langle \mathbf{e}_1, \mathbf{e}_1 \rangle \implies c_1 = \langle \mathbf{x}, \mathbf{e}_1 \rangle / \langle \mathbf{e}_1, \mathbf{e}_1 \rangle \tag{3.32}$$

But $\langle \mathbf{e}_1, \mathbf{e}_1 \rangle$, which is the square of the length of $\mathbf{e}_1$, can be computed regardless of $\mathbf{x}$ since $\mathbf{e}_1$ is known. In other words, by using the scalar product operation, we have achieved projecting $\mathbf{x}$ on $\mathbf{e}_1$ since we could find c_1, which was the unknown coordinate scaling factor. If we could do it for $\mathbf{e}_1$, we should be able to do it for the remaining two coefficients as well. This is done without effort by just replacing the index 1 with the desired index, that is:

$$c_i = \langle \mathbf{x}, \mathbf{e}_i \rangle / \langle \mathbf{e}_i, \mathbf{e}_i \rangle \tag{3.33}$$

Eq. (3.33) is the reason for why the scalar product is often referred to as the *projection operator* because, except for the denominator, it is just a scalar product of any member $\mathbf{x}$ of the space, with the known basis vectors $\{\mathbf{e}_i\}_i$. The denominator does not spoil the equivalence of the scalar product to projection since the denominator is simply the square of the length of the basis vectors that can be computed ahead of the time, and in fact also be used to rescale $\{\mathbf{e}\}_i$. For this reason, it is justified to assume that the length of basis vectors are normalized to 1,

$$\hat{\mathbf{e}}_i = \mathbf{e}_i / \sqrt{\langle \mathbf{e}_i, \mathbf{e}_i \rangle} \qquad (3.34)$$

so that $\langle \hat{\mathbf{e}}_i, \hat{\mathbf{e}}_i \rangle = 1$ and the equation (3.33) reduces to

$$c_i = \mathbf{x}(i) = \langle \mathbf{x}, \hat{\mathbf{e}}_i \rangle \qquad (3.35)$$

The hat on $\hat{\mathbf{e}}_i$ is a way to tell that this vector has the length 1. Often, the hat is even omitted, when this fact is clear from the context.

Definition 3.4. *The vectors $\mathbf{u}_m$ and $\mathbf{u}_n$ are orthonormal if*

$$\langle \mathbf{u}_m, \mathbf{u}_n \rangle = \delta(m - n) \qquad (3.36)$$

where δ is the Kronecker delta *and is defined as*

$$\delta(m) = \begin{cases} 1, \textit{ if } m = 0; \\ 0, \textit{ otherwise.} \end{cases} \qquad (3.37)$$

In analogy with E_3, even in E_N (as well as in C_N) there must be exactly N vectors in order that they will be just sufficient, neither too many nor too few, to represent any point in E_N. Also in E_N, such a set of vectors is called a *basis*. If the basis vectors are orthogonal the basis is called an orthogonal basis.

3.6 Tensors as Hilbert Spaces

Discussed in linear algebra textbooks (e.g., [147, 210]) matrices are arrays with two indices having scalar (real or complex) elements. Because addition and scaling are well defined for matrices, we already know from Sect. 3.2 that they constitute a vector space. Furthermore, in Sect. 3.2 we saw that actually all arrays with multiple indices are vector spaces, and many of them correspond to real images. We only need a suitable scalar product to make such spaces Hilbert spaces. For simplicity we assume four indices below, but the conclusions are readily generalizable.

Lemma 3.2. *Let $\mathbf{A}$ and $\mathbf{B}$ be two arrays that have the same size. Then, with the definition*

$$\langle \mathbf{A}, \mathbf{B} \rangle = \sum_{k,l,m,n} \mathbf{A}(k, l, m, n)^* \mathbf{B}(k, l, m, n) \qquad (3.38)$$

the space of such arrays is a Hilbert space.

$$\blacklozenge$$

To show that this fulfills the conditions of a scalar product, one can proceed as in Eq. (3.27). This scalar product prompts the following norm:

$$\|\mathbf{A}\|^2 = \langle \mathbf{A}, \mathbf{A} \rangle = \sum_{k,l,m,n} |A(k, l, m, n)|^2 \qquad (3.39)$$

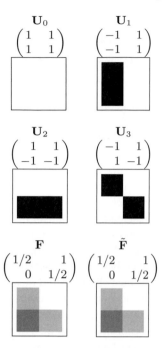

Fig. 3.5. The discrete image $\mathbf{F}$ can be expanded as the sum of the basis vectors $\mathbf{U}_0, \cdots, \mathbf{U}_3$ weighted with the expansion coefficients c_i. The image intensities, representing matrix values, vary uniformly from black to white as the matrix values change from -1 to 1. The matrix $\tilde{\mathbf{F}}$ is obtained by summing $c_i \mathbf{U}_i$ and should be identical to $\mathbf{F}$

Example 3.3. *Assume that we have a discrete image of size* 2×2

$$\mathbf{F} = \begin{pmatrix} 0.5 & 1 \\ 0 & 0.5 \end{pmatrix} \tag{3.40}$$

and the four orthogonal basis vectors as

$$\mathbf{U}_0 = \begin{pmatrix} 1 & 1 \\ 1 & 1 \end{pmatrix}, \quad \mathbf{U}_1 = \begin{pmatrix} -1 & 1 \\ -1 & 1 \end{pmatrix}, \quad \mathbf{U}_2 = \begin{pmatrix} 1 & 1 \\ -1 & -1 \end{pmatrix}, \quad \mathbf{U}_3 = \begin{pmatrix} -1 & 1 \\ 1 & -1 \end{pmatrix}$$

which are illustrated by Fig. 3.5.

We calculate the coefficients c_i *when* $\mathbf{F}$ *is expanded in the basis* $\mathbf{U}_i$ *as:*

$$\mathbf{F} = \sum_i c_i \mathbf{U}_i \tag{3.41}$$

The space of 2×2 *images is a Hilbert space with the scalar product given by Eq. (3.38). The coefficients are obtained according to*

$$c_i = \frac{\langle \mathbf{U}_i, F \rangle}{\langle \mathbf{U}_i, \mathbf{U}_i \rangle} \tag{3.42}$$

yielding:

$$\mathbf{c} = (2/4, 1/4, 1/4, 0)^T \tag{3.43}$$

By summing the basis matrices $\mathbf{U}_i$ weighted with the coefficients c_i, $\mathbf{F}$ can be reconstructed.

Tensors represent physical or geometric quantities, e.g., a force vector field in a solid, that are viewpoint-invariant in that they have invariant representations w.r.t. a coordinate frame. For practical manipulations they need to be represented in a coordinate system, however. The data corresponding to these representations are usually stored as arrays having multiple indices, e.g., as ordinary vectors and matrices of linear algebra, although their representation does not critically depend on the choice of the coordinate system more than a *change of basis*, the *viewpoint transformation*. In other words, the numerical representation of a tensor field given in a coordinate frame should be recoverable from its representation in another known frame. For example, the directions and the magnitudes of the internal forces in a body will be fixed with respect to the body, whereas the body itself and thereby the force field can be viewed in different coordinate systems. Once observed and measured in two different coordinate systems, the two measurements of the force fields will only differ by a viewpoint transformation and *nothing else* because the forces (which stayed fixed relative the body) are not influenced by the coordinate system. The principal property of the tensors is that the influence of the coordinate system becomes nonessential. The tensors look otherwise like ordinary arrays having multiple indices, e.g., matrices.

Geometric or physical quantities are characterized by the degrees of freedom they have. For example, the mass and the temperature are scalars having zero degrees of freedom. A velocity or a force is determined by an array of scalars that can be accessed with one index, a vector. Force and velocity accordingly have one degree of freedom. The polarization and the inertia are quantities that are described by matrices; therefore they require two indices.[3] They have then two degrees of freedom. The elements of arrays representing physical quantities are real or complex scalars and are accessed by using 1, 2, 3, 4, etc. indices. The number of indices or the degree of freedom is called the *order of a tensor*. Accordingly, the temperature, the velocity, and the inertia are tensors with the orders of 0, 1, and 2, respectively. First- and second-order tensors are arrays with 1 and 2 indices, respectively, but an array with 1 or 2 indices is a tensor if the array elements are not influenced by the observing coordinate system by more than a viewpoint transformation.

Tensors consitute Hilbert spaces too since their realizations are arrays. It is most interesting to study the first- and second-order tensors as being *tensor valued pixels* in the scope of this book. That is to say, apart from the space–time indices, which are the points where the tensor elements are either measured or computed, we will have at most two indices representing a tensor as a pixel value. For example, in a color image sequence there are four indices, the last of which represents the "color vector", which is encoded with one index. Further in the book, Sect. 9.6 and Chap.

---

[3] There are quantities that would require more than two indices, i.e., they are to be viewed as generalized matrices.

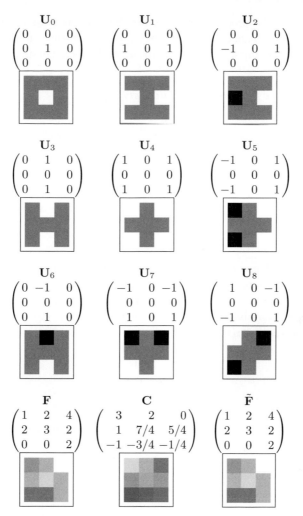

Fig. 3.6. The discrete image $\mathbf{F}$ can be expanded as the sum of the basis vectors $\mathbf{U}_0$, $\mathbf{U}_1$, $\cdots$, and $\mathbf{U}_8$ weighted with the expansion coefficients that are placed in the matrix $\mathbf{C}$ in a row major fashion, i.e., $c_0 = \mathbf{C}(1,2)$, $c_1 = \mathbf{C}(1,3)$, $c_2 = \mathbf{C}(1,3)$, $c_3 = \mathbf{C}(2,1)$ $\cdots$, etc. The image intensities, representing matrix values, vary uniformly from black to white as the matrix values change from -1 to 1, except in the *bottom row,* where the gray range from black to white represents the interval $[-4, 4]$. The matrix $\tilde{\mathbf{F}}$ is obtained by summing $c_i \mathbf{U}_i$ and should be identical to $\mathbf{F}$

10, we will discuss the *discrete local spectrum* and the *structure tensor* at each point of the image. These pixels will have arrays with two indices as values, *tensor values.*

Tensors constitute Hilbert spaces for which addition, scaling, and scalar product rules follow the same rules as arrays with multiple indices. The zeroth and the first-order tensor scalar products are therefore the same as those of the scalars and the vectors of linear algebra. Not surprisingly, the second-order tensors can also be equipped with a scalar product. We restate these results for the first- and second-order tensors.

Lemma 3.3. *With the following scalar products*

$$\langle \mathbf{u}, \mathbf{v} \rangle = \sum_k \mathbf{u}(k)^* \mathbf{v}(k) \tag{3.44}$$

$$\langle \mathbf{A}, \mathbf{B} \rangle = \sum_{kl} \mathbf{A}(k,l)^* \mathbf{B}(kl) \tag{3.45}$$

where k, l run over the components of the tensor indices, the tensors up to the second-order having the same dimension constitute Hilbert spaces.

♦

Example 3.4. *We attempt to expand 3×3 images in an orthogonal basis with matrix elements having ± 1 or 0, in an analogous manner as in Example 3.3. We note, however, that it is now more difficult to guess orthogonal matrices. We suggest "guessing" but only three variables instead of 9, by using the following observation.*
    Consider the vectors $\mathbf{u}(i)$ in E_3:

$$\mathbf{u}_0 = (0, 1, 0)^T$$
$$\mathbf{u}_1 = (-1, 0, 1)^T$$
$$\mathbf{u}_2 = (1, 0, 1)^T$$

which are orthogonal in the sense of the ordinary scalar product of first-order tensors, i.e.,

$$\langle \mathbf{u}_i, \mathbf{u}_j \rangle = \mathbf{u}_i^T \mathbf{u}_j = 0 \qquad \text{for} \qquad i \neq j \tag{3.46}$$

By using the following product based on the ordinary matrix product rule in linear algebra, tensor product, we can construct 3×3 matrices $\mathbf{U}_k$ that are orthogonal second-order tensors in the sense of the above scalar product (Eq. 3.45).

$$\mathbf{U}_k = \mathbf{u}_i \mathbf{u}_j^T \tag{3.47}$$

That the matrices $\mathbf{U}_k$ are orthogonal to each other follows by direct examination:

$$\langle \mathbf{u}_i \mathbf{u}_j^T, \mathbf{u}_{i'} \mathbf{u}_{j'}^T \rangle = \sum_m \sum_n \mathbf{u}_i(m) \mathbf{u}_j(n) \mathbf{u}_{i'}(m) \mathbf{u}_{j'}(n) \tag{3.48}$$

$$= (\sum_m \mathbf{u}_i(m) \mathbf{u}_{i'}(m))(\sum_n \mathbf{u}_j(n) \mathbf{u}_{j'}(n)) \tag{3.49}$$

$$= 0 \tag{3.50}$$

when $(i, j) \neq (i', j')$. Consequently, we can expand all 3×3 real matrices by using the scalar product in Eq. (3.45) and the basis $\mathbf{U}_k$.

In Fig. 3.6, we show the nine basis vectors obtained via Eq. (3.47), and the reconstruction coefficients of the example matrix:

$$\mathbf{F} = \begin{pmatrix} 1 & 2 & 4 \\ 2 & 3 & 2 \\ 0 & 0 & 2 \end{pmatrix} \tag{3.51}$$

The matrix is expanded in this basis and is reconstructed by a weighted summation of the same basis with the weighting coefficients c_i.

We generalize the result of the example as a theorem.

Theorem 3.1. *Let $\mathbf{u}_i \in E_N$ be orthogonal to each other, i.e., $\langle \mathbf{u}_i, \mathbf{u}_j \rangle = \mathbf{u}_i^T \mathbf{u}_j = 0$ for $i \neq j$. Then the (tensor) products of these, $\mathbf{U}_k = \mathbf{u}_i \mathbf{u}_j^T$, are also orthogonal to each other, i.e., $\langle \mathbf{U}_k, \mathbf{U}_l \rangle = \sum_{mn} \mathbf{U}_k(m,n) \mathbf{U}_l(m,n) = 0$ when $k \neq l$.*

We can construct images where image values are not gray values but tensors. A gray image can thus be viewed as a *field* with *zero-order tensors*. A color image can be viewed as a *field* having *first-order tensors* as values. In Sect. 10.3 we will discuss the direction as a *second-order tensor*. An image depicting the local direction will thus be a *field of second-order tensors*. Evidently, even fields of tensors are Hilbert spaces as they are multi-index arrays, which are Hilbert spaces. We restate this property for the sake of completeness.

Lemma 3.4. *With the following scalar products:*

$$\langle \mathbf{A}, \mathbf{B} \rangle = \sum_{ij} \sum_k \mathbf{A}(i,j,k)^* \mathbf{B}(i,j,k) \tag{3.52}$$

$$\langle \mathbf{A}, \mathbf{B} \rangle = \sum_{ij} \sum_{kl} \mathbf{A}(i,j,k,l)^* \mathbf{B}(i,j,k,l) \tag{3.53}$$

(where k, l run over the components of the tensor indices, and i, j run over the discrete points of the image) tensor fields up to the second degree having the same dimension constitute Hilbert spaces.

◆

3.7 Schwartz Inequality, Angles and Similarity of Images

We explore here the concept of "angle" which we already met, at least verbally in connection with the "orthogonality" which has visual conotations with the right angles. For this too, we need the scalar product. First, we present an important inequality, known as the Schwartz inequality.

Theorem 3.2 (Schwartz inequality I). *The* Schwartz inequality,

$$|\langle \mathbf{u}, \mathbf{v} \rangle| \leq \|\mathbf{u}\| \|\mathbf{v}\| \tag{3.54}$$

holds for Hilbert spaces.

◆

Exercise 3.6. *Prove that the Schwartz inequality holds for arrays with one or two indices.*

The theorem says

$$\frac{|\langle \mathbf{u}, \mathbf{v} \rangle|}{\|\mathbf{u}\|\|\mathbf{v}\|} \leq 1 \tag{3.55}$$

which in turn yields that

$$-1 \leq \frac{\langle \mathbf{u}, \mathbf{v} \rangle}{\|\mathbf{u}\|\|\mathbf{v}\|} \leq 1 \tag{3.56}$$

When the dimension of the vector space is large ($N > 3$) the concept of angle is difficult to imagine. The following definition helps to bridge this difficulty. The idea is to use the amplification factor that is needed to turn the Schwartz inequality into an equality to pinpoint the angle between two vectors. This is possible because the amplification factor is always in the interval $[-1, 1]$, as shown above.

Definition 3.5. *The angle φ between the two vectors $\mathbf{u}$ and $\mathbf{v}$ is determined via*

$$\cos \varphi = \frac{\langle \mathbf{u}, \mathbf{v} \rangle}{\|\mathbf{u}\|\|\mathbf{v}\|} \tag{3.57}$$

Angle computation requires a scalar product. If we have a Hilbert space, however, it comes with a method that produces the scalar product.

The Schwartz inequality, Eqs. (3.54-3.57), with a suitable scalar product e.g., (3.44)–(3.45) or (3.52)–(3.53) are extensively used in machine vision. In particular Eq. (3.57) is known as the *normalized correlation*, where $\mathbf{U}$ is a template or an image of an object that is searched for, and $\mathbf{V}$ is a test image which is usually a subpart of an image that is matched with the template. The direction cosine between the template and the test vector is measured by the expression in Eq. (3.57). When the two patterns are vectors that share the same direction, then the system decides for a match. This angular closeness is often tested by checking whether the absolute value of Eq. (3.57) is above a threshold, which is empirically decided e.g., by inspecting the image data manually or by using statistical tools such as a classifier or neural network.

Example 3.5. *Consider the photograph of the text shown in Fig. 3.7. The digital photographs of pages of books or manuscripts such as this can be conveniently stored in computers. Yet, to let computers process the stored data, e.g., to search for a particular word in such images, demands that the letters and words be located and recognized automatically by computers themselves. The goal is to automatically locate and identify all characters of the relevant alphabet in digital pictures of text. This is known as the* optical character recognition (OCR) *problem in image analysis. We use the Schwartz inequality to construct a simple OCR system for the displayed script as summarized next.*

Let $\mathbf{U}$ be the character to be searched for, and $\mathbf{V}$ be the test image that is any subimage of Fig. 3.7 having the same size as $\mathbf{U}$. The angle φ between $\mathbf{U}$ and $\mathbf{V}$ is determined via the equation

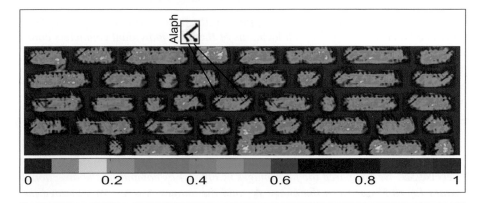

Fig. 3.7. On the *top*, a digital photograph of a text (Aramaic), in which individual characters are to be automatically identified, is displayed. In the *middle*, the directional cosines between the template of a character (alaph) and the local images is shown in color. At *bottom*, the identification result is shown for 3 letters where the same color represents the same letter

$$\cos(\varphi) = \frac{\langle \mathbf{U}, \mathbf{V} \rangle}{\|\mathbf{U}\| \cdot \|\mathbf{V}\|} \tag{3.58}$$

Because the values in the matrices are greater than or equal to zero, the interval for $\cos(\varphi)$ *will be* $[0, 1]$. *In the case when the scalar product* $\langle \mathbf{U}, \mathbf{V} \rangle$ *equals zero, the angle* φ *will be 90 degrees (orthogonal), i.e., least similar. By contrast, if they share direction, then we will have* $\mathbf{V} = \alpha \mathbf{U}$, *where* α *is a positive scalar. Because* $\langle \mathbf{U}, \mathbf{U} \rangle = \|\mathbf{U}\|^2$,

$$\cos(\varphi) = \langle \mathbf{U}, a\mathbf{U} \rangle / (\|\mathbf{U}\| * \|a\mathbf{U}\|) = a\langle \mathbf{U}, \mathbf{U} \rangle / (a\|\mathbf{U}\|^2) = 1, \qquad (3.59)$$

to the effect that **V** *is an ideal match if* $\cos(\varphi)$ *is 1.*

Since all occurrences of **U** in the manuscript must be tested, the norm has to be calculated for all (same size) subimages in the picture of the manuscript. The result of Eq. (3.58) is illustrated in Fig. 3.7 by showing the automatically found positions of three different characters, "alaph", "mim", and "tau". The more "redish" an image, the closer $\cos(\varphi)$ is to 1 at that location (middle). As example, two positions of "Alaph" character are indicated.

With the same threshold of 0.925 applied to the three directional cosine images (only one is shown in Fig. 3.7), all locations of the three individual characters could be found and marked with respective colors. An "alaph" template painted the black pixel values in the original image as red, if the directional cosine corresponding to the template center was above the threshold. A similar procedure was applied to other two templates to obtain the result at the bottom.

	$\mathbf{A}_2$	$\mathbf{A}_3$	$\mathbf{A}_2$
$\mathbf{A}_1$	0.844	0.778	0.826

Exercise 3.7. *We compared the four images in Fig. 3.4 by computing the directional cosines Eq. (3.57) between the image* $\mathbf{A}_1$ *and the images* $\mathbf{A}_2$-$\mathbf{A}_4$. *We obtained the directional cosines, listed above. Would you prefer the directional cosines (i.e., the Schwartz inequality ratios) to the triangle inequality ratios to measure similarities between images?*
HINT: Use average relative similarties in your judgement.

In summary, if we have a Hilbert space then we have a vector space on which we have a scalar product. The existence of a scalar product always allows us to derive a useful norm, the $\mathcal{L}^2$ norm. In turn, this norm allows to measure distances between points in the Hilbert space. Furthermore, we can also measure the angles between vectors by a scalar product.

4

Continuous Functions and Hilbert Spaces

We can reconstruct as well as decompose a discrete signal by means of a set of discrete signals that we called the basis in the previous chapter. Just as a vector in E_3 can be expressed as a weighted sum of three orthogonal vectors, so can functions defined on a continuous space be expressed as a weighted sum of orthogonal basis functions with some coefficients. This will be instrumental when modeling nondiscrete images. Examples of nondiscrete images are the images on photographic papers or films. Tools manipulating functions defined on a continium rather than on a discrete set of points are necessary because without them, understanding and performing many image processing operations would suffer, e.g., to enlarge or reduce the size of a discrete image. In this chapter we deliberately chose to be generic about functions as vector spaces. We will be more precise about the function families that are of particular interest for image analysis in the subsequent chapters. First, we will need to introduce functions, which are abstractions that define a rule, as points in a vector space. Then, we define addition and scaling to establish that functions are vector spaces. To make a Hilbert space of the function vector space we need a scalar product, the generic form of which will be presented in the subsequent section. Finally, we present orthogonality and the angle concept for nondiscrete images.

4.1 Functions as a Vector Space

If one scrutinizes the theory of *Hilbert space* that we have studied so far, one concludes that there are few difficult assumptions hindering its extension to cover quantities other than arrays of discrete scalars. Continuous functions are quantities that we will study in the framework of vector spaces. A major difference between the thus far discussed arrays and the function spaces is that the latter are abstract quantites that are continuous. A photograph is a *continuous function* defined on an ordinary plane. In that sense it is called a *continuous image*. Its discrete version, a digital image, is defined on a set of points on the same 2D plane. The continuous photograph assumes many more values because it is defined even between the set of points where the discrete image is undefined.

As in the case of E_3, we should get used to the idea that a function, e.g., $f(t)$, is a point in the function space. This is probably the hardest part of all, since many students do not have experience other than to imagine functions as curves placed over the x-axis or surfaces placed over an (x, y)-plane and so on. So, we need to imagine that the entire graph is a point among its likes in a function space. This idea is transmitted by writing a *function* as f, without its argument(s), which also serves to keep from becoming distracted by the (t) or whatever *function arguments* that usually follow f. If one has a cosine or arctan function, for example, these will be points in an appropriate function space. Observe that we are not changing the fact that a function is an abstract rule that tells how to produce a *function value* (e.g., the number that a function delivers) when coordinate arguments are given to it. When one has a unique rule (a function), then one has a unique point (in a vector space of functions to be discussed below) that corresponds to that function. If every point is a function then the origin of the function space must also be a function, which is perfectly true. The origin of the function space, the null function, is a function that represents a unique rule: "deliver the function value zero for all arguments". What is more, it is contained in all vector spaces of functions!

4.2 Addition and Scaling in Vector Spaces of Functions

The next step is to have a way to obtain a new "point" (function) by a rule of scaling, *function scaling*. This will yield a different function (rule) than the one we started with. Since we are looking for a rule, we have to define it as such and assign a new function symbol to it. Let us call it g:

$$g = \alpha f \tag{4.1}$$

We know g if we know a rule how to obtain its value when an argument is presented to it. Since we know the rule for f, (this is $f(t)$), the rule to obtain the value of g can be simply defined as to multiply the function value of f at t with the scalar α:

$$g(t) = \alpha f(t) \tag{4.2}$$

This is, of course, the way we are used to multiplying functions with scalars from calculus. But now we have additionally the interpretation that this is an operation making the corresponding vector "longer".

We also need a way to add two vectors in the function spaces. *Function addition* is also defined in the old fashion: The new rule h,

$$h = f + g \tag{4.3}$$

is obtained by using the already known rules regarding f and g:

$$h(t) = f(t) + g(t) \tag{4.4}$$

4.3 A Scalar Product for Vector Spaces of Functions

Finally, we need to introduce a *scalar product of functions*:

$$\langle f, g \rangle = \int f^* g \tag{4.5}$$

where the $\cdot^*$ on the top of g represents the complex conjugate operation. This definition resembles the scalar product for C_N, see Eq. (3.27). Instead of a summation over a discrete set, we integrate now over a continuum. The integration is taken over a suitable definition domain of the functions f and g, that is in practice a suitable range of the arguments of f and g. Integral operation can be seen as a "degenerate" form of summation when we have so many terms to sum that they are uncountably infinite. The idea of uncountably infinite needs some more explanation. Integers that run from 0 to ∞ are, as an example, infinite in number, yet they are countable (namable). That is, there is no integer between two successive integers since all integers can be named one after the other. By contrast, the real numbers are infinite and uncountable, because there is always another real number between two real numbers, no matter how close they are chosen in any imaginable process that will attempt to name them one after the other. So if we want to make summations over terms that are generated by real number "indices" (instead of integer indices), we need integration to sum them up because these "indices" are uncountably infinite. It is possible to define other scalar products as well, but for most signal analysis applications Eq. (4.5) will be a useful scalar product.

We can see that if an auto–scalar product of a vector in the Hilbert space is taken then the result is real and nonnegative. To be more precise, $\langle f, f \rangle \geq 0$, and equality occurs if and only if $f = 0$.

4.4 Orthogonality

With a scalar product we have the tool to test if functions are "orthogonal" in the same way as before. Two functions are said to be orthogonal if

$$\langle f, g \rangle = 0 \tag{4.6}$$

We can reconstruct and analyze a signal by means of a set of orthogonal basis signals. Just as a vector in E_3 can be expressed as a weighted sum of three orthogonal vectors, so can a function be expressed as a weighted sum of orthogonal basis functions with some "correctly" chosen coefficients. But in order to do that, we will need to be more precise about the Hilbert space, i.e., about our scalar product and functions that are allowed to be in our function space. In Chap. 5 we will give further details on these matters.

4.5 Schwartz Inequality for Functions, Angles

The Schwartz inequality in Sect. 3.7 was defined for conventional matrices and vectors. An analogue of this inequality for function spaces yields the following result:

Theorem 4.1 (Schwartz inequality II). *The* Schwartz inequality

$$|\langle f, g \rangle| \leq \|f\| \|g\| \tag{4.7}$$

where

$$\langle f, g \rangle = \int f^* g \tag{4.8}$$

holds for functions.

♦

Exercise 4.1. *Prove that the Schwartz inequality holds even for functions.*

Dividing both sides of inequality (4.7) and subsequently removing the magnitude operator yields

$$\frac{|\langle f, g \rangle|}{\|f\| \|g\|} \leq 1 \quad \Leftrightarrow \quad -1 \leq \frac{\langle f, g \rangle}{\|f\| \|g\|} \leq 1 \tag{4.9}$$

Accordingly, we can define an "angle" φ between the functions f and g as follows:

$$\cos(\varphi) = \frac{\langle f, g \rangle}{\|f\| \|g\|} \tag{4.10}$$

because the right-hand side of the equation varies continuously between -1 and 1 as $f/\|f\|$ changes. Notice that $\cos \varphi = 1$ exactly when $f/\|f\|$ equals $g/\|g\|$, whereas it equals -1 when $f/\|f\|$ equals $-g/\|g\|$.

5

Finite Extension or Periodic Functions—Fourier Coefficients

In this section we will make use of Hilbert spaces being a vector space in which a scalar product is defined. We will do this for a class of functions that are powerful enough to include physically realizable signals. The scalar product for this space will be made precise and utilized along with an orthogonal basis to represent the member signals of the Hilbert space by means of a discrete set of coefficients.

5.1 The Finite Extension Functions Versus Periodic Functions

We define the *finite extension functions* and the *periodic functions* as follows:

Definition 5.1. *A function f is a finite extension FE function if there exists a finite real constant T such that*

$$t \notin [-\frac{T}{2}, \frac{T}{2}] \quad \Rightarrow \quad f(t) = 0 \tag{5.1}$$

A function f is a periodic function if there is a positive constant T, called a period, such that $f(x) = f(x + nT)$ for all integers n.

The definitions for higher dimensions are analogous. Examples of finite extension signals include (i) a sound recording in which there is a time when the signal starts and another when it stops; (ii) a video recording which has, additionally, a space limitation representing the screen size; (iii) an (analog) photograph, which has a boundary outside of which there is no image. A periodic signal is a signal, that repeats copies of itself in a basic interval that is given by a fixed interval T. Because of this all real signals that are *finite extension signals* can be considered to be equivalent to *periodic signals* as well. The FE signals can be repeated to yield a periodic function, and periodic functions can be truncated such that all periods but one are set to zero. Accordingly, a study of periodic functions is also a study of FE functions, and vice versa. In this chapter we will use f to mean either finite extension or periodic.

Arguments of periodic functions are often measured or labelled in angles because after one period the function is back at the same point where it started, like in the trigonometric circle that represents the angles from 0 to 2π. This can always be achieved by stretching or shrinking the argument of the periodic function $f(t)$ to yield $t' = \frac{2\pi}{T}t$. In terms of physics, t' becomes a "dimensionless" quantity since it is an angle. The quantity

$$\omega_1 = \frac{2\pi}{T} \tag{5.2}$$

serves the goal that t' obtains values in $[0, 2\pi]$ when t is assigned values in $[0, T]$.

Theorem 5.1. *The space of periodic functions that share the same period is a Hilbert space with the scalar product:*

$$\langle f, g \rangle = \int_{-\frac{T}{2}}^{\frac{T}{2}} f^* g \tag{5.3}$$

♦

This is the scalar product that we have seen before, except that it has been given a precision with respect to function type and the integration domain. It is straightforward to extend it to higher dimensions where the integration domains will be squares, cubes, and hypercubes.

5.2 Fourier Coefficients (FC)

Since there are many types of orthogonal functions that are periodic, one speaks of *orthogonal function families*. A very useful orthogonal function family is the *complex exponentials*:

$$\psi_m(t) = \exp(i\omega_m t) \tag{5.4}$$

where

$$\omega_m = m\frac{2\pi}{T} \qquad \text{and,} \qquad m = 0, \pm 1, \pm 2, \pm 3 \cdots \tag{5.5}$$

In other words for each ω_m, we have a different member of the same family. The function family is then

$$\{\psi_m\}_m = \{\cdots, e^{-i2\omega_1 t}, e^{-i\omega_1 t}, 1, e^{i\omega_1 t}, e^{i2\omega_1 t}, \cdots\} \tag{5.6}$$

This is also called the *Fourier basis*, because the set acts as a basis for function spaces and defines the Fourier transform, Sect. 6.1. Here, we discuss the Fourier series as an intermediary step.

We can verify that the members of the Fourier basis are orthogonal via their mutual scalar products. Assuming that $m \neq n$, we can find the primitive function to the exponential easily and obtain

$$\langle \psi_n, \psi_m \rangle = \int_{-\frac{T}{2}}^{\frac{T}{2}} \exp(-in\omega_1 t) \exp(im\omega_1 t) dt$$

$$= \left[\frac{1}{i(m-n)\omega_1} \exp(i(m-n)\omega_1 t) \right]_{t=-\frac{T}{2}}^{t=\frac{T}{2}}$$

$$= \frac{1}{i(m-n)\omega_1} [\exp(i(m-n)\omega_1 T) - \exp(i0)] = 0 \qquad (5.7)$$

where $m - n \neq 0$. Now we also need to find the norms of these basis vectors, and therefore we assume that $m = n$.

$$\langle \psi_m, \psi_m \rangle = \|\Psi_m\|^2 = \int_{-\frac{T}{2}}^{\frac{T}{2}} \exp(-im\omega_1 t) \exp(im\omega_1 t) dt$$

$$= \int_{-\frac{T}{2}}^{\frac{T}{2}} 1 dt = T \qquad (5.8)$$

The norm of the complex exponential is consequently independent of m. That is, any member of this orthogonal family has the same norm as the others. This result is not obvious, but thanks to the scalar product we could reveal it conveniently. Accordingly, we obtain:

$$\langle \psi_n, \psi_m \rangle = T\delta(m-n) \qquad (5.9)$$

where δ is the Kronecker delta, see Eq. (3.37). Since complex exponentials are orthogonal and we know their norms, any function f that has a limited extension T can be reconstructed by means of them as:

$$f(t) = \sum_m c_m \exp(i\omega_m t) \qquad (5.10)$$

where c_ms are complex-valued scalars. An arbitrary coefficient $F(n)$ can be determined by taking the scalar product of both sides of Eq. (5.10) with $\exp(i\omega_n t)$:

$$\langle \exp(i\omega_n t), f \rangle = \langle \exp(i\omega_n t), \sum_m c_m \exp(i\omega_m t) \rangle$$

$$= \sum_m c_m \langle \exp(i\omega_n t), \exp(i\omega_m t) \rangle$$

$$= \sum_m c_m T\delta(m-n) = c_n T \qquad (5.11)$$

yielding,

$$c_n = \frac{1}{T} \langle \exp(i\omega_n t), f \rangle = \frac{1}{T} \int_{-\frac{T}{2}}^{\frac{T}{2}} f(t) \exp(-i\omega_n t) dt \qquad (5.12)$$

Consequently, c_ns are the *projection coefficients* of f on the Fourier basis. They are also known as the *Fourier coefficients* (FCs). We summarize this result in a slightly different way in the next theorem, to facilitate the derivation of the Fourier transform in Sect. 6.1.

Theorem 5.2 (FC I). *There exists a set of scalars $F(\omega_m)$ which can synthesize a function $f(t)$ having the finite extension T,*

$$F(\omega_m) = \frac{1}{2\pi} \int_{-\frac{T}{2}}^{\frac{T}{2}} f(t) \exp(-i\omega_m t)dt \qquad \text{(Analysis)} \qquad (5.13)$$

such that

$$f(t) = \frac{2\pi}{T} \sum_m F(\omega_m) \exp(i\omega_m t) \qquad \text{(Synthesis)} \qquad (5.14)$$

using $\omega_m = m\frac{2\pi}{T}$ and $m = 0, \pm 1, \pm 2, \pm 3 \cdots$.

♦

Exercise 5.1. *Prove the theorem.*
HINT: Expand the function $\frac{T}{2\pi}f$ in the Fourier basis.

While this is nothing but a convenient way of computing projection coefficients in a function space where the functions are interpreted as points, few formulas have had as much practical impact on science as these two, including signal analysis, physics, and chemistry. Equations (5.13-5.14) are the famous Fourier[1] series. They will be generalized further below to yield the usual Fourier transform on $[0, \infty]$. These formulas can be seen as a transform pair. In that sense the coefficients:

$$c_m = \frac{2\pi}{T} F(\omega_m) \qquad (5.15)$$

constitute a *unique* representation of f, and vice versa.

It is worth noting that, per construction, the complex exponentials also constitute a "basis" for sine and cosine functions whose periods exactly fit the basic period one or more (integer) times. We can sometimes find the *projection coefficients* even without integration since FCs are unique. We can, for example, write the sine and cosine as weighted sums of complex exponentials by using the Euler formulas:

$$\cos(\omega_1 t) = \frac{1}{2}\exp(i\omega_1 t) + \frac{1}{2}\exp(-i\omega_1 t) = \frac{1}{2}\psi_1 + \frac{1}{2}\psi_{-1} \qquad (5.16)$$

$$\sin(\omega_1 t) = \frac{1}{2i}\exp(i\omega_1 t) - \frac{1}{2i}\exp(-i\omega_1 t) = \frac{1}{2i}\psi_1 - \frac{1}{2i}\psi_{-1} \qquad (5.17)$$

The scalars in front of each orthogonal function ψ. are then the projection coefficients c_ms, very much like the coordinates of the ordinary vectors in E_3. They are also referred to as *coordinates*.

Exercise 5.2. *Show that $\langle\cos(\omega_1 t), e^{i\omega_1 t}\rangle/T = \frac{1}{2}$*

---

[1] J. B. J. Fourier, 1768–1830, French mathematician and physicist.

5.3 (Parseval–Plancherel) Conservation of the Scalar Product

We consider a scalar product between two members of our Hilbert space of periodic functions, $f = \frac{2\pi}{T} \sum_m F(\omega_m)\psi_m$ and $g = \frac{2\pi}{T} \sum_m G(\omega_m)\psi_m$:

$$
\begin{aligned}
\langle f, g \rangle &= \left(\frac{2\pi}{T}\right)^2 \langle \sum_m F(\omega_m)\psi_m, \sum_n G(\omega_n)\psi_n \rangle \\
&= \left(\frac{2\pi}{T}\right)^2 \sum_{mn} F^*(\omega_m)G(\omega_n)\langle \psi_m, \psi_n \rangle \\
&= \left(\frac{2\pi}{T}\right)^2 \sum_{mn} F^*(\omega_m)G(\omega_n)T\delta(m-n) \\
&= \frac{(2\pi)^2}{T} \sum_m F^*(\omega_m)G(\omega_m)
\end{aligned}
\tag{5.18}
$$

To obtain Eq. (5.18), we changed the order of the summation and the integration (of the scalar product). This is allowed for all functions f and g that are physically realizable. The $\delta(m-n)$ is the Kronecker delta whereby we obtained the last equality. Notice that the Kronecker delta reduced the double sum to a single sum by replacing n with m everywhere before it disappeared, a much appreciated behavior of δ under summation. We explain the reason for this "sum-annihilating" property of the Kronecker delta. Assume that we have written down all the terms one after the other,

$$
F^*(\omega_1)G(\omega_1)\delta(1-1), \quad F^*(\omega_1)G(\omega_2)\delta(1-2), \quad F^*(\omega_1)G(\omega_3)\delta(1-3)\cdots
\tag{5.19}
$$

The only terms that are nonzero are those when both indices are equal, $n = m$. Accordingly, we can reach all terms that are nonzero in the sum by using a single index in a single sum. Behaviorally, this is the same as saying that δ "replaces" one of its indices in every other terms such that the argument of the δ becomes zero and it "erases" the sum corresponding to the disappeared index, before vanishing itself.

The set of the discrete Fourier coefficients

$$
\{F(\omega_m)\}_m = (\cdots F(\omega_{-2}), F(\omega_{-1}), F(\omega_0), F(\omega_1), F(\omega_2)\cdots)
\tag{5.20}
$$

can be viewed as a vector with an infinite number of elements. Members constitute a faithful and unique representation of f because of the synthesis formula, Eq. (5.14). Can such infinite dimensional vectors be considered a Hilbert space on their own account? Yes, indeed. Scaling and addition are extended versions of their counterparts in finite dimensional vector spaces:

$$
(\cdots \alpha F(\omega_{-1}), \alpha F(\omega_0), \alpha F(\omega_1), \cdots) = \alpha(\cdots F(\omega_{-1}), F(\omega_0), F(\omega_1), \cdots)
\tag{5.21}
$$

and

$$
\begin{aligned}
&(\cdots F(\omega_{-1}) + G(\omega_{-1}), F(\omega_0) + G(\omega_0), F(\omega_1) + G(\omega_1), \cdots) \\
&= (\cdots F(\omega_{-1}), F(\omega_0), F(\omega_1), \cdots) + (\cdots G(\omega_{-1}), G(\omega_0), G(\omega_1), \cdots)
\end{aligned}
$$

As scalar product, the sum

$$\langle \{F(\omega_m)\}_m, \{G(\omega_n)\}_n \rangle = \sum_m F^*(\omega_m) G(\omega_m) \tag{5.22}$$

which we already obtained from Eq. (5.18), can be utilized. That this is a scalar product is readily verified, given the sum defining the scalar product *converges*. In other words, the more terms used in the sum, the closer it will get to the one and same converged quantity. *Convergence* is not a problem for physically realizable signals. Using Eq. (5.18), we state the *Parseval–Plancherel theorem* for the FC transform:

Theorem 5.3 (Parseval–Plancherel FC). *The scalar products are conserved under the FC transform*

$$\langle f, g \rangle = \frac{(2\pi)^2}{T} \langle \{F(\omega_m\}_m, \{G(\omega_n\}_n \rangle \tag{5.23}$$

◆

As a special case, we obtain the conservation of the norms (Energies).

$$\|f\|^2 = \frac{(2\pi)^2}{T} \sum_m |F(\omega_m)|^2 = \frac{(2\pi)^2}{T} \|\{F(\omega_m)\}_m\|^2 \tag{5.24}$$

This is analogous to the 3D Euclidean space where the angle between two vectors is independent of the particular orthogonal coordinate system in use. We will discuss this further in Sect. 10.3, because in image analysis the interpretation of this theorem has important applicatons.

It can be shown that the function series in Eq. (5.14) always converges for the functions f that we will be interested in (physically realizable functions that have finite extension) so that the sum on the right side always produces the function value on the left side at all points t (*strong convergence*, or *pointwise convergence*). This implies that the Fourier coefficients constitute a decreasing sequence:

$$\lim_{m \to \infty} F(\omega_m) = 0 \tag{5.25}$$

Accordingly, the sum in equation Eq. (5.14) can be truncated after a finite number of terms without much harm to the original signal. This trick is often used in applications such as image enhancement or image compression, since the high-index Fourier coefficients correspond to complex exponentials (sinusoids) with high spatial frequencies that are likely to be noise. We will even use it to justify the theory of a fully discrete version of the FC transform, DFT, in Sect. 6.3

5.4 Hermitian Symmetry of the Fourier Coefficients

The Fourier series equations in Eqs. (5.14) and Eq. (5.13) are also valid for functions f that are complex-valued, since our scalar product can cope with complex functions from the beginning. However, the signals of the real world are real, even if humans put two real signals together and interpret them as the real and the imaginary parts of a complex-valued signal, e.g., the electromagnetic signals. Yet the Fourier coefficients of real signals as well as complex signals are complex-valued. Since there is much more information in a complex signal than in a real signal, there must be some redundancy in the Fourier coefficients of the real signals. Here we will bring further precision to this redundancy.

A real signal is the complex conjugate of itself since the imaginary part is zero,

$$f = f^* \tag{5.26}$$

The Fourier series reconstruction of the signal f yields

$$f(t) = \sum_m F(\omega_m) \exp(im\omega_1 t) \tag{5.27}$$

so that its complex conjugate becomes

$$f(t)^* = \sum_m F^*(\omega_m) \exp(-im\omega_1 t) \tag{5.28}$$

Replacing the index m with $-m$, we obtain

$$f(t)^* = \sum_{m=\infty}^{-\infty} F^*(\omega_{-m}) \exp(im\omega_1 t) \tag{5.29}$$

In doing so, notice that the summation order of m has been reversed, i.e., m runs now from positive to negative integers. But because the relative signs of m in the summands are unchanged, i.e., m in $F(\omega_m)$ with respect to the one in $\exp(im\omega_1 t)$, we have the same terms in the sum as before. However, summing in one direction or the other does not change the total sum, so we are allowed to reverse the direction of the summation back to the conventional direction:

$$f(t)^* = \sum_{m=-\infty}^{\infty} F^*(\omega_{-m}) \exp(im\omega_1 t) \tag{5.30}$$

Accordingly, the Fourier coefficients of f^* will fulfill

$$f^* = \{F^*(\omega_{-m})\}_m \tag{5.31}$$

Now using Eq. (5.26) yields $f = f^*$ to the effect that

$$F(\omega_m) = F^*(\omega_{-m}) \tag{5.32}$$

because the Fourier basis ψ_m is an orthogonal basis set yielding unique FCs. This property of the FCs, which is only valid for real functions f, is called *Hermitian symmetry*. Because of this, the Fourier coefficients of a real signal contains redundancy. If we know a coefficient, say $F(\omega_{17}) = 0.5 + i0.3$, then the mirrored coefficient is locked to its conjugate, i.e., $F(\omega_{-17}) = 0.5 - i0.3$.

The coefficient $F(\omega_m)$ (as well as $F^*(\omega_{-m})$) is complex and can be written in terms of its real and imaginary parts

$$F(\omega_m) = \Re(F(\omega_m)) + i\Im(F(\omega_m)) \tag{5.33}$$

so that

$$F^*(\omega_{-m}) = \Re(F^*(\omega_{-m})) + i\Im(F^*(\omega_{-m})) = \Re(F(\omega_{-m})) - i\Im(F(\omega_{-m}))$$

Thus, Eq. (5.32) can be rewritten as

$$\Re(F(\omega_m)) = \Re(F(\omega_{-m})), \quad \text{and} \quad \Im(F(\omega_m)) = -\Im(F(\omega_{-m})). \tag{5.34}$$

We summarize our finding on the redundancy of the FC coefficents of f as follows.

Theorem 5.4. *The Fourier coefficients of a real signal have* Hermitian symmetry:

$$F(\omega_m) = F^*(\omega_{-m}) \tag{5.35}$$

so that the real and imaginary parts of their FCs are even and odd, respectively, whereas their magnitudes are even.

♦

6

Fourier Transform—Infinite Extension Functions

In the previous chapter we worked with finite extension functions and arrived at useful conclusions for signal analysis. In this chapter we will lift the restriction on finiteness on the functions allowed to be in the Hilbert space. As a result, we hope to obtain similar tools for a wider category of signals.

6.1 The Fourier Transform (FT)

If f is not a finite extension signal, we still would like to express it in terms of the complex exponentials basis via the synthesis Eq. (5.14), and the analysis Eq. (5.13) formulas. To find a workaround, we will express a part of such a function in a finite interval and then let this interval grow beyond all bounds. To be precise, we will start with the finite interval $[-T/2, T/2]$, and then let T go to infinity (Fig. 6.1) Synthesizing the function f on a finite interval by complex exponentials amounts to setting the function values to zero outside the interval, knowing that we will only be able to reconstruct the signal inside of it. Outside of $[-T/2, T/2]$, the synthesis delivers a repeated version of the (synthesized) function inside the interval. We call the function restricted to the finite interval f_T, to put forward that it is constructed from a subpart of f, and that it converges to f when T approaches, ∞:

$$f = \lim_{T \to \infty} f_T \tag{6.1}$$

The symmetric interval is a trick that will enable us to control both ends of the integration domain with a single variable T. By sending T to infinity both ends of the integration domain will approach infinity while the function f_T will be replaced by f.

We first restate our scalar product, Eq. (5.3), for convenience:

$$\langle f, g \rangle = \int_{-T/2}^{T/2} f^* g \tag{6.2}$$

The vector space of periodic functions or the space of FE functions having the extension T constitutes a Hilbert space, with the above definition of the scalar product.

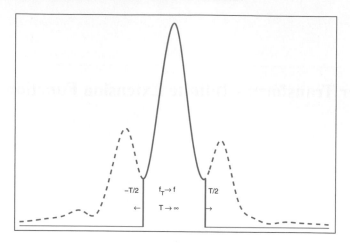

Fig. 6.1. The graph drawn *solid* illustrates f_T. As T grows beyond every bound, f_T will include the *dashed graph* and will equal f

Exercise 6.1. *Show that (i) complex exponentials are orthogonal under any scalar product taken over an interval with the length of the basic period. (ii) The norms of the complex exponentials are not affected by the shift of the interval over which the scalar product is taken. (iii) Show the results in (i) and (ii) for 2D functions, i.e.,* $C \exp(i\omega_x x + i\omega_y y)$.

Remembering that

$$\omega_m - \omega_{m-1} = m\frac{2\pi}{T} - (m-1)\frac{2\pi}{T} = \frac{2\pi}{T} \qquad (6.3)$$

we restate the synthesis formula of Eq. (5.14) as

$$f_T(t) = \sum_m F(\omega_m) \exp(i\omega_m t)(\omega_m - \omega_{m-1}) \qquad (6.4)$$

and the analysis formula of Eq. (5.13) as

$$F(\omega_m) = \frac{1}{2\pi} \int_{-T/2}^{T/2} f_T(t) \exp(-i\omega_m t)dt \qquad (6.5)$$

Passing to the limit with $T \to \infty$, we observe that $\omega_m - \omega_{m-1} = \frac{2\pi}{T}$, which is the distance between two subsequent elements of the equidistant *discrete grid*:

$$\cdots \omega_{-2}, \omega_{-1}, \omega_0, \omega_1, \omega_2 \cdots , \qquad (6.6)$$

approaches to zero,

$$\frac{2\pi}{T} = \omega_m - \omega_{m-1} = \Delta\omega \to 0 \qquad (6.7)$$

Under the passage to the limit, the discrete sequence of $F(\omega_m)$ will approach a function $F(\omega)$ that is continuous in ω almost everywhere[1]. This is because at any fixed ω, $F(\omega)$ can be approximated by $F(\omega_m)$ as (i) for some m, the difference between $\omega - \omega_m$ will be negligible, and (ii) the analysis formula, Eq. (5.13), exists almost everywhere when ω_m changes. Thus, the synthesis formula, Eq. (6.4), approximates the "measure/area" (integral) of the limit function $F(\omega)\exp(i\omega t)$ better and better, to the effect that $\sum$ will be replaced by $\int$ while $\Delta\omega$ will be replaced by $d\omega$. The analysis formula remains as an integral, but the integration domain approaches the entire real axis, $[-\infty, \infty]$. In consequence, f_T will approximate f, the function that we started with, ever better. Because f is integrable, the analysis formula will converge as T increases. As mentioned, the value ω_m is a constant in the analysis integral and it can therefore be replaced by the fixed point ω, yielding $F(\omega)$. We formulate these results in the following theorem:

Theorem 6.1 (FT). *Provided that the integrals exist, the integral transform pair*

$$\mathcal{F}(f)(\omega) = F(\omega) = \frac{1}{2\pi} \int_{-\infty}^{\infty} f(t)\exp(-i\omega t)dt \qquad \text{(Forward) FT} \qquad (6.8)$$

$$\mathcal{F}^{-1}(F)(t) = f(t) = \int_{-\infty}^{\infty} F(\omega)\exp(i\omega t)d\omega \qquad \text{(Inverse) FT} \qquad (6.9)$$

defines the Fourier transform *(FT) and the* Inverse Fourier transform, *which relate a pair of complex-valued functions $f(t)$ and $F(\omega)$, both defined on the domain of the real axis. The function F is called the Fourier Transform of the function f.*

$\blacklozenge$

As their counterparts in the FC transform, the FT and inverse FT equations are also referred to as *analysis formula* and *synthesis formula*, respectively. The pair (f, F) is commonly referred to as the FT pair. Because the forward FT is the same as the inverse FT except for a reflection in the origin, i.e., $\mathcal{F}(f)(\omega) = \mathcal{F}^{-1}(f)(-\omega)$, the forward FT can be used to implement the inverse FT in practice. Reflection can be implemented by reordering the result. Notwithstanding its simplicity, this observation is useful when practicing the FT and results in the following convenience.

Lemma 6.1 (Symmetry of FT). *If*

$$(f(t), F(\omega)) \qquad (6.10)$$

is an FT pair, so is

$$(F(\omega), f(-t)) \qquad (6.11)$$

The conclusions and principles established assuming the forward direction are also valid in the inverse direction.

$\blacklozenge$

---

[1] Almost everywhere means everywhere except for a set of points with zero measure. See [226] for further reading on measure and integral.

Fig. 6.2. The image on the *right* is obtained by applying the forward FT to the image on the *left* two times. The *red lines* are added for comparison convenience

Example 6.1. *First, we illustrate the symmetry of FT by showing how one would need a simple rearrangement to make the ordinary FT a replacer of the inverse FT. Subsequently, we use the same lemma to establish a third version of the FC theorem.*

- *Figure 6.2 represents a forward FT applied to an image twice. The second FT acts as an inverse FT, except for a reflection. The FT has been applied to all three color components, in RGB.*

Because of the symmetry of FT, the conclusions on FT remain valid even if the roles of $f(t)$ and $F(\omega)$ are interchanged. The FC theorem can therefore be formulated for limited frequency functions.

Theorem 6.2 (FC II). *There exists a set of scalars $f(t_m)$ that can synthesize a function $F(\omega)$ having the finite extension of Ω,*

$$F(\omega) = \frac{1}{\Omega} \sum_m f(t_m) \exp(-it_m\omega), \qquad (Synthesis) \qquad (6.12)$$

where

$$f(t_m) = \int_{-\frac{\Omega}{2}}^{\frac{\Omega}{2}} F(\omega) \exp(it_m\omega)d\omega, \qquad (Analysis) \qquad (6.13)$$

using $t_m = m\frac{2\pi}{\Omega}$.

♦

Exercise 6.2. *Prove theorem 6.2.*
HINT: Expand $\Omega F(\omega)$ in $\exp -it_m\omega$.

6.2 Sampled Functions and the Fourier Transform

Via theorem 5.2, we established that there exists a unique set of values capable of reconstructing any $f(t)$ by means of complex exponentials, provided that f has an extension T that is finite. This technique relies heavily on the fact that T is finite, and therefore we can make copies of f to obtain a periodic function. This function can in turn be expanded in Fourier series by use of a scalar product defined on a period, e.g., $[0, T]$. In Eq. (5.13), we obtained the discrete set of values used in the reconstruction. We viewed these values as an equidistant *sampling* of a continuous function, although we did not speculate how this function of "imagination" behaved between the grid points. As long as the function values turn out to be the correct values on the grid, we can always take this view, by letting the values of F between the discrete points $\omega_m = m\frac{2\pi}{T}$ vary according to some rule of our choice.

However, theorem 6.1 (FT) states that we can reconstruct f even if it is not a finite extension signal, albeit that the reconstruction now cannot be achieved by using a discrete sequence of values $F(\omega_m)$, but instead a continuous function $F(\omega)$. By using the FT theorem we are now, in fact, capable of reconstructing f even *without* periodizing it. Assuming that the signal f is nonzero in the symmetric interval $[-T/2, T/2]$, the FT of f is:

$$F(\omega) = \frac{1}{2\pi} \int_{-\infty}^{\infty} f(t) \exp(-i\omega t)dt = \frac{1}{2\pi} \int_{-T/2}^{T/2} f(t) \exp(-i\omega t)dt \qquad (6.14)$$

Eq. (6.14) tells us that not only the values of F are unique on the grid ω_m, but also in the continuum between the grid points! The conclusion must be that there exists a continuous function $F(\omega)$ that can be represented via its discrete values, also referred to as *discrete samples* or *samples*, to such an extent that even the function values between the discrete grid points are uniquely determined by the simple knowledge of its discrete values on the grid ω_m. In other words, the discrete samples $F(\omega_m)$ constitute not only an exact representation of the periodized original signal f, but also an exact representation of $F(\omega)$, which is in turn an exact representation of the original (unperiodized) function f. Notice that $F(\omega)$ is the Fourier transform of the limited extension function f (without periodization).

How do we estimate (interpolate) $F(\omega)$ when we only know its values at discrete points? The answer to this is obtained by substituting Eq. (5.14) in Eq. (6.14).

$$
\begin{aligned}
F(\omega) &= \frac{1}{2\pi} \int_{-T/2}^{T/2} f(t) \exp(-i\omega t)dt \\
&= \frac{1}{2\pi} \int_{-T/2}^{T/2} \left(\sum_m F(\omega_m)\frac{2\pi}{T} \exp(i\omega_m t) \right) \exp(-i\omega t)dt \\
&= \sum_m F(\omega_m)\frac{2\pi}{T}\frac{1}{2\pi} \int_{-T/2}^{T/2} \exp(-i(\omega - \omega_m)t)dt \\
&= \sum_m F(\omega_m)\frac{1}{T} \int_{-T/2}^{T/2} \exp(-i(\omega - \omega_m)t)dt \\
&= \sum_m F(\omega_m)\mu(\omega - \omega_m) \qquad (6.15)
\end{aligned}
$$

where

$$\mu(\omega) = \frac{1}{T} \int_{-T/2}^{T/2} \exp(-i\omega t) dt \qquad (6.16)$$

Clearly $F(\omega)$ is reconstructed from its samples $F(\omega_m)$ by using these as weight functions for a series of a displaced versions of μ, the *interpolation function*. We can identify μ in Eq. (6.16) as the Fourier transform of a piecewise constant function $\chi_T(t)$, which is constant inside the interval where f is nonzero, and zero elsewhere.

$$\chi_T(t) = \begin{cases} 1, & \text{if } t \in [-\frac{T}{2}, \frac{T}{2}]; \\ 0, & \text{otherwise.} \end{cases} \qquad (6.17)$$

We will refer to χ_T as the *characteristic function* of the interval $[-\frac{T}{2}, \frac{T}{2}]$. The relationship in Eq. (6.16) along with the definition in Eq. (6.17) reveals that

$$\mu(\omega) = \frac{1}{T} \int_{-\infty}^{\infty} \chi_T(t) \exp(-i\omega t) dt = \frac{2\pi}{T} \mathcal{F}(\chi_T)(\omega) \qquad (6.18)$$

Computing the integral in Eq. (6.16) in a straightforward manner yields

$$\mu(\omega) = \frac{1}{T} \left[\frac{\exp(-i\omega t)}{-i\omega} \right]_{t=-T/2}^{t=T/2} \qquad (6.19)$$

$$= \frac{1}{T} \frac{1}{\omega} 2 \sin(\omega \frac{T}{2}) \qquad (6.20)$$

In consequence, defining the *sinc function* as

$$\text{sinc}(\omega) = \frac{\sin \omega}{\omega} \qquad (6.21)$$

yields the interpolator function

$$\mu(\omega) = \text{sinc} \left(\frac{T}{2} \omega \right) \qquad (6.22)$$

which, using Eq. (6.18), results in

$$\mathcal{F}(\chi_T)(\omega) = \frac{T}{2\pi} \text{sinc} \left(\frac{T}{2} \omega \right) \qquad (6.23)$$

A characteristic function and its corresponding interpolation function is illustrated in Fig. 6.3. Before we discuss the properties of the sinc functions further, we present a few observations and elucidate the concept of functions with finite frequency extension, also known as band-limited functions.

- The sampling interval for F is given by $\omega_m - \omega_{m-1} = \frac{2\pi}{T}$, so that if the distance between the samples defines the unit measurement, $\omega_m - \omega_{m-1} = 1$, then T's value is locked to $T = 2\pi$. If another measurement unit is used to quantify the length of the interval that samples F, alternatively the length of the period of f, an ordinary scaling of the arguments will restore the correspondence. Because of this, it is customary to assume that a finite extension 1D function is nonzero in $[-\pi, \pi]$, and the sampling interval of its FT is equal to unit length.

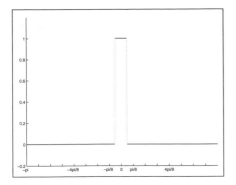 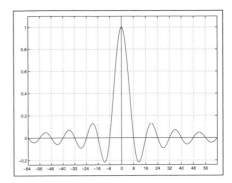

Fig. 6.3. The graph shows the finite extension of one domain that matches a particular interpolation function in the other domain where the domains are, interchangeably, the time and the frequency domains. Notice that the interpolator (*sinc function*) crosses zero only at every eighth integer, while its total extension (in the other domain) is $\frac{2\pi}{8}$

- The function $F(\omega)$ can be represented exactly by means of its samples on an integer grid, if $f(t)$ vanishes outside an interval having length $T = 2\pi$.

Functions with Finite Frequencies or Band-limited Functions

By using the symmetry of the Fourier transform, we can switch the roles of F and f while retaining the above results. In other words, we assume F to have a limited extension Ω, outside of which it will vanish (and therefore can be periodized). We will refer to such functions as finite frequency (FF) functions, which are also known as *band-limited functions*. Consequently, the continuous function $f(t)$ can be synthesized faithfully by means of its samples. This combination, limited extension F in the ω-domain and discrete f in the t-domain, is the most commonly encountered case in signal analysis applications. Evidently, the roles of μ and χ are also switched to the effect that μ now interpolates the samples of f whereas χ defines the extension of F. As discussed previously, the sampling interval of f equals 1 if the extension of F is $\Omega = 2\pi$.

Properties of Sinc Functions

We note that the *sinc function*,

$$\mathrm{sinc}(t) = \frac{\sin t}{t} \tag{6.24}$$

is a continuous function, even at the origin, where it attains the maximum value 1. In fact, it is not only continuous at the origin, but it is also *analytic* there (and everywhere), meaning that all orders of its derivatives exist continuously yielding a "smooth" function. As has been observed already, a scaled version of sinc yields the interpolation function μ

$$\mu(t) = \text{sinc}(\pi t) = \frac{\sin(\frac{\Omega}{2}t)}{\frac{\Omega}{2}it}. \tag{6.25}$$

If Ω, the extension or the support of $F(\omega)$ where it is nonzero, is an integer share of 2π, i.e., $\Omega = 2\pi/\kappa$ for some integer κ, then the interpolator μ is a sinc function in the t-domain with zero crosses at every κth grid point. If $\kappa = 1$, the zero crosses will occur at every integer (except at the origin). We illustrate a χ, μ pair with $\kappa = 8$ in Fig. 6.3. As a consequence of the previous discussion, shifting the above sinc function yields a series of functions μ_m,

$$\mu_m(t) = \mu(t - t_m) = \text{sinc}(\frac{\Omega}{2}(t - t_m)) = \frac{\sin(\frac{\Omega}{2}(t - t_m))}{\frac{\Omega}{2}(t - t_m)} \tag{6.26}$$

which are capable of reconstructing any band-limited function f via

$$f(t) = \sum_m f(t_m)\mu_m(t) \tag{6.27}$$

However, it can also be shown[2] that these functions $\{\mu_m\}_{m=-\infty}^{\infty}$ are orthogonal to each other:

$$\langle \mu_m, \mu_n \rangle = \frac{2\pi}{\Omega}\delta(m - n) \tag{6.28}$$

under the scalar product,

$$\langle f, g \rangle = \int_{-\infty}^{\infty} f^* g \tag{6.29}$$

Thus, the band-limited functions constitute a Hilbert space with the above scalar product.[3] Accordingly, any band-limited signal f can be reconstructed via an orthogonal expansion and the scalar product from Eq. (6.29)

$$f(t) = \sum_m F(\omega_m)\mu_m(t) \tag{6.30}$$

where

$$F(\omega_m) = \langle f, \mu_m \rangle / \langle \mu_m, \mu_m \rangle \tag{6.31}$$

In consequence, the coefficients $F(\omega_m)$ in Eq. (6.30) must equal to $f(t_m) = f(m\frac{2\pi}{\Omega})$ appearing in Eq. (6.27)

$$F(\omega_m) = \frac{\langle \mu_m, f \rangle}{\langle \mu_m, \mu_m \rangle} = \frac{\langle \mu_m, f \rangle}{(\frac{2\pi}{\Omega})} = f(m\frac{2\pi}{\Omega}) \tag{6.32}$$

In other words, the *projection coefficients* of a band-limited signal f onto the sinc functions basis are the samples of the signal on the grid:

---

[2] See Exercise 7.2.

[3] This is not too surprising because just another representation of the same functions, i.e., their FTs, constitutes the space of limited extension functions. We have already seen that such functions constitute a Hilbert space with a scalar product enabling their reconstruction via FCs and the Fourier series.

$$\langle \mu_m, f \rangle = \frac{2\pi}{\Omega} f(t_m) \tag{6.33}$$

which uses no integrals, this result is a fairly simple tool to find projection coefficients.

Example 6.2. *We illustrate sampling and periodization by Fig. 6.4.*

- In the top, left graph, we have drawn a band-limited signal $f(t)$. The very definition of f as band-limited implies that $F(\omega)$ has a limited extension in the spectrum, which is drawn in the top, right graph. The width of F is Ω.
- In the middle-left graph, we have drawn f after discretization. In the middle right graph, we have drawn the ω-domain after discretization. The discretization in one domain is equivalent to a periodization in the other domain. Also, the discretization step is inversely proportional to the period of the other domain.
- In the bottom, left graph we have shown the sampled f, when we periodize F, with a larger period than the extension of F. This periodization of F, using twice as large an Ω as compared to the extension of F, is shown in the bottom, right graph.

We summarize the results of this section by the *Nyquist theorem.*

Theorem 6.3 (Nyquist). *Let f and F be a Fourier transform pair. Then sampling of either of the functions is equivalent to periodization of the other function. To be more precise, the following two statements are valid:*

- *(Time-limited signals) If f is nonzero only in the interval $[-\frac{T}{2}, \frac{T}{2}]$, then F can be sampled without loss of information provided that the sampling period is $\frac{2\pi}{T}$ or less.*
- *(Band-limited signals) If F is nonzero only in the interval $[-\frac{\Omega}{2}, \frac{\Omega}{2}]$ then f can be sampled without loss of information provided that the sampling period is $\frac{2\pi}{\Omega}$ or less.*

♦

When working with band-limited signals, the distance between the samples in the t-domain can be assumed to be equal to 1, for convenience. Then we obtain the critical frequency domain parameter $\Omega = 2\pi$, so that the critical frequency in the theorem yields:

$$\frac{\Omega}{2} = \pi \quad \Leftrightarrow \quad -\pi < \omega < \pi \tag{6.34}$$

In signal processing literature, $\frac{\Omega}{2}$ is often referred to as the Nyquist frequency. Even its normalized version, π, is called the Nyquist frequency, because a scaling of ω and t-domains is always achievable. The basic interval $-pi < \omega < \pi$ is sometimes called the *Nyquist period*, or the *Nyquist block*.

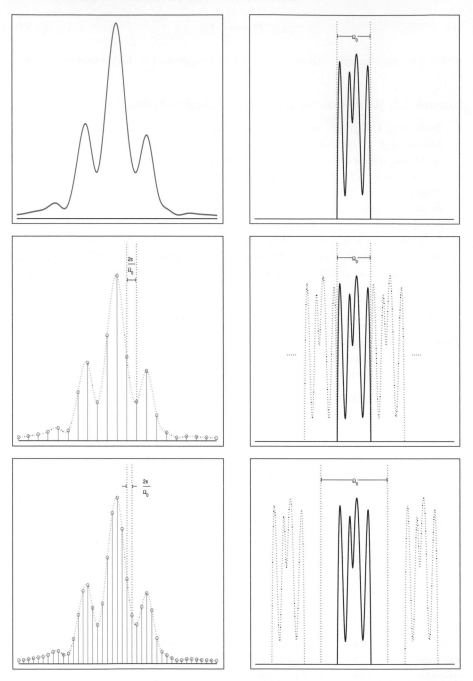

Fig. 6.4. The effect of sampling on a band-limited signal is a repetition of its finite extension spectrum. On the *left*, the graphs are in the t-domain, on the *right*, they are in the ω-domain

6.3 Discrete Fourier Transform (DFT)

When f is a finite extension function with the extension T, Eq. (5.14) states that f can be periodized and synthesized by means of FCs, $F(n\frac{2\pi}{T})$, which are samples of an infinite extension function F. Albeit theoretically infinite in number, the discrete function values, $F(n\frac{2\pi}{T})$, decrease since they are FCs. As such, they must decrease, or else the synthesis formula will not converge to the periodized version of f. Accordingly,

$$\lim_{n\to\infty} F(n\frac{2\pi}{T}) \to 0 \tag{6.35}$$

is a fact for real signals and from Fourier series we obtain a periodic function which approximates the periodized $f(t)$ well. Consequently, when synthesizing f from its FCs as in Eq. (5.14), there exists N such that we can ignore the $F(m\frac{2\pi}{T})$ for which $m \geq N$, and still obtain a reasonably good approximation of the periodized $f(t)$.

In accordance with the above, we will attempt to bring the concept of the FT closer to applications, by assuming in this section that not only f has a finite extension T, but that the corresponding FCs are also finite in number, i.e., only FCs

$$\{F(n\omega_n)\}_{n=0}^{N-1} \qquad \text{with} \qquad \omega_n = n\frac{2\pi}{T} \tag{6.36}$$

are nonzero. According to Eq. (5.14), we can synthesize $f(t)$

$$f(t) = \sum_{m=0}^{N-1} \frac{2\pi}{T} F(\omega_m) \exp(i\omega_m t) \tag{6.37}$$

by means of the following N coefficients:

$$F(\omega_m) = \frac{1}{2\pi} \int_{-\frac{T}{2}}^{\frac{T}{2}} f(t) \exp(-i\omega_m t) dt \tag{6.38}$$

where $m \in 0, \cdots N - 1$. Because f has a finite extension, any sampling of it will yield a finite number of function values. If we choose to sample f at a set of t_n in such a way that the sampling interval corresponds to what the Nyquist theorem affords us, it should be possible not to lose information. The largest sampling step suggested by the Nyquist theorem is $2\pi(N\frac{2\pi}{T})^{-1} = \frac{T}{N}$, where $N\frac{2\pi}{T}$ represents the extension of $\{F(n\omega_n)\}_{n=0}^{N-1}$. We proceed accordingly, and substitute t_n in Eq. (6.37)

$$f(t_n) = \sum_{m=0}^{N-1} \frac{2\pi}{T} F(\omega_m) \exp(i\omega_m t_n) \quad \text{with} \quad t_n = n\frac{T}{N} \tag{6.39}$$

Notice that there are exactly N samples of $f(t_n)$. Thus, it should be possible to compute $F(\omega_m)$ appearing in Eq. (6.39) directly from $f(t_n)$, without passing through a continuous reconstruction of $F(\omega)$ (and then sampling it) in Eq. (6.38). We observe that the sequence of complex numbers $\exp(it_n\omega_m)$ appearing in Eq. (6.39) constitutes an N-element vector[4] $\boldsymbol{\psi}_n$, i.e.,

$$\boldsymbol{\psi}_n = (1, \exp(it_n\omega_1), \exp(it_n\omega_2) \cdots \exp(it_n\omega_{N-1}))^T \qquad (6.40)$$

or

$$\psi_n(m) = \exp(it_n\omega_m), \qquad \text{with} \qquad m \in 0 \cdots N - 1. \qquad (6.41)$$

There are N such vectors $\boldsymbol{\psi}_n$ because $n \in 0 \cdots N - 1$. Furthermore, $\boldsymbol{\psi}_n$ are orthogonal to each other under the conventional vector scalar product of Eq. (3.27). That is,

$$\langle \boldsymbol{\psi}_n, \boldsymbol{\psi}_m \rangle = \sum_{l=0}^{N-1} \exp(i(m-n)\frac{l}{N}) = N\delta(m-n) \qquad (6.42)$$

where δ is the Kronecker delta, holds.

Exercise 6.3. *Show that ψ_m are orthogonal.*
HINT: Identify $\langle \boldsymbol{\psi}_m, \boldsymbol{\psi}_n \rangle$ as a geometric series.

To obtain the kth element of the vector $F(\omega_k)$, we compose the scalar product between the vector $(f(t_0, \cdots, f(t_{N-1}))^T$ and $\boldsymbol{\psi}_k$, i.e., we multiply both sides of Eq. (6.39) with $\exp(-i\omega_k t_n)$ and sum over the index n:

$$\sum_{n=0}^{N-1} f(t_n) \exp(-i\omega_k t_n) = \sum_{n=0}^{N-1} \exp(-i\omega_k t_n) \sum_{m=0}^{N-1} \frac{2\pi}{T} F(\omega_m) \exp(i\omega_m t_n)$$

$$= \sum_{n=0}^{N-1} \sum_{m=0}^{N-1} \frac{2\pi}{T} F(\omega_m) \exp(-i\omega_k t_n) \exp(i\omega_m t_n)$$

$$= \sum_{m=0}^{N-1} \frac{2\pi}{T} F(\omega_m) \sum_{n=0}^{N-1} \exp(-i\omega_k t_n) \exp(i\omega_m t_n)$$

$$= \sum_{m=0}^{N-1} \frac{2\pi}{T} F(\omega_m) \sum_{n=0}^{N-1} \exp\left(-ink\frac{2\pi}{N}\right) \exp\left(inm\frac{2\pi}{N}\right)$$

$$= \sum_{m=0}^{N-1} \frac{2\pi}{T} F(\omega_m) N\delta(k-m)$$

$$= N\frac{2\pi}{T} F(\omega_k) \qquad (6.43)$$

where we have used $t_n\omega_m = nm\frac{2\pi}{N}$ and the orthogonality of ψ_m as given by Eq. (6.42). Consequently, we obtain

$$F(\omega_m) = (N\frac{2\pi}{T})^{-1} \sum_{n=0}^{N-1} f(t_n) \exp(-i\omega_m t_n) \qquad (6.44)$$

---

[4] Note the typographic difference that now $\boldsymbol{\psi}_n$ is in boldface because it is an array or a conventional vector, in contrast to ψ_n, which represents a function.

This equation, together with Eq. (6.39), establishes f and F as a discrete transform pair when sampled on the grids of t_m and ω_n, respectively. We state the result as a lemma.

Lemma 6.2. *Let $f(t)$ be a finite extension function with the extension T and have N nonzero FCs. The samples of f*

$$f(t_n) = \sum_{m=0}^{N-1} \frac{2\pi}{T} F(\omega_m) \exp(i\omega_m t_n), \quad \text{with} \quad t_n = n\frac{T}{N}, \qquad (6.45)$$

and the samples of its FT,

$$F(\omega_m) = (N\frac{2\pi}{T})^{-1} \sum_{n=0}^{N-1} f(t_n) \exp(-i\omega_m t_n), \quad \text{with} \quad \omega_m = m\frac{2\pi}{T}, \qquad (6.46)$$

constitute a discrete transform pair.

$\blacklozenge$

Using an analogous reasoning and theorem 6.2 we can restate this result for band-limited functions.

Lemma 6.3. *Let $f(t)$ be a band-limited function, i.e., $F(\omega)$ has the finite extension Ω, that has at most N nonzero samples $f(t_n)$ with $t_n = n\frac{2\pi}{\Omega}$. The samples of f*

$$f(t_n) = \sum_{m=0}^{N-1} \frac{2\pi}{TN} F(\omega_m) \exp(i\omega_m t_n), \quad \text{with} \quad t_n = n\frac{2\pi}{\Omega}, \qquad (6.47)$$

and the samples of its FT,

$$F(\omega_m) = (\frac{2\pi}{T})^{-1} \sum_{n=0}^{N-1} f(t_n) \exp(-i\omega_m t_n), \quad \text{with} \quad \omega_m = m\frac{\Omega}{N}, \qquad (6.48)$$

constitute a discrete transform pair.

$\blacklozenge$

The two lemmas can be simplified further because we are free to define $T = 2\pi$, or $\Omega = 2\pi$. In either case, the transform pair can be interpreted as a finite discrete sequence, even without reference to the sampling distance between t and ω variables. Accordingly, both lemmas can be reduced to a dimensionless form that only differs with where we place the transform constant $1/N$. We choose to reduce the first lemma and give it as a theorem.

Theorem 6.4 (DFT). *The discrete Fourier transform (DFT) for arrays with N elements, defined as*

$$\mathbf{F}(n) = \frac{1}{N} \sum_{m=0}^{N-1} \mathbf{f}(m) \exp\left(-imn\frac{2\pi}{N}\right) \qquad (6.49)$$

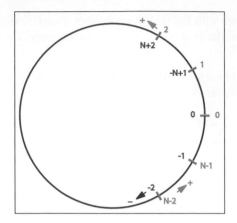

Fig. 6.5. The circular topology of DFT indices illustrated for N points. An index j represents the same point as $j + nN$, where n is an arbitrary integer

has an inverse transform

$$\mathbf{f}(m) = \sum_{n=0}^{N-1} \mathbf{F}(n) \exp\left(imn\frac{2\pi}{N}\right) \tag{6.50}$$

♦

The DFT or its two variants in this section are special cases of the FT. In 1961 Cooley and Tukey, [48], published an algorithm that could quickly compute the discrete Fourier transform. Since then this and other algorithms for quickly computing DFTs are collectively called FFT (Fast Fourier Transform) algorithms although there are several variants to choose from. The FFT has found many applications because of its efficiency. In many cases, it is faster to do two FFTs, a multiplication of the resulting functions, and then to do the inverse FFT, than doing shift-invariant computations, e.g., convolutions with arbitrary filters.

6.4 Circular Topology of DFT

We note that m and n on the right-hand side of the equations in theorem 6.4, do not necessarily have to be one of the integers in $0 \cdots N-1$. We will investigate this issue along with a practical way of dealing with it next.

In arrays with N elements that are DFT pairs, there can be at most N different $\mathbf{f}(m)$ or $\mathbf{F}(n)$ values, although m, n can be any integer according to the DFT lemma. It can be easily shown that if for any integer m, n, and k the equation

$$n = m + kN \tag{6.51}$$

holds, then

$$\mathbf{f}(n) = \mathbf{f}(m), \qquad \text{and} \qquad \mathbf{F}(n) = \mathbf{F}(m). \tag{6.52}$$

Hence, the translation of indices in the time, or frequency domain must be done by means of a *circular addition*, which means that $n+m$ and $n+m+kN$ are equivalent. This results in a *circular topology of DFT* where the first array element labelled as 0 is the nearest neighbor of the one labelled as 1 and $N - 1$.

Although it is feasible to find neighbors using visual workarounds for discrete functions defined on 1D-domains (Fig. 6.5), it is definitely a challenge to "visually" find the neighbor of a point close to grid boundaries in higher dimensions. The arithmetic modulo function, here abbreviated as $\mathrm{Mod}(m, N)$, is handy in practical applications of DFT to avoid the practical difficulties associated with its circular topology. The function computes the unique remainder of m after division by N; that is, the result is always one of the integers $0 \cdots N - 1$, e.g.,

$$\mathrm{Mod}(3, 17) = \mathrm{Mod}(-14, 17) = \mathrm{Mod}(20, 17) = 3 \tag{6.53}$$

Ordinary addition using the Mod function, i.e., $\mathrm{Mod}(m + n, N)$, defines a so-called finite group so that integers that differ with integer multiples of N are equivalent to each other. This allows a seamless interpretation of *all* translations on rectangular discrete grids, in all dimensions, which is precisely what is needed to interpret the indices of DFT correctly.

For example, when $N = 17$, the integers $3, -14, 20$ represent the same coefficient

$$\mathbf{f}(3) = \mathbf{f}(-14) = \mathbf{f}(20) \qquad \text{and} \qquad \mathbf{F}(3) = \mathbf{F}(-14) = \mathbf{F}(20) \tag{6.54}$$

although -14 and 20 are obviously not possible to use as the indices of an array having 17 elements, which has index labels $0 \cdots 16$. After an application of the modulo function, e.g., Eq. (6.53), the resulting integers can be used as indices of arrays holding the discrete values of f and F, Eq. (6.54).

A systematic use of the Mod function in all index references, additions, and subtraction, known also as *modulo arithmetic*, enables *circular translation*, i.e., translations that hit no grid boundaries. This can be used to "walk" around in the DFT arrays conveniently, which is particularly important for a correct implementation and interpretation of the convolution operation by DFT, as we will study in Chap. 7. The circular interpretation of the DFT indices on 1D domains is illustrated by Fig. 6.5 using N points (*taps*). Note that the first index is 0 and the last index is $N - 1$, and the indices are periodic with the period N in the modulo N arithmetic.

7

Properties of the Fourier Transform

In this chapter we study the FT to reveal some of its properties that are useful in applications. As has been shown previously, the FT is a generalization of both FC and DFT. There is an added value, to use FT indifferently both to mean FC and to mean DFT. To do that, however, we need additional results. The first section contributes to that by introducing the Dirac distribution [212]. This allows us to Fourier transform a sinusoid, or a complex exponential, which is important to virtually all applications of signal analysis in science and economics. Fourier transforming a sinusoid is not a trivial matter because a sinusoid never converges to zero. In Sect. 7.2, we establish the invariance of scalar products under the FT. This has many implications in the practice of image analysis, especially in the direction estimation and quantification of spectral properties. Finally, we study the concept of convolution in the light of FT, and close the circle by suggesting the Comb tool to achieve formal sampling of arbitrary functions. This will automatically yield a periodization in the frequency domain via a convolution. With the rise of computers, convolution has become a frequently utilized tool in signal analysis applications.

7.1 The Dirac Distribution

What is the area or the integral of the sinc signal that was defined in Eq. (6.21)? The answer to this question will soon lead us to construct a sequence of functions that will produce a powerful tool when working with the Fourier transform.

We answer the question by identifying the characteristic function of an interval as the inverse FT of a sinc function via Eq. (6.23):

$$\chi_T(t) = \begin{cases} 1, & \text{if } t \in [-\frac{T}{2}, \frac{T}{2}]; \\ 0, & \text{otherwise.} \end{cases}$$

$$= \mathcal{F}^{-1}\left(\frac{T}{2\pi}\text{sinc}(\omega\frac{T}{2})\right)(t) = \int \frac{T}{2\pi}\text{sinc}\left(\omega\frac{T}{2}\right)\exp(i\omega t)d\omega \quad (7.1)$$

Evaluating $\chi_T(0)$ yields the integral of a sinc function:

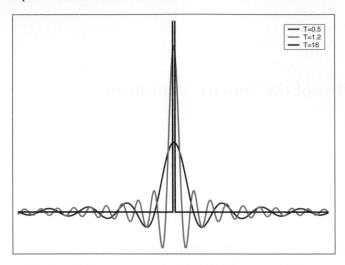

Fig. 7.1. The sinc function sequence B_T given by Eq. (7.3) for some T values. In *blue* the (capped) function indicates that B_T approaches the Dirac distribution with "∞ value" at the origin

$$\int \frac{T}{2\pi} \mathrm{sinc}\left(\omega\frac{T}{2}\right) d\omega = 1 \qquad (7.2)$$

Obviously, the "area under the curve",

$$B_T(\omega) = \frac{T}{2\pi} \mathrm{sinc}\left(\omega\frac{T}{2}\right) \qquad (7.3)$$

is 1, independent of T, as long as T is finite! In fact for every T we have a different function B_T, so that we can fix a sequence of Ts and study B_T. This is what we will do next.

The B_Ts were obtained by Fourier transforming a function that is constant in a finite and symmetric interval $[\frac{-T}{2}, \frac{T}{2}]$. What happens to sinc when T approaches infinity? In other words, what is the Fourier transform of an "eternal" constant? We restate Eq. (6.23) and study it when T increases.

$$\mathcal{F}(\chi_T)(\omega) = B_T(\omega) = \frac{1}{2\pi}T \cdot \mathrm{sinc}\left(T\frac{\omega}{2}\right) \qquad (7.4)$$

Any function contracts when we multiply its argument with a large constant T. The function

$$\mathrm{sinc}(T\frac{\omega}{2}) \qquad (7.5)$$

is accordingly a contracted version of $\mathrm{sinc}(\omega)$ as T increases.

The sinc function is a smooth function that is both continuous and has continuous derivatives everywhere, including at the origin. When we increase T, the sinc function contracts in the horizontal direction only. In particular, the maximum of the sinc function remains 1, and this maximum is attained at the origin for all Ts. However, when T increases, the values of the functions $\{B_T(\omega)\}_T$ at $\omega = 0$ increase with T, since $\mathrm{sinc}(T \cdot 0) = 1$. As a result, $B_T(\omega)$ approaches infinity at $\omega = 0$, whereas the effective width of the B_T functions shrink to zero. Thus, the $B_T(\omega)$ approaches to something which is zero everywhere except at the origin and yet has an area which is always 1!

That something is the so-called *Dirac distribution* represented by δ

$$\lim_{T \to \infty} B_T(\omega) = \lim_{T \to \infty} \frac{1}{2\pi} T \cdot \mathrm{sinc}(\omega \frac{T}{2}) = \delta(\omega) \tag{7.6}$$

which is shown in Fig. 7.1. Other names for the Dirac distribution are the *delta function*, *unit impulse*, and the *Dirac function*, although, $\delta(\omega)$ is not very useful when considered strictly as a function. This is because the $\delta(\omega)$ interpreted as a function, will only deliver the value zero, except at the origin, where it will not even deliver a finite value, but delivers the value "infinity".

In mathematics, the limit of the B_T function is not defined in terms of what it does to the points ω, its argument, but instead what it does to a class of other functions under integration. That is also where its utility arises, namely, it consistently delivers a value as a result of the integration. Such objects are called generalized functions, or distributions. To obtain a hint on how we should define the limit of B_T, we should thus study the behavior of B_T when it is integrated with arbitrary (integrable) functions f:

$$\langle B_T, f \rangle = \int B_T f(\omega) d\omega$$

$$= \int (\frac{1}{2\pi} \int \chi_T(t) \exp(-i\omega t) dt) f(\omega) d\omega$$

$$= \int (\frac{1}{2\pi} \int \exp(-i\omega t) f(\omega) d\omega) \chi_T(t) dt$$

$$= \int F(t) \chi_T(t) dt$$

Thus, in the limit we will obtain

$$\lim_{T \to \infty} \langle B_T, f \rangle = \int F(t) dt = f(0) \tag{7.7}$$

because χ_T approaches to the constant function 1 with increasing T. Also, changing the order of the integration and the limit operator is not a problem for physically realizable functions. This shows that as T grows beyond any bound, B_T will consistently "kick out" the value of its fellow integrand f at origin, no matter the choice of f. Therefore, we give the following precision to δ, the Dirac distribution.

Definition 7.1. *The Dirac-δ distribution is defined as*

$$\langle \delta, f \rangle = \int f(\omega) \delta(\omega) d\omega = f(0) \tag{7.8}$$

The argument "ω" in $\delta(\omega)$ is there to mark the "hot point" of the delta distribution, not to tempt us to conclude that δ is a function delivering a useful value at the argument. For example, $\delta(\omega - \omega')$ becomes infinity when $\omega = \omega'$ (whereas it vanishes elsewhere). In fact, taking a close look at the rule that defines the Dirac distribution, we note that it is not based on real numbers as arguments but on functions, represented by f. Whereas functions are rules that deliver numbers for arguments that are numbers, the Dirac distribution is a rule that delivers numbers for "arguments" that are functions appearing in scalar products, Eq. (7.1). In plain English, this is what the Dirac distribution does. No matter what the integration domain is, as long as the domain contains the "hot-point" of the δ-distribution, the distribution kicks out the value of its "fellow integrand" at the hot point. Example behaviors include

$$\int f(x)\delta(x - y)dx = f(y) \tag{7.9}$$

and

$$\iint f(x)\delta(x - y)dxdy = \int f(y)dy. \tag{7.10}$$

Theorem 7.1. *The Fourier transform of the constant function "1" is a Dirac distribution:*

$$\mathcal{F}(1)(\omega) = \frac{1}{2\pi} \int \exp(-i\omega t)dt = \delta(\omega) \tag{7.11}$$

♦

Exercise 7.1. *The proof of the theorem is obtained by Egs. (7.4) and Eq. (7.6). In particular, provide the following:*

  i) *Prove that the Fourier transform of the complex exponential* $\exp(i\omega t)$ *is a shifted delta distribution.*
  ii) *Prove that the Fourier Transform of a Dirac-δ is a constant.*
 iii) *Create a δ in a 2D image. Compute the real part of its Fourier transform (use DFT on a large image). Compute the imaginary part of its Fourier transform. What are the parameters that determine the direction and the frequencies of planar waves?*

7.2 Conservation of the Scalar Product

In the case of Fourier series, we were able to show that the scalar products were conserved between the time domain and the Fourier coefficients. Now we show that *scalar product conservation* is valid even for the Fourier transform. The scalar products are defined as

$$\langle f, g \rangle = \int_{-\infty}^{\infty} f(t)^* g(t)dt \tag{7.12}$$

and

$$\langle F, G \rangle = \int_{-\infty}^{\infty} F(\omega)^* G(\omega) d\omega \tag{7.13}$$

We use the inverse FT to transfer the scalar product in the time domain to the frequency domain:

$$
\begin{aligned}
\langle f, g \rangle &= \langle \mathcal{F}^{-1}(F), \mathcal{F}^{-1}(G) \rangle \\
&= \int [\int F(\omega) \exp(i\omega t) d\omega][\int G(\omega') \exp(i\omega' t) d\omega']^* dt \\
&= \int \int \int F(\omega) \exp(i\omega t)[G(\omega') \exp(i\omega' t)]^* d\omega d\omega' dt \\
&= \int \int F(\omega) G(\omega')^* [\int \exp(i\omega t) \exp(-i\omega' t) dt] d\omega d\omega' \\
&= \int \int F(\omega) G(\omega')^* [\int \exp(-i(\omega - \omega')t) dt] d\omega d\omega' \\
&= \int \int F(\omega) G(\omega')^* 2\pi \delta(\omega - \omega') d\omega d\omega' \\
&= 2\pi \int F(\omega) G(\omega) d\omega = 2\pi \langle F, G \rangle \tag{7.14}
\end{aligned}
$$

This establishes the *Parseval–Plancherel theorem* for the FT.

Theorem 7.2 (Parseval–Plancherel). *The FT conserves the scalar products:*

$$\langle f, g \rangle = \int_{-\infty}^{\infty} f(t)^* g(t) dt = 2\pi \int_{-\infty}^{\infty} F(\omega)^* G(\omega) d\omega = 2\pi \langle F, G \rangle \tag{7.15}$$

♦

As a consequence of this theorem, we conclude that FT preserves the norms of functions.

Exercise 7.2.

  i) *Show that the shifted sinc functions, see Eq. (6.26), that interpolate band-limited signals constitute an orthonormal set under the scalar product, Eq. (6.29)*
  HINT: Apply the Parseval–Plancherel theorem to the scalar products.
 ii) *There are functions that are not band-limited. What does the projection of such a signal on sinc functions correspond to?*
  HINT: Decompose the signal into a sum of two components corresponding to portions of the signal inside and outside of Ω.
iii) *Can the sinc functions of Eq. (6.26) be used as a basis for square integrable functions?*
  HINT: Are the square integrable functions band-limited?

7.3 Convolution, FT, and the δ

We define first the convolution operation, which is bilinear.

Definition 7.2. *Given two signals f and g, the* convolution *operation, denoted by $*$, is defined as*

$$h(t) = (f * g)(t) = \int_{-\infty}^{\infty} f(t - \tau)g(\tau)d\tau \tag{7.16}$$

It takes an ordinary variable substitution to show that the convolution is commutative:

$$h(t) = (f * g)(t) = \int_{-\infty}^{\infty} f(t - \tau)g(\tau)d\tau = \int_{-\infty}^{\infty} f(\tau)g(t - \tau)d\tau = (g * f)(t) \tag{7.17}$$

Now we study the FT of the function $h = f * g$.

$$
\begin{aligned}
h(t) = \mathcal{F}(f * g)(\omega) &= \frac{1}{2\pi} \int \exp(-i\omega t)[\int f(t - \tau)g(\tau)d\tau]dt \\
&= \frac{1}{2\pi} \int g(\tau)[\int f(t - \tau)\exp(-i\omega t)dt]d\tau \\
&= \frac{1}{2\pi} \int g(\tau)[\int f(t)\exp(-i\omega(t + \tau))dt]d\tau \\
&= \frac{1}{2\pi} \int g(\tau)\exp(-i\omega\tau)d\tau[\frac{2\pi}{2\pi} \int f(t)\exp(-i\omega t)dt] \\
&= 2\pi F(\omega)G(\omega)
\end{aligned}
\tag{7.18}
$$

Consequently we have the following result, which establishes the behavior of *convolution under the Fourier transform.*

Theorem 7.3. *The bilinear operator $*$ transforms as $\cdot$ under the FT*

$$h = f * g \quad \Leftrightarrow \quad H = 2\pi F \cdot G \tag{7.19}$$

$\blacklozenge$

This is a useful property of FT in theory and applications. Below we bring further precision as to how it relates to sampled functions. Before doing that, we present the *Fourier transform of $\delta(x)$*, the *Dirac distribution*, as a theorem for its practical importance.

Theorem 7.4. *The Dirac-δ acts as the element of unity (one) for the convolution operation:*

$$\delta * f = f * \delta = f \tag{7.20}$$

$\blacklozenge$

Convolution and the FC

To work with discrete signals, we assume now f and g are two band-limited functions, both with the same frequency extension of Ω. With this assumption, $f(t_m)$ and
$g(t_m)$ are discrete versions of f and g which can also synthesize them faithfully. To
obtain relation Eq. (7.19), we did not need to make a restrictive assumption on the
extensions of F and G. In consequence, relation (7.19) holds even for band-limited
functions. The right-hand side is easy to interpret because F and G are finite extension functions that get multiplied pointwise. However, the left-hand side requires
that we are more precise on the definition of the convolution. To obtain a discrete
definition of the convolution operator that is consistent with its continuous analogue,
we pose the question as follows. What is $h(t)$, if $H = 2\pi FG$ is given? To answer
this, we synthesize F and G according to Eq. (6.12)

$$F(\omega) = \left(\frac{1}{\Omega}\right) \sum_m f(t_m) \exp(-it_m\omega) \tag{7.21}$$

$$G(\omega) = \left(\frac{1}{\Omega}\right) \sum_m g(t_m) \exp(-it_m\omega) \tag{7.22}$$

$$\tag{7.23}$$

Evidently, H will have the extension Ω because F and H have the extension Ω. We
can then proceed to compute the FCs of H according to Eq. (6.13)

$$h(t_k) = 2\pi(\frac{1}{\Omega})^2 \int_{-\frac{\Omega}{2}}^{\frac{\Omega}{2}} \sum_{mn} f(t_m)g(t_n) \exp[-i(t_m+t_n)\omega] \exp(it_k\omega)d\omega$$

$$= 2\pi(\frac{1}{\Omega})^2 \sum_{mn} f(t_m)g(t_n) \int_{-\frac{\Omega}{2}}^{\frac{\Omega}{2}} \exp[i(t_k-t_m-t_n)\omega]d\omega$$

$$= 2\pi(\frac{1}{\Omega})^2 \sum_{mn} f(t_m)g(t_n)\delta(t_k-t_m-t_n)\Omega$$

$$= \frac{2\pi}{\Omega} \sum_n f(t_k-t_n)g(t_n) \tag{7.24}$$

Using the standard assumption $\Omega = 2\pi$ results in $t_k - t_{k-1} = 1$ for any k, and yields
the following definition:

Definition 7.3. *Given two discrete signals $f(k)$ and $g(k)$, with k being integers, the
convolution operation for such signals is denoted by $*$, and is defined by:*

$$h(k) = (f * g)(k) = \sum_n f(k-n)g(n) \tag{7.25}$$

With this definition at hand, we summarize our finding in Eq. (7.24).

Lemma 7.1. *The discrete convolution defined by Eq. (7.25) is a special case of its
counterpart given by Eq. (7.16) in that theorem 7.3 holds when the discrete signals*

$f(k)$ and $g(k)$ are samples of band-limited signals $f(t)$ and $g(t)$. Likewise, theorem 7.4 also holds if δ is interpreted as the Kronecker delta.

♦

Convolution and DFT

It is logical to use the same definition of convolution on an array holding N scalars as the one given for an infinitely large grid containing samples of a function.

$$\mathbf{h}(k) = (\mathbf{f} * \mathbf{g})(k) \sum_{n=0}^{N-1} \mathbf{f}(k-n)\mathbf{g}(n) \qquad (7.26)$$

However, the translated function $\mathbf{f}(k-n)$ needs to be more elaborately defined. The index $k-n$ must refer to well-defined array elements. This is because even if k and n are within $0 \cdots N-1$, their difference $k-n$ is not necesserily in that range, which is a problem that an infinite grid does not have. Remembering that the DFT was derived from lemma 6.2 and that both $\mathbf{f}(t)$ and $\mathbf{F}(\omega)$ have been extended to be periodic, it is clear that the translation $k-n$ on the grid is equivalent to the *circular translation* discussed in Sect. 6.4. In other words,

$$\mathbf{f}(m) \leftarrow \mathbf{f}(\mathrm{Mod}(m, N)) \qquad (7.27)$$

as well as

$$\mathbf{f}(k-n) \leftarrow \mathbf{f}(\mathrm{Mod}(k-n, N) \qquad (7.28)$$

should be used in practice. Using the above definition for convolution together with circular translation is also called *circular convolution*.

Upon this interpretation of convolution, we now investigate how a circular convolution transforms under DFT. Let the symbol $\mathcal{F}$ represent the DFT, and let $\mathbf{F}$, $\mathbf{G}$ and $\mathbf{H}$ be DFTs of the discrete function values $\mathbf{f}$, $\mathbf{g}$, and $\mathbf{h}$ on a grid. The functions $\mathbf{f}$, $\mathbf{g}$, and $\mathbf{h}$ are assumed to have been defined on a finite grid of t_m, while the functions $\mathbf{F}$, $\mathbf{G}$, and $\mathbf{H}$ are defined on a grid of ω_m. Both grids have the same number of grid points, N, as generated by integers $m \in 0, 1, 2 \cdots N-1$.

$$\mathcal{F}(\mathbf{f} * \mathbf{g})(k) = \frac{1}{N} \sum_l \left(\sum_j \mathbf{f}(j)\mathbf{g}(l-j) \right) \exp\left(-ik\frac{2\pi l}{N}\right)$$

$$= \frac{1}{N} \sum_l \left\{ \sum_j \left[\sum_m \mathbf{F}(m) \exp\left(ij\frac{2\pi m}{N}\right) \right] \right.$$

$$\left. \times \left[\sum_n \mathbf{G}(n) \exp\left(i(l-j)\frac{2\pi n}{N}\right) \right] \right\} \exp\left(-ik\frac{2\pi l}{N}\right) \quad (7.29)$$

We change the order of summations so that we can obtain a "delta" function that helps us to annihilate summations. However, we first shift the order of the sums with indices l, j with those with the indices n, m:

$$\mathcal{F}(\mathbf{f} * \mathbf{g})(k) = \frac{1}{N} \sum_{n} \sum_{m} \mathbf{F}(m)\mathbf{G}(n) \times$$

$$\sum_{l} \sum_{j} \exp\left(ij\frac{2\pi m}{N}\right) \exp\left(i(l-j)\frac{2\pi n}{N}\right) \exp\left(-ik\frac{2\pi l}{N}\right)$$

$$= \frac{1}{N} \sum_{n} \sum_{m} \mathbf{F}(m)\mathbf{G}(n) \sum_{l} \exp\left(il\frac{2\pi(n-k)}{N}\right) \sum_{j} \exp\left(ij\frac{2\pi(m-n)}{N}\right)$$

$$= \frac{N^2}{N} \sum_{m} \sum_{n} \mathbf{F}(m)\mathbf{G}(n)\delta(n-k)\delta(m-n)$$

$$= \frac{N^2}{N} \sum_{n} \mathbf{F}(n)\mathbf{G}(n)\delta(n-k)$$

$$= N\mathbf{F}(k)\mathbf{G}(k) \tag{7.30}$$

We note here that the two Kronecker delta functions are obtained by recognizing sums (the ones with l and j) as scalar products between complex exponentials, see Eq. (6.42), and that these deltas annihilated[1] two summations (the ones with m and n) before they disappeared. Accordingly, we give the definition for circular convolutions.

Definition 7.4. *Given two finite discrete signals* $\mathbf{f}(m)$ *and* $\mathbf{g}(m)$, *with* m *being integer in* $0, \cdots N-1$, *the circular convolution operation for such signals is denoted by* $*$, *and is defined as*

$$\mathbf{h}(k) = (\mathbf{f} * \mathbf{g})(k) = \sum_{n=0}^{N-1} \mathbf{f}(k-n)\mathbf{g}(n) \tag{7.31}$$

All references to indices, including additions, are in $\mathrm{Mod}N$ *arithmetic.*

Accordingly, $\mathbf{h}(m)$, the outcome of the discrete convolution, is periodic with the period N, just as the component sequences $\mathbf{f}(m)$ and $\mathbf{h}(m)$. We summarize the result on *DFT of convolution* as a lemma.

Lemma 7.2. *Given the discrete circular convolution defined by Eq. (7.31), it transforms as multiplication under the DFT*

$$\mathbf{h}(m) = (\mathbf{f} * \mathbf{g})(m) \quad \Leftrightarrow \quad \mathbf{H}(k) = N\mathbf{F}(k) \cdot \mathbf{G}(k) \tag{7.32}$$

Likewise, theorem 7.4 holds if δ *is interpreted as the Kronecker delta.*

♦

---

[1] The δs are very efficient "summation annihilators". They erase summations twice: The first time when they are created, and the second time when they disappear. The Dirac distributions are the analogous "integration annihilators".

7.4 Convolution with Separable Filters

The 1D convolution, Eqs. (7.25) and (7.31), and the FT results are also valid in multiple dimensions. Specially, when the filter size is large, it is more efficient to perform the convolution in the frequency domain for general filters. However, there is an exception to this rule if filters have a special form.

Here we will discuss how to implement a special class of filters in multiple dimensions, separable filters, that can be implemented very quickly. This class covers a frequently used type of filters used in bandpass or scale space filtering, e.g., to change the size of an image or to compress it. We discuss it for 2D continuous functions here, but the results are also valid for both types of discrete convolutions discussed above, and for even higher dimensions than 2.

Definition 7.5. *A function* $g(x, y)$ *is called a* separable function *if*

$$g(x, y) = g_1(x)g_2(y) \tag{7.33}$$

Assume that $g(x, y)$ is a separable filter function with which we wish to convolve the function $f(x, y)$. Using separability of g, we can rewrite the convolution:

$$h(x, y) = (g * f)(x, y) = \int\int g(\tau_x, \tau_y)f(x - \tau_x, y - \tau_y)d\tau_x d\tau_y \tag{7.34}$$

$$= \int g_1(\tau_x)\left(\int g_2(\tau_y)f(x - \tau_x, y - \tau_y)d\tau_y\right)d\tau_x$$

$$= \int g_1(\tau_x)[\tilde{h}(x - \tau_x, y)]d\tau_x \tag{7.35}$$

$$= (g1 * \tilde{h})(x, y) \tag{7.36}$$

where $\tilde{h}$ introduced in Eq. (7.35) is a function that represents the intermediary result:

$$\tilde{h}(x, y) = \int g_2(\tau_y)f(x, y - \tau_y)d\tau_y = (g_2 * \tilde{h})(x, y) \tag{7.37}$$

This is in fact a convolution that involves only one of the argument dimensions of $f(x, y)$, i.e., y. It is computed by integrating along the y-axis of the image with the 1D filter $g_2(y)$. The operation that follows is also a pure 1D convolution, but applied to the *result* of the first convolution and along the remaining dimension, x, so that we can write the entire operation as a cascade of 1D convolutions.

$$h(x, y) = (g_1(x) \cdot g_2(y)) * f(x, y) = g_1(x) * (g_2(y) * f) \tag{7.38}$$

Assuming that, in practice, the filter as well as the image are discrete, and the integrations have to be implemented as summations, we can estimate the cost of convolution as the number of arithmetic operations[2] per image point. If the filter g is $m \times m$, then

---

[2] One arithmetic operation in computational cost estimation corresponds to one multiplication plus one addition.

the filter g_1 will be of size $1 \times m$, whereas the filter g_2 will be of size $m \times 1$. Implementing the convolution directly via a 2D filter, as in Eq. (7.34), will require m^2 operations/pixel. Implementing it as in Eq. (7.38) demands first m operations/pixel to compute $\tilde{h} = g_2 * f$, then another m operations/pixel to compute the final h, totaling to $2m$ operations/pixel. It should be stressed that both ways of implementing the filtering will yield results that are identical. Assuming a typical filter with $m = 20$, the gain in terms of arithmetic operations is, however, a factor of 10. This is an appreciable difference for most applications because it translates to a speed-up of the same amount in conventional computation environments. For this reason, separable filters are highly attractive filters, in image and other multidimensional signal processing applications.

7.5 Poisson Summation Formula, the Comb

We elaborate in this section on how to sample a function formally. We will use displaced dirac distributions to achieve this. For a quick initiation, the theorems and lemmas can be directly read without proofs.

Assume that we have an integrable function $F(\omega)$, which is not necessarily a finite frequency function. Then we define the periodized version of it $\tilde{F}(\omega)$ as

$$\tilde{F}(\omega) = \sum_n F(\omega + n\Omega) \tag{7.39}$$

Evidently, $\tilde{F}(\omega)$ is periodic with the period Ω, and we can expand it in terms of FCs, see Eq. (6.12)

$$\tilde{F}(\omega) = \frac{1}{\Omega} \sum_m \tilde{f}(t_m) \exp(-it_m\omega) \tag{7.40}$$

where we obtain $\tilde{f}(t_m)$ through Eq. (6.13), if we keep in mind that $t_m = m\frac{2\pi}{\Omega}$,

$$
\begin{aligned}
\tilde{f}(t_m) &= \int_{-\frac{\Omega}{2}}^{\frac{\Omega}{2}} \tilde{F}(\omega) \exp(it_m\omega) d\omega \\
&= \int_{-\frac{\Omega}{2}}^{\frac{\Omega}{2}} \sum_n F(\omega + n\Omega) \exp(it_m\omega) d\omega \\
&= \sum_n \int_{-\frac{\Omega}{2}}^{\frac{\Omega}{2}} F(\omega + n\Omega) \exp(it_m\omega) d\omega
\end{aligned}
$$

Here, we changed the order of summation and integration which is permitted for physically realizable functions. Accordingly, $\tilde{f}(t_m)$ is obtained as:

$$\tilde{f}(t_m) = \sum_n \int_{n\Omega - \frac{\Omega}{2}}^{n\Omega + \frac{\Omega}{2}} F(\omega) \exp(it_m(\omega - n\Omega)) d\omega$$

$$= \exp(-it_m n\Omega) \sum_n \int_{n\Omega - \frac{\Omega}{2}}^{n\Omega + \frac{\Omega}{2}} F(\omega) \exp(it_m \omega) d\omega$$

$$= \exp(-imn2\pi) \int_{-\infty}^{\infty} F(\omega) \exp(it_m \omega) d\omega$$

$$= f(t_m) \tag{7.41}$$

Thus, $\tilde{F}(0)$ according to the formula in Eq. (7.40) yields

$$\tilde{F}(0) = \frac{1}{\Omega} \sum_m f(m \frac{2\pi}{\Omega}) \tag{7.42}$$

On the other hand, using the definition of $\tilde{F}(\omega)$ in Eq. (7.39) we obtain the same $\tilde{F}(0)$ as

$$\tilde{F}(0) = \sum_n F(n\Omega) \tag{7.43}$$

so that we can write the *Poisson summation* as a theorem.

Theorem 7.5 (Poisson summation). *If f and F are a FT pair, then their sample sums fulfill*

$$\sum_n F(n\Omega) = \frac{1}{\Omega} \sum_m f(m \frac{2\pi}{\Omega}) \tag{7.44}$$

♦

The power of this formula is appreciable because we did not require F to have finite frequencies nor f to have a finite extension. It says that if we know an FT pair, then we can deduce their discrete sums from one another.

The *Comb distribution* is defined as a train of δ distributions:

$$\mathrm{Comb}_T(t) = \sum_n \delta(t - nT) \tag{7.45}$$

It is a convenient analytic tool when sampling physical functions, as it relates sampling in one domain with the periodization in the other domain, elegantly. However, we need to derive its behavior under the FT before exploiting it.

The left side of Eq. (7.44) is

$$\sum_n F(n\Omega) = \langle \sum_n \delta(\omega - n\Omega), F \rangle \tag{7.46}$$

whereas the right side of Eq. (7.44) yields

$$\frac{1}{\Omega} \sum_m f(m\frac{2\pi}{\Omega}) = \frac{1}{\Omega} \langle \sum_m \delta(t - m\frac{2\pi}{\Omega}), f \rangle$$

$$= \frac{1}{\Omega} \int \sum_m \delta\left(t - m\frac{2\pi}{\Omega}\right) \left(\int F(\omega) \exp(it\omega) d\omega\right) dt$$

$$= \frac{1}{\Omega} \int F(\omega) \left[\int \sum_m \delta\left(t - m\frac{2\pi}{\Omega}\right) \exp(it\omega) dt\right] d\omega$$

by changing the order of the integrations and the summation. Because $\delta(t) = \delta(-t)$, we can perform a variable substitution in the inner integral,

$$\frac{1}{\Omega} \sum_m f(m\frac{2\pi}{\Omega}) = \frac{1}{\Omega} \int F(\omega) \left[\int \sum_m \delta\left(t - m\frac{2\pi}{\Omega}\right) \exp(-it\omega) dt\right] d\omega$$

$$= \frac{1}{\Omega} \int F(\omega) \mathcal{F}\left[\sum_m \delta\left(t - m\frac{2\pi}{\Omega}\right)\right](\omega) d\omega$$

$$= \langle \frac{1}{\Omega} \mathcal{F}\left[\sum_m \delta\left(t - m\frac{2\pi}{\Omega}\right)\right], F \rangle \qquad (7.47)$$

and identify it formally as a FT of a Comb. In fact, because δ is a distribution and F was an arbitrary function, the object

$$\mathcal{F}\left(\sum_n \delta(\omega - n\Omega)\right) \qquad (7.48)$$

is also a distribution which, thanks to Eqs. (7.44), (7.46), and (7.47),

$$\sum_n F(n\Omega) = \langle \sum_n \delta(\omega - n\Omega), F \rangle = \langle \frac{1}{\Omega} \mathcal{F}\left[\sum_m \delta\left(t - m\frac{2\pi}{\Omega}\right)\right], F \rangle \ (7.49)$$

acts in a well-defined manner on arbitrary functions under integration. Accordingly, the last equation delivers the FT of a Comb, which we state in the following theorem.

Theorem 7.6 (FT of a Comb). *The FT of a Comb distribution yields a Comb distribution*

$$\mathcal{F}\left[\sum_m \delta\left(t - m\frac{2\pi}{\Omega}\right)\right] = \Omega \sum_n \delta(\omega - n\Omega) \qquad (7.50)$$

♦

Up to now, we did not impose any restriction on $F(\omega)$. We now assume that F is zero outside of $-\frac{\Omega}{2}, \frac{\Omega}{2}$. Then, we can, by applying the right-hand of the theorem, see that the periodized F is obtained as a convolution:

$$\tilde{F}(\omega) = (\Omega \sum_n \delta(\omega - n\Omega)) * F$$

$$= \Omega \sum_n \delta(\omega - n\Omega) * F$$

$$= \Omega \sum_n F(\omega - n\Omega) \qquad (7.51)$$

As there is no overlap, the information relative to F is still intact after the periodization of F. Since convolution in the ω-domain is equivalent to a multiplication in the t-domain Eq. (7.51) is equivalent to a multiplication in t-domain,

$$\left[\sum_m \delta \left(t - m\frac{2\pi}{\Omega} \right) \right] f(t) = \sum_m f \left(m\frac{2\pi}{\Omega} \right) \delta \left(t - m\frac{2\pi}{\Omega} \right) \qquad (7.52)$$

where Eq. (7.50) has been used. However, a periodization of the ω-domain corresponds to a sampling in the t-domain. Accordingly, we conclude that a multiplication by a Comb distribution yields the mathematical representation of sampled function. The following lemma, summarizes this and restates the Nyquist theorem:

Lemma 7.3. *Let f be a finite frequency function with F vanishing outside of $-\frac{\Omega}{2}, \frac{\Omega}{2}$. Then the sampled f is given by the distribution*

$$Comb_T(t) f(t) = \sum_m f(mT)\delta(t - mT) \qquad (7.53)$$

with $T < \frac{2\pi}{\Omega}$. A lossless reconstruction of f is achieved by convolving this distribution with the inverse FT of the characteristic function χ_Ω.

7.6 Hermitian Symmetry of the FT

The Hermitian symmetry of FCs were discussed in Sect. 5.4. There is an analogous result even for FT. We suggest the reader to study it in the frame of the following exercises.

Exercise 7.3. *Show that the $F(\omega)$ that is the Fourier transform of $f(t)$ is Hermitian:*

$$F(\omega) = F(-\omega)^* \qquad (7.54)$$

Do the same in 2D, i.e., when $F(\omega_x, \omega_y)$ is the FT of $f(x, y)$, show that

$$F(\omega_x, \omega_y) = F(-\omega_x, -\omega_y)^* \qquad (7.55)$$

Exercise 7.4. *Show that even DFT coefficients are Hermitian. Which coefficients represent $F(-2, -2)$ in a 256×256 image (the first element is labelled as 0 in both dimensions)?*

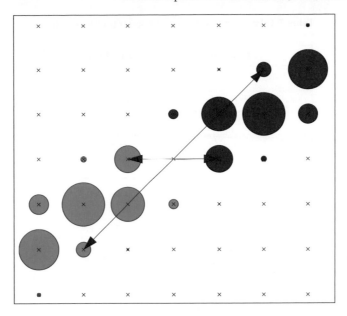

Fig. 7.2. FT of a real function is Hermitian, i.e., the points mirrored through the origin have the same absolute value, shown as the *sizes of circles*, but negative arguments, shown as the *red* hue and its negative hue, *cyan*

Exercise 7.5. *Take a digital image and apply the DFT to it. Put all strictly negative horizontal frequencies to zero, i.e., half the frequency plane. Inverse Fourier transform this, and display the real part and the imaginary part. The complex image you obtained is called the* analytic *signal in signal processing literature. Comment on your result.*

Exercise 7.6. *Now fill the points in the half-plane with zero values with values from the other half-plane according to Eq. (7.55). Comment on your result.*

7.7 Correspondences Between FC, DFT, and FT

We started with the Fourier series and derived from it the FC transform, theorem 5.2. By theorem 6.1, we showed that a generalization of the FCs leads to FT if the finite extension restriction imposed on functions is removed. In that we also established that not only the t-domain but also the ω-domain can be continuous. This led to the idea of reformulating the FC theorem for the finite frequency functions, yielding theorem 6.2. From both of the theorems FC I and FC II, we could derive a fully discrete transform pair, DFT, where the number of FCs synthesizing the current finite extension function is finite, theorem 6.4.

Table 7.1. Frequently referenced Fourier transform pairs

Function	FT of function
$f(t)$	$F(\omega)$
$f(t + t')$	$\exp(-it'\omega)F(\omega)$
$\frac{\mathrm{d}f(x)}{\mathrm{d}x}$	$i\omega F(\omega)$
$(f * g)(x)$	$2\pi F(\omega)G(\omega)$
$\delta(t)$	$\frac{1}{2\pi}$
$\sum_m \delta(t - mT)$	$\frac{2\pi}{T} \sum_n \delta(\omega - n\frac{2\pi}{T})$
$\chi_T(t) = \begin{cases} 1, & \text{if } t \in [-\frac{T}{2}, \frac{T}{2}]; \\ 0, & \text{otherwise.} \end{cases}$	$\frac{T}{2\pi} \frac{\sin(\frac{T}{2}\omega)}{(\frac{T}{2}\omega)}$
$\frac{1}{\sqrt{2\pi\sigma^2}} \exp(-\frac{t^2}{2\sigma^2})$	$\exp(-\frac{\omega^2}{2(\frac{1}{\sigma})^2})$

The same essential properties hold true for FTs, FCs and DFTs. For this reason and to keep the clutter of variables and indices to a minimum, it is natural to work with FTs, at least at the design stage of applications. We will summarize the important aspects of FT pair interrelationship by using only the forward and inverse FT integrals. In table 7.1, we list some commonly used properties of the transform, and useful FT pairs in signal analysis studies. Here, we recall that lemma 6.1 is convenient to use to obtain some other entries for the table.

Exercise 7.7. *What is the FT of* $\delta(t - t_0)$ *?*
HINT: Study the FT of $f(t - t')$.

Exercise 7.8. *What is the FT of* $\cos(t)$ *?*
HINT: Use $\cos(t) = \frac{1}{2}(\exp(it) + \exp(-it))$.

By interpreting the symbols according to the *FT correspondence table* (Table 7.2), the FT pair can represent both variants of the FC transform and the DFT. However, it should be noted that the table can not be used for translation of FT pairs. For example, the FT of a Gaussian is a Gaussian, but the FC and DFTs of a Gaussian are strictly speaking not sampled Gaussians, although for many purposes this is a good approximation.

Exercise 7.9. *What are the FT, FC and DFT of* $\cos(t)$ *on the appropriate parts of the real axis?*
HINT: Apply FT, FC, and DFT to translated versions of δ *while interpreting the latter as Dirac and Kronecker-δs.*

Exercise 7.10. *Plot the FT, FC and DFT of a Gaussian by using a numerical software package. How should we proceed to approximate a true Gaussian by the results better and better?*

Table 7.2. Correspondence table for forward and inverse FT, FC, and DFT

$$F(\omega) = \mathcal{F}(f)(\omega) = \frac{1}{P} \int_{t \in \mathcal{D}} f(t) \exp(-i\omega t) dt \qquad \textbf{(Forward)}$$

$$f(t) = \mathcal{F}^{-1}(F)(t) = \int_{\omega \in \mathcal{D}} F(\omega) \exp(it\omega) d\omega \qquad \textbf{(Inverse)}$$

	FT	FC I	FC II	DFT
FORWARD				
Symbol	$\mathcal{F}$	$\mathcal{F}$	$\mathcal{F}$	DFT
Constant	$P = 2\pi$	$P = 2\pi$	$P = 1$	$P = N$
exp	$\exp(-i\omega t)$	$\exp\left(-in\frac{2\pi}{T}t\right)$	$\frac{1}{\Omega}\exp\left(-im\frac{2\pi}{\Omega}\omega\right)$	$\exp(-in\frac{2\pi}{N}m)$
Sum rule	$\int$	$\int$	$\sum_m$	$\sum_m$
$\mathcal{D}$	$t \in [-\infty, \infty]$	$t \in [\frac{-T}{2}, \frac{T}{2}]$	$m \in Z$	$m \in \{0, \cdots, N-1\}$
Translation	$t + t'$	$\mathrm{Mod}(t + t', T)$	$m + m'$	$\mathrm{Mod}(m + m', N)$
δ	Dirac-δ	Dirac-δ	Kronecker-δ	Kronecker-δ
f	integrable f	FE f	FF f	FEF f
INVERSE				
Symbol	$\mathcal{F}^{-1}$	$\mathcal{F}^{-1}$	$\mathcal{F}^{-1}$	IDFT
exp	$\exp(i\omega t)$	$\frac{2\pi}{T}\exp\left(in\frac{2\pi}{T}t\right)$	$\exp\left(im\frac{2\pi}{\Omega}\omega\right)$	$\exp\left(in\frac{2\pi}{N}m\right)$
Sum rule	$\int$	$\sum_n$	$\int$	$\sum_n$
$\mathcal{D}$	$\omega \in [-\infty, \infty]$	$n \in Z$	$\omega \in [\frac{-\Omega}{2}, \frac{\Omega}{2}]$	$n \in \{0, \cdots, N-1\}$
Translation	$\omega + \omega'$	$n + n'$	$\mathrm{Mod}(\omega + \omega', \Omega)$	$\mathrm{Mod}(n + n', N)$
δ	Dirac-δ	Kronecker-δ	Dirac-δ	Kronecker-δ

FE: Finite extension functions
FF: Finite frequency functions (band-limited functions)
FEF: Finite extension and frequency functions

8

Reconstruction and Approximation

Often we only know gray values of an image on a discrete set of points. This is increasingly the case due to the ever more affordable digital imaging equipment, e.g., consumer still and motion picture cameras. Even when the original image is on film, it must often be digitized for further processing on digital equipment, e.g., when analog processing is not available. In this chapter we study the techniques used to obtain some discrete image processing schemes for the most common mathematical operations.

We discuss first the interpolation function and the characteristic function as this concept has a significant impact on how one implements approximation of functions and operators. Then we study the important class of linear operators and illustrate them by computing the partial derivatives of images, and affine coordinate transformations, e.g., rotation and zooming, of images. Subsequently, we contrast the linear operators with nonlinear operators and discuss "square of an image", and "multiplication of two images", which are nonlinear.

The basic approach to approximation of mathematical operators is to reconstruct the continuous image from its samples and then to apply the operator followed by an appropriate sampling of the result. While discussing the typical pitfalls, our goal is to generate discrete signal processing schemes using discrete data to approximate continuous operators.

8.1 Characteristic and Interpolation Functions in N Dimensions

Definition 8.1. *The* characteristic function *on a limited volume $\mathcal{D}$ in N dimensions (ND) is defined as*

$$\chi_D(\mathbf{r}) = \begin{cases} 1, & \textit{if } \mathbf{r} \in \mathcal{D}; \\ 0, & \textit{otherwise}. \end{cases} \tag{8.1}$$

This function is also known as the *indicator function* because it is used to indicate two volumes, one in which the function is zero, and the complementary volume in which it is not zero.

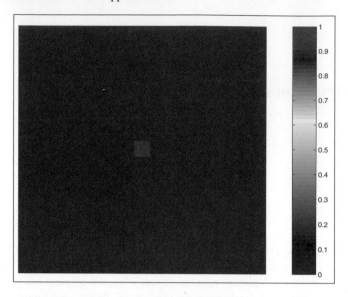

Fig. 8.1. A characteristic function that corresponds to a square in 2D.

We now multiply f with the characteristic function $\chi_{\mathcal{D}}(\mathbf{r})$ to obtain a limited extension function from a periodic function, $\chi_{\mathcal{D}}(\mathbf{r}) f(\mathbf{r})$. We assume that $\mathcal{D}$ fits entirely in a central hypercube having the vertices $\pi(\pm 1, \cdots, \pm 1)^T$, the *Nyquist block*. The Nyquist block defines one period of the function f which is repeated in all possible coordinate axis directions to cover the entire ND space. The boundaries of the Nyquist block have the distance to the origin, π, implying that the distances between samples are integer values in directions parallel to the coordinate axes. We wish to study how the finite extension function $\chi_{\mathcal{D}}(\mathbf{r}) f(\mathbf{r})$ is Fourier transformed. This problem is the same as the one discussed previously, except that we now have multiple dimensions.

$$F(\boldsymbol{\omega}) = \frac{1}{2\pi} \int_{-\infty}^{\infty} \chi_{\mathcal{D}}(\mathbf{r}) f(\mathbf{r}) \exp(-i\boldsymbol{\omega}^T \mathbf{r}) d\mathbf{r} \tag{8.2}$$

$$= \frac{1}{2\pi} \int_{-\infty}^{\infty} \chi_{\mathcal{D}}(\mathbf{r}) (\sum_m F(\omega_m) \exp(i\boldsymbol{\omega}_m^T \mathbf{r}) \exp(-i\boldsymbol{\omega}^T \mathbf{r}) d\mathbf{r} \tag{8.3}$$

Here we have used the fact that a periodic function can be expanded in its Fourier coefficients, $F(\omega_m)$, given by an ND version of theorem 5.2. By changing the order of integration and summation we obtain

$$F(\boldsymbol{\omega}) = \sum_m F(\omega_m) \frac{1}{2\pi} \int_{-\infty}^{\infty} \chi_{\mathcal{D}}(\mathbf{r}) \exp(i\boldsymbol{\omega}_m^T \mathbf{r}) \exp(-i\boldsymbol{\omega}^T \mathbf{r}) d\mathbf{r} \tag{8.4}$$

so that

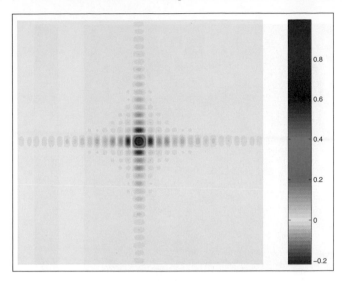

Fig. 8.2. The interpolation function corresponding to a square region. The cross section of this function is periodically negative, as is shown in Fig. 8.4

$$F(\boldsymbol{\omega}) = \sum_m F(\boldsymbol{\omega}_m)\frac{2\pi}{T}\frac{1}{2\pi}\int_{-\infty}^{\infty}\chi_{\mathcal{D}}(\mathbf{r})\exp(-i(\boldsymbol{\omega}-\boldsymbol{\omega}_m)^T\mathbf{r})d\mathbf{r} \qquad (8.5)$$

$$= \frac{2\pi}{T}\sum_m F(\boldsymbol{\omega}_m)\mathcal{F}(\chi_{\mathcal{D}})(\boldsymbol{\omega}-\boldsymbol{\omega}_m) \qquad (8.6)$$

Eq. (8.6) states that the Fourier transform of a limited extension function is a discrete sum of shifted continuous functions weighted with some coefficients. These continuous functions can be identified as interpolation functions in ND. They are, as the equation indicates, determined entirely by the characteristic function $\chi_{\mathcal{D}}$, and $\mathcal{D}$ is in turn determined by the volume $\mathcal{D}$, outside of which the function f vanishes.

The characteristic function on a (limited) symmetric interval in 1D is given as

$$\chi_T(x) = \begin{cases} 1, & \text{if } x \in [\frac{-T}{2}, \frac{T}{2}]; \\ 0, & \text{otherwise.} \end{cases} \qquad (8.7)$$

The FT of this characteristic function yields

$$\mathcal{F}(\chi_T)(\omega) = \frac{T}{2\pi}\text{sinc}(\omega\frac{T}{2}), \qquad (8.8)$$

a fact already established in Eq. (6.22).

The repetition of an ND volume in the $\mathbf{r}$-domain, which is also called a tessellation, is always achievable by use of the suggested Nyquist block. This corresponds to a hypercubic sampling in the $\boldsymbol{\omega}$-domain. Conversely, we can switch the roles of $\mathbf{r}$ and $\boldsymbol{\omega}$ and arrive at the conclusion that a band-limited signal $f(\mathbf{r})$ can always be

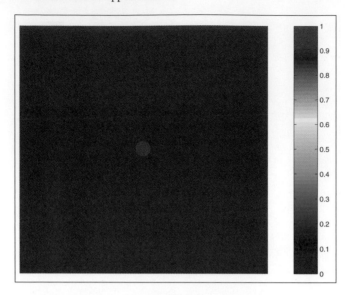

Fig. 8.3. The circular region defining a characteristic function

sampled at the vertices or at the centers of a hypercube. This is done by using a sufficiently large hypercube that entirely contains the volume $\mathcal{D}'$ of the characteristic function, the Nyquist block, which is now defined in the ω-domain. Fortunately, hypercubic sampling is utilized to obtain discrete images from imaging devices in most applications, e.g., a scanner, a digital camera, a digital video camera, magnetoresonance camera, and so on. However, other repeatable volumes, such as repeating a hexagonal region in a hexagonal manner in 2D, are also possible. The corresponding interpolation functions can be identified by Fourier transformation of the repeated volume containing the characteristic function.

The information that encodes the shape of the boundary of the characteristic function is sufficient to determine the interpolation function in the "other" domain which can be represented accurately by sampling. In 1D we cannot talk about the shape of the boundaries because the boundaries are points, which results in the sinc functions as interpolators when $\mathcal{D}$ is a contiguous interval. If it consists of several juxtaposed intervals, then the interpolation function is expressed as sums and differences of sinc functions.

In 2D, an extension of a central interval to a square interval results in a simple characteristic function too. Assuming that it is the ω-domain that contains the limited extension function F, this leads to the characteristic function

$$\mathrm{sinc}(\pi x) \cdot \mathrm{sinc}(\pi y) \tag{8.9}$$

in the **r**-domain, which is illustrated by Fig. 8.1. Likewise, in 3D the analogous interpolation function is defined by

$$\mathrm{sinc}(\pi x) \cdot \mathrm{sinc}(\pi y) \cdot \mathrm{sinc}(\pi z) \tag{8.10}$$

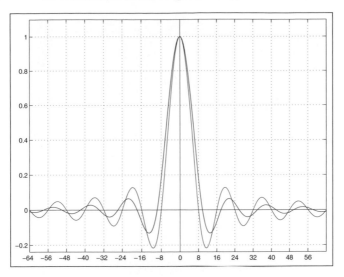

Fig. 8.4. The cross sections (along the horizontal axes) of the interpolators shown in Fig. 8.5 and 8.2. Note that both of the latter 2D interpolators have bandwidth of $\frac{\pi}{8}$ (horizontally). The circular region interpolator, the *Besinc function*, is shown in *blue*, whereas the quadratic region interpolator, the sinc function, is shown in red

It is straightforward to show that both in 2D as well as higher dimensions, characteristic functions defined on such rectangular volumes always lead to products of sinc functions of the cartesian variables in the "other" domain.

Admittedly, the sinc functions are suggested by the theory as the ideal interpolation functions for 1D band-limited signals, and they can be easily extended to higher dimensions. However, from this, one should not hasten to conclude that the sinc functions are the ideal (and the best) interpolation functions in practice. This is far from the truth because of the various shortcomings of the sinc functions. To start with, the sinc functions are not ideal interpolators for 2D and higher dimensions even in theory, since it is most probable that natural ND signals to be processed will have characteristic functions (in the $\boldsymbol{\omega}$-domain) that do not have direction preference, e.g., the average camera images in 2D are best described by characteristic functions with boundaries that are circles. In 2D, such a circular characteristic function results in an interpolation function that is obtained by substituting $\pi\|\boldsymbol{\omega}\| = \sqrt{\omega_x^2 + \omega_y^2}$ into a 1D function, as below:

$$\text{Besinc}(\|\boldsymbol{\omega}\|) = \frac{2J_1(\pi\|\boldsymbol{\omega}\|)}{\pi\|\boldsymbol{\omega}\|} \tag{8.11}$$

where J_1 is a *Bessel function* (of the first-order of the first kind). Since it is the analogue of the sinc function, we call Eq. (8.11) *Besinc function*. The interpolator function is illustrated by Fig. 8.3, whereas its profile through the origin, the Besinc function, is given in Fig. 8.4.

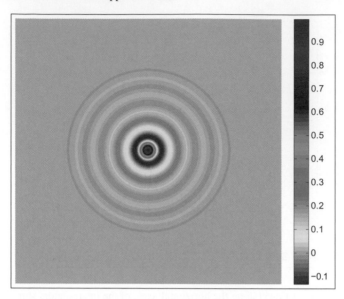

Fig. 8.5. The interpolation function corresponding to a circular characteristic function. A cross section of this function is shown in Fig. 8.4

The most important reasons why sinc functions fall out of grace in practice are, however, as follows:

 i) They decrease as $1/\|\mathbf{r}\|$, which is too slow, yielding very large filters.
 ii) They are not direction isotropic, meaning that in 2D and higher dimensions, sharp changes of image intensities with well-defined directions are not treated by a sinc interpolator in the same manner for all directions, creating artifacts. We show this in Sect. 9.4 via Example 9.3.
 iii) They introduce ringing near sharp changes of image intensities, creating artifacts.

Point i) is a sufficient reason to look for other interpolation functions in 1D. In 2D and higher dimensions, the subsequent points are so imposing that sinc functions are almost never the first choice.

Consider the functions that look like tents with the maximum 1

$$\mu(t) = \begin{cases} 1 - t, & \text{if} \quad 0 < t \le 1 \\ 1 + t, & \text{if} \quad -1 \le t \le 0 \\ 0, & \text{otherwise;} \end{cases} \tag{8.12}$$

as illustrated by Fig. 8.6. This is the linear interpolation function, which, because of its appearance, is sometimes called a *tent function*. Equation (8.15) suggests the weighting of the shifted versions of the continuous interpolation function, μ, with the known image values f on the grid $\mathbf{r}_l$. When these amplified functions are summed, the continuous signal $f(\mathbf{r})$ is obtained from its discrete values $f(\mathbf{r}_l)$. This process is called *synthesis* or *reconstruction*, and is illustrated by Fig. 8.6 for tent functions. Since the sum of two linear functions is another linear function, and because the interpolation functions are exact on the grid (integers of the x-axis, in the figure), the approximation using Eq. (8.12) results in a piecewise linear function. Continuous by construction, the approximating function is delivered by the weighted tent summation, which "automatically" joins the values of the original function on a set of discrete points. Thus, the approximation is error free if the function to be approximated is piecewise linear. If the constant "1" is to be approximated, then the entire set of interpolation functions is summed and the result is error free as shown in the figure. However, the linear interpolator suffers from the following drawbacks:

i) Extending them to 2D via $\mu(x)\mu(y)$ will make them anisotropic because this 2D function has isocurves that are not circles (but biased by squares). This means that certain directions of edges will be artificially favored over the others.

ii) It is not differentiable everywhere. As a result we cannot construct all partial derivatives easily everywhere. We discuss the partial derivative operator in the next section.

Another function family, called *B-splines*, can be obtained by successive convolution of the linear interpolator by itself. The result is a piecewise polynomial interpolator that is smooth. Because they do not suffer from the second point above, B-splines are often preferred over the linear interpolator [123]. They have found many applications because of their speed thanks to their separability in x and y. However, they too suffer from being anisotropic.

A nonpolynomial interpolator, that we will study further in Sect. 9.2 is the *Gaussian* interpolator, which can be shown to be the asymptotic limit of B-splines [221]. In Fig. 8.7 we show a set of shifted Gaussians that are used to illustrate a synthesis process. A Gaussian is isotropic when extended to 2D via $\mu(x)\mu(y)$, and it is infinitely differentiable, everywhere. In the next section, it suffices to know that there exists an interpolation function that has a localized support (it decreases rapidly to zero), and that the choice is not restricted to the sinc functions.

8.2 Sampling Band-Preserving Linear Operators

We will be concerned here with *continuous operators* acting on vector spaces, in particular, function spaces. Such operators act on functions and deliver elements that stay in the same function space, as a result. The continuity of an operator means that a small change of its argument (a function) results in a small change of its result (also a function). Derivation, convolving with a particular filter, and taking the square root of a function are examples of operators. We start by defining linear operators.

Definition 8.2. *An operator* $\mathcal{T}$ *is a* linear operator *if it satisfies*

$$\mathcal{T}(f + g) = \mathcal{T}(f) + \mathcal{T}(g) \tag{8.13}$$
$$\mathcal{T}(\lambda f) = \lambda \mathcal{T}(f) \tag{8.14}$$

where λ *is a scalar.*

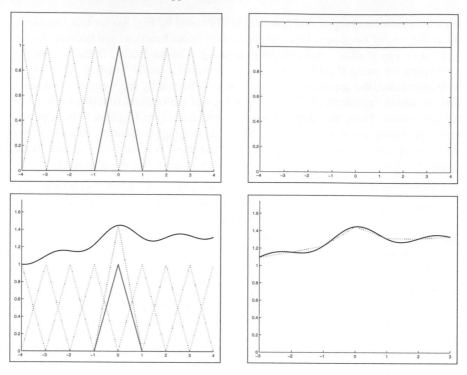

Fig. 8.6. (*Top, left*) The graph illustrates the set of linear interpolators on a discrete set of points. (*Top, right*) Summing up the interpolators, approximates the constant function 1. (*Bottom, left*) A function to be approximated along with an interpolator amplified by its function sample. (*Bottom, right*) The green curve is the result of the approximation, the sum of the amplified interpolators

Assume that we are given a set of function values $f(\mathbf{r}_l)$, where $\mathbf{r}_l$ is a discrete set of points that we will call *grid*. From this, the continuous $f(\mathbf{r})$ can be synthesized by using Eq. (8.6)

$$f(\mathbf{r}) = \sum_l f(\mathbf{r}_l)\mu(\mathbf{r} - \mathbf{r}_l) \tag{8.15}$$

with μ being an interpolation function. This continuous *reconstruction* is merely a mental process that is done to implement continuous linear operators discretely. Ultimately, we wish to be able to implement every continuous operator discretely. However, there is no effective formula that will fit all types of operators. Instead, we aim here to illustrate how to implement some frequently occurring linear continuous operators, in particular, those that do not change the bandwidth of the functions. The latter is an important restriction that raises the hope of success because we can always represent every function with finite frequencies well enough by its samples. We illustrate the technique by showing how to implement partial derivation and noninteger shifts on a discrete grid.

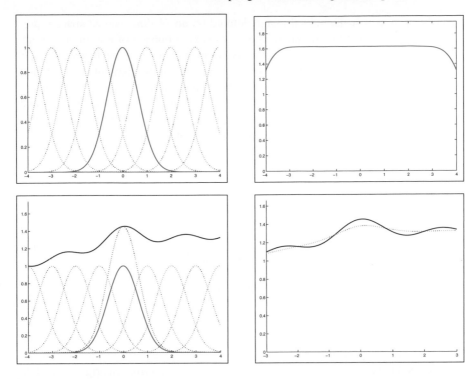

Fig. 8.7. (*Top, left*) The graph illustrates a set of Gaussians, with $\sigma = 0.65$, that are used as differentiable interpolators. (*Top, right*) Summing up the interpolators approximates a constant function. (*Bottom, left*) A function to be approximated (the same as in Fig. 8.6) along with an interpolator amplified by its function sample. (*Bottom, right*) The *green curve* is the result of the approximation, the normalized sum of the amplified interpolators

Partial Derivatives

In image analysis, the partial derivative of a function, which is only discretely available, is a frequently demanded operation that ideally should also result in a discrete image. Ideally, the computations performed on the discrete samples of the image should result in a discrete version of the result delivered by the continuous operator applied to the continuous image. To take the partial derivative of a function is evidently linear because

$$\frac{\partial}{\partial x}(f + g) = \frac{\partial}{\partial x}(f) + \frac{\partial}{\partial x}(g), \quad \text{and} \quad \frac{\partial}{\partial x}(\lambda f) = \lambda \frac{\partial}{\partial x}(f). \quad (8.16)$$

The partial derivative is one of the most frequently used operators in image processing, e.g., to extract edges, lines, direction, curvature, and texture properties. In particular, arbitrary *partial derivatives* of a differentiable scalar function $f(\mathbf{r})$, where $\mathbf{r} = (x_1, x_2, \cdots x_N)^T$, are represented by $\frac{\partial f(\mathbf{r})}{\partial x_j}$ with $j = 1, 2, \cdots N$. The function

f can be seen as a scalar-valued image defined on an N-dimensional space represented by $\mathbf{r}$. If f has finite frequencies, i.e., F is zero outside of a limited volume, $\frac{\partial f}{\partial x_j}$ will also have finite frequencies. This is because the latter is equivalent to $iw_j F$ in the frequency domain, which will evidently be zero in the same places (volume) as F is zero. Thus, the partial derivatives of the continuous function can be faithfully represented by sampling these functions on the same grid that samples the original image. To compute the partial derivatives, the operator $\frac{\partial}{\partial x_i}$ is applied to Eq. (8.15), and it is moved past the sum because partial derivation is a linear operator:

$$\frac{\partial f(\mathbf{r})}{\partial x_i} = \sum_l f(\mathbf{r}_l) \frac{\partial \mu(\mathbf{r} - \mathbf{r}_l)}{\partial x_i}, \qquad i, j : 1 \cdots N \qquad (8.17)$$

At this point, we must require that the interpolation function μ is differentiable. Assuming this, the linear operator can be absorbed by the interpolator function, and the resulting continuous function can be discretized too. Because $\frac{\partial f}{\partial x}$ has the same finite frequencies as f, we can also sample it on the same grid $\mathbf{r}_k$, yielding

$$\frac{\partial f(\mathbf{r}_k)}{\partial x_i} = \sum_l f(\mathbf{r}_l) \frac{\partial \mu(\mathbf{r}_k - \mathbf{r}_l)}{\partial x_i}, \qquad i, j : 1 \cdots N \qquad (8.18)$$

This is a discrete scalar product of two functions, the first of which is the original discrete image. The second function is the discretized interpolation function after it has absorbed the partial derivation operator. As can be seen from the right-hand side of Eq. (8.18), the approximation can be realized by a convolution when the values of $\frac{\partial f}{\partial x_i}$ at all points of the grid have to be computed.

Arbitrary Translation or Shifting

Translating an image $f(\mathbf{r})$ with $\Delta\mathbf{r}$ is another frequently needed operation. Here, $\Delta\mathbf{r}$ does not have to be a multiple of the sampling period that is assumed to be 1. For example the translation could be $\Delta\mathbf{r} = (0.25, 0.75)^T$ in 2D. In the continuous domain, this is equivalent to a coordinate transformation,

$$\mathbf{r}' = \mathbf{r} + \Delta\mathbf{r} \qquad (8.19)$$

so that the translated version of $f(\mathbf{r})$ is given by $f(\mathbf{r} - \Delta\mathbf{r})$. Because the FT of the translated image is $\exp(-i\Delta\mathbf{r}^T \boldsymbol{\omega}) F(\boldsymbol{\omega})$, translation is a band-preserving operator. Without loss of generality, one can assume that all components of the vector $\Delta\mathbf{r}$ are in the open interval $]0, 1[$ because otherwise the integer translation is performed first. Integer translations are conveniently implemented as a permutation. Using Eq. (8.15), the remaining noninteger translation is implemented by sampling the reconstructed and shifted signal at the grid points as

$$f(\mathbf{r}_j - \Delta\mathbf{r}) = \sum_l f(\mathbf{r}_l)\mu(\mathbf{r}_j - \Delta\mathbf{r} - \mathbf{r}_l) \qquad (8.20)$$

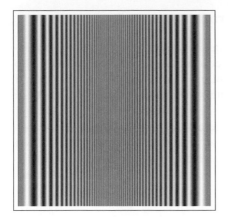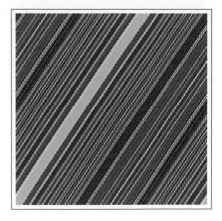

Fig. 8.8. A 1D function (a line) containing slow and rapid gray changes is translated with zero shifts (*left*) and 0.67 points (noninterpolated) shifts (*right*) between successive lines. Each color represents a unique gray level of the original, which is given in color in Fig. 12.4 along with the interpolated shifts. The *vertical axis* is time

The result is a scalar product between the shifted, sampled interpolation function, which easily lends itself to be implemented as a convolution.

If this scalar product is not implemented, translation will result in aliasing effects. This is illustrated for 1D functions by Fig. 8.8, where we show 1D functions as lines of an image. The image on the left is obtained by repeating the same line (the 1D function), i.e., the translation is zero between the successive lines. We applied 0.67 pixels (cyclical) translation to the same 1D function to obtain the successive lines, by rounding off $0.67j$ where j is the horizontal index, of the image on the right. Apart from the jaggedness of the lines, we also see erroneous gray levels in the high-frequency parts, where both aliasing problems are due to the straightforward implementation of the translation. This should be compared to Fig. 12.4 (top and middle), where we show the translation performed according to Eq. (8.20), which contains interpolation.

Note, however, when one produces a (higher dimensional) *motion image* by successive application of a translation to a static image f, such as the one in the example, the resulting image will potentially have a large *temporal* frequency extent as well. This extent depends on the size of the shift as well as on the frequency content of the static pattern f. Because the temporal frequencies must obey the law of sampling to avoid the errors of discretization, for every temporal sampling frequency there is a maximum speed/shift that should not be exceeded. The implications of motion on the spectrum are discussed in further detail in Sect. 12.7.

In conclusion, a linear band-preserving continuous operator can be discretized by discretizing the result of the operator applied to the interpolation function in analogy with the partial derivative operator.

8.3 Sampling Band-Enlarging Operators

Here we discuss how to discretely carry out certain operations that will finitely en-large the volume in which a band-limited function is non-zero, i.e., the result is also a band-limited function, albeit with a different band limitation. There is no general formula for nonlinear operators, although with some care many can be discretized as long as they do not enlarge the frequency support to infinity, i.e., the volume of frequencies in which F is zero diminishes to zero after the operation.

Multiplication of two images. Assume that we know the discrete values of two images f and g on the same grid $\mathbf{r}_l$ and we need the discrete values of fg. This is usually referred to as a (pointwise) *multiplication of two images*. An intuitive ap-proximation is

$$(fg)(\mathbf{r}_l) = f(\mathbf{r}_l)g(\mathbf{r}_l). \tag{8.21}$$

However, the use of the original grid $\mathbf{r}_l$ poses a problem because the resulting dis-crete function on the right-hand side of Eq. (8.21) may not represent the continuous $(fg)(\mathbf{r})$ faithfully enough, although $f(\mathbf{r}_l)$ and $g(\mathbf{r}_l)$ may do so with their respec-tive functions $f(\mathbf{r})$ and $g(\mathbf{r})$. The problem is traced to the fact that the continuous fg has a Fourier spectrum that is wider than each of its constituent functions. The multiplication fg in the spatial domain is equivalent to the convolution $F * G$ in the frequency domain to the effect that the result will be "wider" than each of F and G. Particularly undesirable effects may result if the sum of the bandwidths of F and G is larger than π in one or more of their dimensions, $(x_1, x_2, \cdots x_N)^T$. Discretizing such a product of images at the sampling rate of the original grid will yield aliasing errors because of the violation of the fundamental sampling conditions given by the Nyquist theorem. Thus, reconstructing fg from $f(\mathbf{r}_l)g(\mathbf{r}_l)$ will contain unacceptable errors.

To avoid a violation of the Nyquist theorem, discrete images containing very high frequencies should not be multiplied with each other directly first, but they should be up-sampled with a factor 2 to force the highest frequency components to be less than $\pi/2$. Multiplying such images is then risk-free if the product is sampled at the rate of the new, finer grid. Although the continuous product has a larger bandwidth than its constituents, the bandwidth is not as large as it can violate the Nyquist theorem upon sampling on the new grid. A lossy, but faster alternative to this procedure is, prior to multiplication, to apply a lowpass filtering to the image to make sure that no significant frequency components exist outside of the central square having the width and height π in the spectrum. Squaring, being a special case of multiplication of two digital images, is subject to the same reasoning, i.e., no significant power of high-frequency components above the central square with height and width of π should exist upon squaring. The reasoning is extended to ND in a straightforward fashion.

Because $\frac{\partial f}{\partial x_i}$ has the same bandwidth as f, products and squares of the partial derivative images must not contain components outside of the central square with height and width of π.

Rotation of an image. The *rotation of an image* with the arbitrary angle of θ rep-resents a linear operator. It can be achieved by replacing $\mathbf{r}$ in Eq. (8.6) or equivalently in Eq. (8.15), with $\mathbf{Q}\mathbf{r}'$, where $\mathbf{Q}$ is a *rotation matrix*:

$$f(\mathbf{Q}\mathbf{r}') = \sum_l f(\mathbf{r}_l)\mu(\mathbf{Q}\mathbf{r}' - \mathbf{r}_l) \tag{8.22}$$

As an example of rotation matrices, we mention

$$\mathbf{Q} = \begin{pmatrix} \cos\theta \ \sin\theta \\ -\sin\theta \ \cos\theta \end{pmatrix} \tag{8.23}$$

which is the matrix used to rotate 2D images. The new coordinates $\mathbf{r}'$ can be sampled at a desired rate yielding $\mathbf{r}'_k$. If the image f is band-limited so is $f(\mathbf{Q}\mathbf{r}')$ because a rotation merely rotates the spectrum:

$$f(\mathbf{Q}\mathbf{r}'_k) = \sum_l f(\mathbf{r}_l)\mu(\mathbf{Q}\mathbf{r}'_k - \mathbf{r}_l) \tag{8.24}$$

Inspecting this equation shows that rotation is achieved by a scalar product between the image values on the original grid and the interpolation function rotated via $\mathbf{Q}\mathbf{r}'_k$ to align the axes of the new coordinates $\mathbf{r}'$, but sampled at the original grid points. The result of the scalar product is placed in the new grid at the location $\mathbf{r}'_k$. In other words, here too, the rotation operation is absorbed by the interpolation function. The interpolator is aligned to the new grid coordinates, but it is sampled at the old grid points to generate the scalar product coefficients. A scalar product between the latter and the function samples on the old grid delivers the rotated function value on the new grid. The sampling rate on the new grid must be appropriately chosen so it does not lose some spectral components. This is because a 2D image f can be faithfully sampled on a rectangular grid only if F is confined to the square $[-\pi, \pi] \times [-\pi, \pi]$, (and to the hypercube $[-\pi, \pi] \times [-\pi, \pi] \times ...[-\pi, \pi]$ for an ND image), according to the Nyquist theorem. Rotating f, and thereby F, will require a denser sampling rate on the new grid if all spectral components are to be retained.

In Fig. 8.9 we illustrate the frequency behavior of two example rotations. Using the original density will result in a *rotation aliasing* caused by certain components, see the "magenta" zones, which may not be acceptable to some applications. For example, a rotation with $\theta = \pi/4$ may require a sampling grid that is $\sqrt{2}$ times denser than the old grid if all frequency components are to be retained. For square-shaped 2D images, a frequency band magnification with a factor $\sqrt{2}$ is the largest factor for any rotation. Alternatively, the area between the blue circle and the boundaries of the red square, representing the frequencies that risk causing aliasing, can be simply suppressed by a lowpass filtering as this is an area which could be accepted as noise by the application at hand. Such a filtering can be incorporated directly into the rotation operator and amounts to having a larger kernel to reconstruct f from its samples. Similarly, in ND images having the same size in all coordinate axes, the maximum density magnification due to rotation is $\sqrt{N}$.

Affine warping of an image. Evidently, the same signal processing principles and reasoning used to discretize rotation apply to discretizing *affine coordinate transformations*, also known as *affine warping*. The old coordinates are replaced by

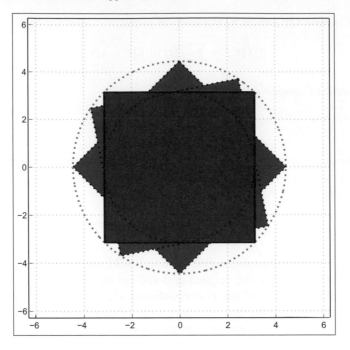

Fig. 8.9. The *red square* shows the basic area $[-\pi, \pi] \otimes [-\pi, \pi]$ that is preserved (and repeated) by a rectangular sampling grid. The *two squares behind* show the same frequency content but rotated with $\frac{\pi}{16}$ and $\frac{\pi}{4}$ radians. The *magenta regions* show the areas that should be suppressed if the original grid density is retained

$$\mathbf{r} = \mathbf{A}\mathbf{r}' + \mathbf{r}_0 \qquad (8.25)$$

where $\mathbf{A}$ is a constant invertible matrix, and $\mathbf{r}_0$ is a constant translation vector. Even affine warping is a linear operator. Because a translation with the constant vector $\mathbf{r}_0$ corresponds to a multiplication with $\exp(-i\mathbf{r}_0^T\mathbf{k})$ in the frequency domain, the frequency support of F is unchanged. Accordingly, only the effect of the matrix multiplication, i.e., $\mathbf{r}_0 = 0$ above, on the spectral support is of relevance when discussing sampling of an affine deformation of an image. Just like in the rotation transformation,

$$\mathbf{r} = \mathbf{A}\mathbf{r}' \quad \Rightarrow \quad f(\mathbf{A}\mathbf{r}') \qquad (8.26)$$

the result of affine warping is a band-limited function, albeit the boundary of the characteristic function of F now undergoes a more general linear transformation instead of a simple rotation,

$$\mathbf{r} = \mathbf{A}\mathbf{r}' \quad \Leftrightarrow \quad \mathbf{k}' = \mathbf{A}^{-1}\mathbf{k} \qquad (8.27)$$

An affine coordinate transformation is therefore absorbed by the interpolation function, and the result is a value of the image at the new grid coordinates but computed via a scalar product on the old grid. The translation part can be implemented according to Eq. (8.20).

9

Scales and Frequency Channels

Blurring a function by a linear filter is equivalent to suppressing high-frequency components of its spectrum. Having void information at high frequencies suggests in turn that there might be a redundancy in the representation which can be avoided. This observation can be utilized in several ways in signal analysis, notwithstanding image analysis, including the following:

1. The amount of data representing an image can be reduced, e.g., image compression.
2. The image can be efficiently filtered to contain only certain frequency components, e.g., recognition of image contents.
3. The image can be resized, e.g., user interfaces and animation.

9.1 Spectral Effects of Down- and Up-Sampling

We consider first the band-limited 1D function $f(t)$ to investigate the effects of up- and down-sampling on $F(\omega)$, the spectrum.

Let the maximum frequency of the band-limited signal f be $\Omega_0/2$ i.e., $F(\omega)$ is effectively zero outside of the interval $[-\frac{\Omega_0}{2}, \frac{\Omega_0}{2}]$. Then f can be sampled with the distance $T_0 = \frac{2\pi}{\Omega_0}$ between the samples, without loss of information. What does happen in the frequency domain when we sample f tighter and tighter using smaller discretization steps than T_0? If information is not lost with the discretization step T_0, then it will definitely not be lost when sampling f with a smaller sampling step, T, with $T < \frac{2\pi}{\Omega_0}$. Sampling in one domain with the sampling period T corresponds to a periodization with the basic period of $\Omega = \frac{2\pi}{T}$ in the other domain.

Conversely, when periodizing F with a period Ω that is larger than the bandwidth of the signal, we pad zeros after the highest frequency of the signal. This has the effect that the part of the void band in which F is practically zero increases with increased sampling rate, i.e., with smaller T. Both up- and down-sampling can be achieved via continuous reconstruction as follows:

1. Confine F to the basic period around the origin. This operation has the purpose to hinder F from being periodic because if F is not periodic but limited, then f is continuous. This is achieved by multiplying F with a suitable characteristic function, $\chi_{\mathcal{D}}(\omega)$. Such a multiplication is equivalent to a convolution in the spatial domain, leading to a reconstruction of the continuous signal from its sampled signals, Eq. (8.6):

$$f(t) = \sum_m f(t_m)\mu_0(t - t_m) \tag{9.1}$$

where $\mu_0 = \mathcal{F}^{-1}(\chi_{\mathcal{D}})$.

2. Resample the continuous signal at the desired rate, which at all circumstances should be done with a finer step size than what the highest frequency of the signal content allows (Nyquist theorem). Effectively, this sampling of Eq. (9.1) amounts to, sampling the interpolation function at the desired points. The equation itself becomes an ordinary discrete scalar product between the filter samples and the old function samples, $f(t_m)$:

$$f(t'_n) = \sum_m f(t_m)\mu_0(t'_n - t_m) \tag{9.2}$$

In Sects. 9.2 and 9.3 we will discuss further the choice of μ_0 in practice. To establish the principles of how the interpolation functions are used when performing up and down-sampling of discrete signals, which we do next, it is sufficient to imagine them as functions that look like tents for now.

Down-sampling

Here we will *down-sample* a given discrete signal $f(mT_0)$ with an integer factor of κ by destroying as little information as possible. In that, we follow the procedure outlined in items and 1 and 2, above.

We reconstruct the continuous signal $f(t)$ and obtain

$$f(t) = \sum_m f(mT_0)\mu_0(t - mT_0) \tag{9.3}$$

where μ_0 is the interpolation function, the effective width of which is inversely proportional to Ω_0. The highest frequency content of the signal $f(t)$ is $\frac{\Omega}{2}$ and the grid is a regular grid given by $t_m = mT_0$ where[1] $T_0 = \frac{2\pi}{\Omega_0}$.

Before sampling $f(t)$ with the new step size $T > T_0$, we must make sure that the continuous f does not contain frequencies outside the new interval $[-\frac{\pi}{T}, \frac{\pi}{T}]$. Some of the high frequencies that were possible to represent with the smaller step size T_0 are no longer possible to represent with the coarser step, T. This can be achieved first in the mind, by convolving the continuous $f(t)$ with a filter, μ_1 having a frequency extension that is the same as the new frequency interval, yielding:

---

[1] For the sake of simplicity we have made this choice, although the argumentation still holds if we choose T_0 as $T_0 \le \frac{2\pi}{\Omega_0}$.

$$\tilde{f}(t) = \sum_m f(mT_0)\mu(t - mT_0) \tag{9.4}$$

where μ is the new interpolation function obtained as a continuous convolution between two lowpass filters, $\mu = \mu_1 * \mu_0$. The interpolation function μ can be assumed to be equal to μ_1 for the following reasons. In the frequency domain this convolution will be realized as a multiplication between two characteristic functions. Ideal characteristic functions[2] assume, by construction, only values 1 and 0 to define intervals, regions, volumes, and so on. In the case of down-sampling, μ will be equal to μ_1 because a large step size implies a characteristic function with a pass region that is narrower than that of a smaller step size. After the substitution $t = nT$ we obtain:

$$\tilde{f}(nT) = \sum_m f(mT_0)\mu(nT - mT_0) \tag{9.5}$$

We can now restate this result by the substitution $T = \kappa T_0$, where κ is a positive integer:

$$\tilde{f}(n\kappa T_0) = \sum_m f(mT_0)\mu(n\kappa T_0 - mT_0) = \sum_m f(mT_0)\mu((n\kappa - m)T_0) \tag{9.6}$$

Here we used the values of f on the original fine *grid* mT_0 to form the scalar product with the filter μ sampled on the same fine grid. We note, however, that the values $\tilde{f}(n\kappa T_0)$ are to be computed at a coarser grid, i.e., at every κth point of the fine grid. As compared to a full convolution, a reduction of the number of arithmetic operations by the factor κ is possible by building the scalar products only at the new grid points, i.e., at every κth point of the original grid. In other words, the number of arithmetic operations per new grid point is M/κ, where M is the size of the filter μ sampled at the original grid positions. However, the size of the discrete filter is directly proportional to the step size $T = \kappa T_0$. Consequently, the number of arithmetic operations per grid point does not change when changing κ.

Up-sampling

Up-sampling with a positive integer factor of κ works in nearly the same way as down-sampling.

First, the continuous signal $f(t)$ is obtained as in Eq. (9.3):

$$f(t) = \sum_m f(mT_0)\mu(t - mT_0) \tag{9.7}$$

where μ is the interpolation function associated with the original *grid* having the period T_0.

---

[2] In practice where computational efficiency, numerical, and perceptional trade-offs must be made simultaneously, the interpolation filters will be inverse FTs of smoothly decreasing functions.

We will up-sample $f(t)$ with the finer step size of T, i.e., $T = \frac{T_0}{\kappa}$. There is ideally no risk of destroying information with a finer step size, which should be contrasted to down-sampling, cancelling the highest frequency contents that must be void for the best fidelity. As before, sampling is achieved by replacing t with nT:

$$f(nT) = \sum_m f(mT_0)\mu(nT - mT_0) = \sum_m f(mT_0)\mu((\frac{n}{\kappa} - m)T_0) \qquad (9.8)$$

Here too we can identify the sum as a scalar product between the two discrete sequences: the values of f on the original coarse grid mT_0 and the continuous interpolation function μ sampled with the original step size T_0. This is because the summation index m generates the interpolator samples, whereas the quantity nT is an offset that remains unchanged as m changes. We note that the values f will be delivered at a finer grid when changing n, i.e., at κ fractions of the original, coarser grid. This will require as many arithmetic operations as the size of the filter μ, sampled at the step of T_0. This size, which we denote by M, can be determined by truncating the filter when it reaches a sufficiently low value at the boundary as compared to its maximum value, typically at 1%. The filter size of μ is proportional to the sampling step T_0. In up-sampling we make T smaller than T_0, but we neither change μ nor T_0 so that the number of arithmetic operations *per point in the up-sampled signal* is M and remains constant for any κ. We note in the last expression of Eq. (9.8) that a change of n results in a shift of μ, but only as much as an integer multiple of T_0/κ, where κ itself is a positive integer. After κ consecutive changes of n, the original grid is obtained cyclically.

The function $f(nT)$ can be generated as κ convolutions, each using its own fractionally shifted up-sampling filter, μ. This is because in Eq. (9.8) the filter coefficients used to generate $f(nT)$ are the same as those generating $f(n'T)$ only if $\text{Mod}\,(n - n', \kappa) = 0$. An alternative is to view $\mu(nT - m\kappa T)$ as a continuous function μ sampled with the step T, but every κth point is retained when forming the scalar product. This is equivalent to filling $\kappa - 1$ zeros between the available values of the discrete signal to be up-sampled, i.e., $f(nT_0)$, while retaining all points of μ:

$$f(nT) = \sum_m f(m\kappa T)\mu(nT - m\kappa T) = \sum_l \tilde{f}(nT)\mu((n - l)T) \qquad (9.9)$$

where

$$\tilde{f}(mT) = \begin{cases} f(mT), & \text{if } m \text{ Mod}\kappa = 0; \\ 0, & \text{else.} \end{cases} \qquad (9.10)$$

Effectively, Eq. 9.9 implements up-sampling by first generating a sparse sequence $\tilde{f}$ and then smoothing this by the filter μ, *sampled with the step size T*. This process is simpler to remember and to implement, but one might argue that it is less efficient because it involves $\kappa - 1$ multiplications with zeros. This is indeed the case, if these are truly computed. On the other hand, multiplications with zeros occur cyclically, meaning that they can be eliminated with some care in the implementation.

Example 9.1. *We illustrate, in Fig. 9.1, the down-sampling process in 1D for $\kappa = 2$, i.e., we wish to reduce the number of samples of a discrete signal by half keeping as much descriptive power in the resulting samples as possible.*

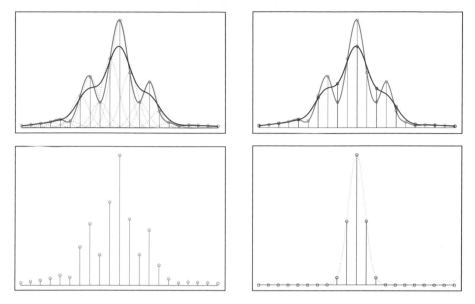

Fig. 9.1. The graphs in the *top row* illustrate the reconstruction of a 1D function and down-sampling by the factor 2. The *bottom row* illustrates the data used that achieve this: the original discrete signal and the sampled filter

1. *Figure 9.1 top, left illustrates a sampled discrete signal (green stems) along with the original (green solid). The dashed magenta curves show the interpolation functions that can be generated by amplifying the interpolation function with the corresponding function values. With the magenta solid curve we represent the reconstructed signal, i.e., the signal that is obtained from the samples by summing up the dashed magenta curves. The reconstructed signal (magenta curve) is a version of the original which lacks rapid variations. We wish to sample the original at a larger step size than the distance between the shown samples. However, a larger step size means smaller repetition period in the frequency domain, meaning that high frequencies must be deleted before the repetition takes place (i.e., sampling in the time domain). If we do not suppress the high frequencies, we will introduce undesired artifacts to the resulting discrete signal. To achieve this implicit smoothing effect, the interpolation function must be chosen twice as wide as it needs to be to reconstruct the original signal.*
2. *On the top, right of the same figure, we show with magenta samples the desired values, which represent the samples of the lowpass filtered reconstructed signal using twice as large a discretization step as the original discretization.*
3. *On the bottom we show how this down-sampling is achieved by processing only discrete signal samples. On the left we show the discrete signal to be down-sampled. On the right we show the sampled interpolation function. The down-sampled signal (shown as the magenta samples in top, right) is obtained by con-*

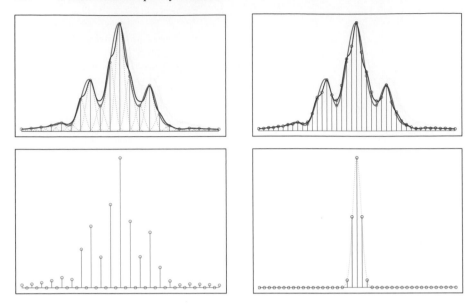

Fig. 9.2. The *top row* illustrates the reconstruction of a 1D function (as in Fig. 9.1) and its up-sampling by the factor 2. The *bottom row* illustrates the data used that achieves this: the original discrete signal and the sampled filter

volving the two discrete signals at the bottom and deleting every second. The deleted samples do not need to be computed if efficiency is cared for in the implementation.

Example 9.2. *In this example, Fig. 9.2, we illustrate the up-sampling of a 1D signal with the factor 2. The goal is to keep the same descriptive power in the resulting samples, because we will double the number of the samples compared to the original sampling.*

1. *The top, left graph illustrates the same sampled discrete signal (green stems) along with the original (green solid) as shown in Fig. 9.1. However, this time the dashed magenta curves show that the interpolation functions that are narrower (magenta curve) are able to reproduce the original signal more faithfully. We wish to sample the original at a smaller step size than the distance between the given samples, and we should therefore not lose the existing richness of the variations.*

2. *On the top, right of the same figure we show with magenta samples the desired values, which represent the samples of the reconstructed signal using twice as small discretization step as the one shown on top, left.*

3. *On the bottom we show how this up-sampling can be implemented by processing only discrete signal samples. On the left we show a slightly different version of the discrete signal to be up-sampled. The difference consists in that midway between the original signal sample positions we have inserted zeros. On the right*

we show the sampled interpolation function to be used in the up-sampling. The up-sampled signal (shown as the magenta samples in top, right) is obtained by convolving the two discrete signals shown at the bottom. Evidently, multiplications with zeros do not need to be executed if efficiency is cared for in the implementation.

9.2 The Gaussian as Interpolator

When resampling a discrete signal, its continuous version is resampled. The continuous signal is in turn obtained by placing the interpolation functions at the grid points and summing them up with weights which are the corresponding discrete signal values. Although this happens in the experimenter's imagination, the implications of it in practice are sizeable because this reasoning influences the discrete filters to be used. Here it is important to note that we generally do not know the original signal but only know its samples. Because of this, any reconstruction is a guess translates to making an assumption on, and a motivation for the interpolation function.

The interpolation function that is commanded by the band-limited signal theory in conjunction with Fourier transform theory is the $\text{sinc}(t)$ function, Eq. (6.27). As pointed out in Sect. 8.1, this choice brings us to the frontier where practical interest conflicts with purely theoretical reasoning. A reconciliation can be reached by not insisting on the strict "limitedness" in the frequency domain in exchange for smaller and direction-isotropic filters (in 2D and higher dimensions). We keep the idea of displaced tentlike functions as a way to reconstruct the original function, but these functions are subjected to a different condition than strict band-limited requirement. Instead, the latter will be replaced by smoothness and compactness in both domains, i.e., we will require from the interpolation function μ_0 that not only μ_0 but also its FT is "compact", in addition to being "smooth".

We will suggest a *Gaussian* to be used as an interpolator, primarily for reasons called for by 2D and higher dimensional image analysis applications. However, to elucidate why the Gaussian family is interesting, we will use 1D in the discussions, for simplicity.

Definition 9.1. *With $0 < \sigma^2 < \infty$, the functions :*

$$g(x) = \frac{1}{\sqrt{2\pi\sigma^2}} \exp(-\frac{x^2}{2\sigma^2}) \tag{9.11}$$

and their amplified versions $A \cdot g(x)$, where A is a constant, constitute a family of functions, called Gaussians.

The Gaussian in Eq. (9.11) is positive and has an area of 1, guaranteed by the constant factor $\frac{1}{\sqrt{2\pi\sigma^2}}$. These qualities make it a *probability distribution function*, the Normal distribution.

A Gaussian with a fixed[3] σ is a decreasing function that converges to zero with increased x. A property of the Gaussian family is that it is invariant under the Fourier

---

[3] With increasing σ a Gaussian converges functionally to the constant A.

transform,

$$\mathcal{F}\{g(x)\} = G(\omega) = \exp\left(-\frac{\omega^2}{2(1/\sigma)^2}\right) \tag{9.12}$$

meaning that the result is still a Gaussian, albeit with an inverted σ and a different constant factor. It is a function that has a "minimal width", even after Fourier transformation. To clarify what this "minimal width" means, we need to be precise about what width means. To that end we first recall that the graph of any function (not just Gaussians) $f(x)$ *shrinks* compared to $f(ax)$ with regard to the x-axis if a is real and $1 < a$. Conversely, it will dilate if $0 < a < 1$. Second, if we have the Fourier transform pair $f(x), F(\omega)$ then the function pair $f(ax), F(\frac{1}{a}\omega)$, with a being real and nonzero, is also a Fourier transform pair. In other words, if f shrinks with regard to the x-axis, its Fourier transform will dilate with regard to ω.

In case f is the uniform distribution, e.g., the characteristic function in Fig. 6.3 left, we know how to interpret the width; we just need to measure the length between the two "ends" of the graph as we measure the length of a brick. The problem with the widths of general functions is that they may not have clearly distinguishable ends. The not-so-farfetched example is the Gaussian family. Therefore one needs to bring further precision to what is meant by the width of a function.

Assuming that the function is integrable, one way to measure the *width of a function $f(x)$*, is via the square root of its variance

$$\Delta(f) = \left(\int_{-\infty}^{\infty} (x - m_0)^2 \tilde{f}(x) dx\right)^{\frac{1}{2}} \tag{9.13}$$

where

$$\tilde{f}(x) = |f(x)|/\int_{-\infty}^{\infty} |f(x)| dx \tag{9.14}$$

and

$$m_0 = \int_{-\infty}^{\infty} x \tilde{f}(x) dx \tag{9.15}$$

The "~" applied to f sees to it that $\tilde{f}$ becomes a probability distribution function, by taking the magnitude of f and normalizing its area to 1. Because f can be negative or complex in some applications, the magnitude operator is needed to guarantee that the result will be positive or zero. After this "conversion", the integral measures the ordinary variance of a probability distribution function.

We can see by inspection that $\Delta(f(\alpha x))$ and $\Delta(F(\frac{\omega}{\alpha}))$ will be inversely proportional to each other to the effect that their product will equal to a constant γ that is independent of α:

$$\gamma = \Delta[f(\alpha x)] \cdot \Delta\left[F\left(\frac{\omega}{\alpha}\right)\right] \tag{9.16}$$

The exact value of γ depends on f, and thereby also on F. As a direct consequence of Cauchy–Bunyakovski–Schwartz inequality, it can be shown that

$$1 \le \gamma = \Delta(f) \cdot \Delta(F) \tag{9.17}$$

In 1920s, when investigating the nature of matter in quantum mechanics, it was discovered via this inequality that joint measurement accuracies of certain physically observable pairs, e.g., position and momentum, time and energy, are limited [103]. Later, Gabor used similar ideas in information science and showed that the joint time–frequency plane is also granular and cannot be sampled with smaller areas than the above inequality affords [77]. The principle obtained by interpreting the inequality came to be known as *Heisenberg uncertainty*, the *uncertainty principle* or even as the *indeterminacy principle*. The principle is a cornerstone of both quantum mechanics and information science.

In the case of a Gaussian, we have $\Delta(g) = \sigma$ because Eq. (9.11) is already a known probability distribution function having the variance σ^2. Likewise, $\tilde{G}$ is a normal distribution with variance $1/\sigma^2$, yielding $\Delta(g) = 1/\sigma$. Consequently, the uncertainty inequality is fulfilled with equality by the Gaussian family. For signal analysis, this outcome means that the Gaussian pair is the most compact Fourier transform pair. Because of this, extracting a local signal around a point by multiplying with a "window" signal will always introduce high-frequency components to the extracted signal. The Gaussian windows will produce the most compact local signal with the least amount of contributions from the high-frequency components. This is also useful when integrating (averaging) a local signal. Using another function than Gaussian with the same width will yield an average that is more significantly influenced by the high frequencies.

9.3 Optimizing the Gaussian Interpolator

Although we have not been precise about the interpolator we used, in Sect. 9.1 we showed the principles of how the continuous interpolators perform up- and down-sampling of the discrete signals. Furthermore, we suggested using the Gaussians as interpolators in the previous section, but we did not precisely state how the variance σ^2 should be chosen as a function of κ, which we will discuss next.

In down-sampling, the Gaussian will not only be used as an interpolator, but at the same time also it will be used as a lowpass filter to reduce the high frequencies while retaining the low frequencies as intactly as possible. Therefore, a Gaussian must mimic or approximate the *characteristic function*:

$$\chi(\omega) = \begin{cases} 1, & -\frac{B}{2} < \omega < \frac{B}{2} \\ 0, & \text{otherwise.} \end{cases}$$

To fix the ideas, we choose $B/2$, determining the symmetric band of interest around the DC component, as $\frac{B}{2} = \frac{\pi}{2}$. The quantity B is the bandwidth of interest that we wish to retain, which includes also the negative frequencies. The particular choice of B above allows a size reduction of the signal with the factor $\kappa = \frac{2\pi}{B} = 2$. In Fig. 9.3 the correponding characteristic (box) function is shown. The Gaussian

$$G(\omega) = \exp\left(-\frac{\omega^2}{2(\sigma')^2}\right) \tag{9.18}$$

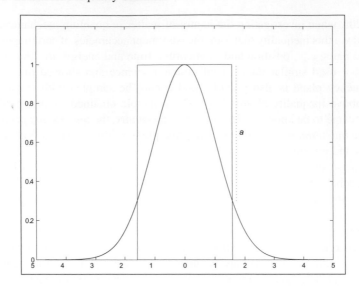

Fig. 9.3. A Gaussian function (*black*), plotted with the χ function (*red*) with $\frac{B}{2} = \frac{\pi}{2}$. *Blue lines* indicate the distance a used when estimating the $||E||_\infty$ norm

that will approximate χ is given in the same graph. Both χ and G have ω as input variables to show that they are in the frequency domain. The value at the origin $G(0) = 1$ guarantees that subsampling of at least the constant signal will be perfectly accomplished, in that the signal value will be unchanged. We will vary σ' to obtain the function G that is least dissimilar to χ.

To obtain a dissimilarity measure, we define first the error signal:

$$e(\omega) = G(\omega) - \chi(\omega)$$

Because the width of the function G depends on the constant σ', the norm of the error signal will depend on σ'. We will use the $\mathcal{L}^k$ norm:

$$\|f\|_k = \left(\int_{-\infty}^{\infty} |f(x)|^k dx \right)^{1/k}$$

to minimize the dissimilarity

$$\min_{\sigma'} \|e\|_k = \min_{\sigma'} \|G - \chi\|_k \qquad (9.19)$$

The higher the value of k the higher the impact of extremes will be on $\|f\|_k$. If we take k towards the infinity, then only the maximum of $|f|$ will influence $\|f\|_k$, yielding:

$$\|f\|_\infty = \max |f| \qquad (9.20)$$

For this reason $\| \cdot \|_\infty$ is also called the max norm.

Using this norm to measure the amount of dissimilarity, we obtain

$$\|e\|_\infty = \max_\omega |G(\omega) - \chi(\omega)| = \max(a, 1 - a) \tag{9.21}$$

where a is in the interval $]0, 1[$ and is defined geometrically as in Fig. 9.3. The figure illustrates a Gaussian with a specific σ', demonstrating that the $\|e\|_\infty$ is determined by the intersection of the line $\omega = \frac{\pi}{2}$ with the graph of $G(\omega)$ because this is where the highest absolute values of the difference will occur. The norm equals the maximum of the pair a and $1 - a$, as depicted in the figure. By varying σ', the intersection point can be varied with the expectation that we will find a low value for $\max(a, 1 - a)$. The lowest possible norm is evidently obtained for σ_0', producing $a = 0.5$ i.e.,

$$G(\frac{\pi}{2}) = \exp\left(-\frac{(\frac{\pi}{2})^2}{2(\sigma')^2}\right) = \frac{1}{2} \tag{9.22}$$

so that the analytic solution,

$$\sigma_0' = \frac{\pi}{\sqrt{\log(256)}} \approx 1.3 \tag{9.23}$$

is the solution for the nonlinear minimization problem given in Eq. (9.19) when k approaches ∞. In the spatial domain this corresponds to a Gaussian filter with

$$\sigma_0 = \frac{\sqrt{\log(256)}}{\pi} \approx 0.75 \tag{9.24}$$

Accordingly, the Gaussian with maximum 1 that approximates the constant function 1, the best in the $\mathcal{L}_\infty$ norm, is the one placed in the middle and that attenuates 50% at the characteristic function boundaries.

The used norm affects the optimal parameter selection. Had we chosen another value for k in the $\mathcal{L}^k$ norm, we would obtain another value for σ_0', yielding a different Gaussian. This is one of the reasons why there is not a unique down-sampling (or up-sampling) filter: because the outcome depends on what norm one chooses to measure the dissimilarity with.

We study here just how much the choice of the error metrics influences the σ_0' that determines the width of the Gaussian. The minimization of Eq. (9.19) can be achieved for finite k numerically. The values of σ_0' that yield minimal dissimilarities measured in different norms are listed in Table 9.1 when the cutoff frequency is $\frac{B}{2} = \frac{\pi}{2}$. Column σ_0 represents the corresponding standard deviations of the filters in the spatial domain, with $\sigma_0 = 1/\sigma_0'$. The table shows that σ_0 varies in the interval $[1, 1.3]$, where the upper bound is given by Eq. (9.24). This suggests that there is a reasonably large interval from which σ_0' can be picked up to depreciate the high frequencies sufficiently while not depreciating the low frequencies too severely for down-sampling. The σ_0' determines the filter G in the frequency domain. The corresponding filter in the spatial domain is given by the $g(x)$ in Eq. (9.11), where the optimal σ_0 is the inverse of the σ_0'. Table 9.1 also lists the σ_0 values that determine

Table 9.1. Gaussians approximating an ideal characteristic function

	Cutoff frequency: $\frac{\pi}{2}$	
Norm	σ_0 [x-domain]	σ_0' [ω-domain]
$\mathcal{L}^1$	0.98	1.0
$\mathcal{L}^2$	0.82	1.2
$\mathcal{L}^3$	0.78	1.3
$\mathcal{L}^\infty$	0.75	1.3

the spatial domain filters. Drawn in the frequency domain, Fig. 9.4 illustrates the corresponding Gaussians and the characteristic function they approximate.

It can be shown that for cases when B has values other than π, corresponding to a size reduction of $\frac{2\pi}{B}$, we obtain the critical frequency domain standard deviation parameter as

$$\sigma_0' \in \left[0.75\frac{B}{\sqrt{\log(256)}}, \frac{B}{\sqrt{\log(256)}}\right] \approx \left[0.65\frac{B}{2}, 0.85\frac{B}{2}\right] \quad \Rightarrow \quad \sigma_0' \lesssim 0.85\frac{B}{2}$$

The interval above is an upper bound for σ_0', for the loosest sampling afforded by the bandwidth B. The critical upper bound on the right originates from the upper bound of the interval, which is obtained by using the $\mathcal{L}_\infty$ norm and observing that one has the option to sample denser than the loosest allowable sampling. Accordingly, the lower bound for the spatial filter parameter for the same size reduction factor yields:

$$\sigma_0 \in \left[0.37\frac{2\pi}{B}, 0.5\frac{2\pi}{B}\right] = \left[0.75\frac{T}{2}, \frac{T}{2}\right] \quad \Rightarrow \quad 0.75\frac{T}{2} \lesssim \sigma_0$$

Here $T = \frac{2\pi}{B}$ is the critical sampling period, which is also the critical size reduction factor. This is because, throughout, we assumed that the original sampling period is 1, implying that the maximum frequency (Nyquist) is normalized to π.

9.4 Extending Gaussians to Higher Dimensions

Gaussians are valuable to image analysis for a number of reasons. In the previous sections, we studied how well it is possible to fit the Gaussian to the characteristic function in the frequency domain and why the Gaussian family should be used instead of the sinc function family. Although in 1D we suggested Gaussians, there are many other function families that are used in practice for a myriad of reasons,[4]. To appreciate the use of the Gaussian family of functions in signal analysis, it is more appropriate to study them in higher dimensions than 1. We first extend their definition to include *Gaussians in N dimensions*:

---

[4] Some frequently encountered 1D filter families include Chebychev, Butterworth, and co-sine functions.

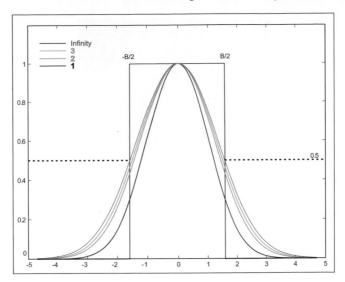

Fig. 9.4. The characteristic function (*red*) plotted with Gaussians optimally approximating it using the norms $\mathcal{L}_1$ (*blue*), $\mathcal{L}_2$(*cyan*), $\mathcal{L}_3$ (*green*), and $\mathcal{L}_\infty$ (*black*). The bandwidth is $B = \pi$

$$g(\mathbf{x}) = a \exp\left(-\frac{1}{2}\mathbf{x}^T\mathbf{C}^{-1}\mathbf{x}\right)$$

where $\mathbf{C}$ is an $N \times N$ symmetric, positive, definite[5] (covariance) matrix, and $\mathbf{x}$ is the N-dimensional coordinate vector. The one-dimensional Gaussian is a special case of this function, with the major difference being that we now have more variance parameters (one σ for each dimension and also covariances), which are encoded in $\mathbf{C}$. Two properties in particular speak in favor of Gaussians: separability and directional indifference (isotropy).

Separability

The separability property stems from the fact that ND Gaussians can always be factored out into N one-dimensional Gaussians, in the following manner:

$$g(\mathbf{x}) = a \prod_{i=1}^{N} \exp\left(-\frac{(\mathbf{x}^T\mathbf{v}_i)^2}{2\sigma_i^2}\right) = a \prod_{i=1}^{N} \exp\left(-\frac{(y_i)^2}{2\sigma_i^2}\right) \qquad (9.25)$$

where $\mathbf{v}_i$ is the ith unit-length eigenvector of $\mathbf{C}$, and (σ_i^2) is the ith eigenvalue of $\mathbf{C}$. The vector $\mathbf{y} = (y_1, y_2, \cdots, y_N)^t$ is given by $\mathbf{y} = \mathbf{Q}\mathbf{x}$. Convolving an ND image $f(\mathbf{x})$ with the ND Gaussian can therefore be achieved by first rotating the image:

---

[5] Positive definite symmetric matrices can always be decomposed as $\mathbf{C} = \mathbf{Q}\Sigma\mathbf{Q}$, where $\mathbf{Q}$ is an orthogonal matrix containing the unit-length eigenvectors of $\mathbf{C}$ in its columns, and Σ is the diagonal matrix containing the eigenvalues, which are all positive. Orthogonal matrices fulfill $\mathbf{Q}^T\mathbf{Q} = \mathbf{I}$, which expresses the orthogonality of the eigenvectors.

Fig. 9.5. FMTEST image, which consists of frequency-modulated planar waves in all directions

$$f(\mathbf{x}) = f(\mathbf{Q}^T \mathbf{y}) = \tilde{f}(\mathbf{y}) \tag{9.26}$$

and then convolving the rotated image separably. In this case, the 1D Gaussians having the arguments y_i, the right-hand side of Eq. (9.25), as discussed in Section 7.3.

Directional Isotropy

It is easy to equip the Gaussian filter with another desirable property, isotropy. From Eq. (9.25) one can conclude that if $\sigma_1 = \sigma_2 = \cdots = \sigma_N = \sigma$ then we can write:

$$
\begin{aligned}
g(\mathbf{x}) &= a \prod_{i=1}^{N} \exp\left(-\frac{(\mathbf{x}^T \mathbf{v}_i)^2}{2\sigma}\right) \\
&= a \prod_{i=1}^{N} \exp\left(-\frac{(y_i)^2}{2\sigma_i^2}\right) = a \exp\left(-\frac{\|\mathbf{y}\|^2}{2\sigma^2}\right)
\end{aligned} \tag{9.27}
$$

Consequently, Gaussians can not only be made separable in ND, but they can also be made fully invariant to rotations, i.e., the function values depend only on the distance to the origin. Gaussians are the only functions that are both separable and rotationally invariant, in 2D and higher dimensions. Used as filters, this property allows us to identically weight all frequencies having the same distance to the origin in the $\boldsymbol{\omega}$-domain. This is significant to image analysis because there is no reason to systematically disfavor a direction compared to the others.

Example 9.3. *We illustrate isotropy in image analysis operators by using an omnidirectional test image consisting of frequency-modulated planar waves in Fig. 9.5. We compare Gaussians with sinc functions when they are used as lowpass filters.*

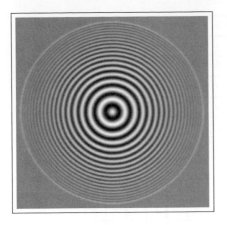 

Fig. 9.6. Lowpass filtering on a test image using both a Gaussian and a sinc. *Left*: filtering using a 2D Gaussian, with $\sigma = 0.75$, corresponding to the black graph in Fig. 9.4. *Right*: filtering using a 2D Sinc, corresponding to the red graph in Fig. 9.4

Such an operation is frequently needed in image analysis, e.g to change the size of an image. Both filter families are separable in that they can be implemented as a cascade of 1D convolutions. However, as illustrated by Fig. 9.6, the sinc filter does not treat all directions equally when it suppresses high frequencies, introducing artifacts. By contrast, the Gaussian filter, being the only filter that is both separable and rotation-invariant, does not suffer from this.

Example 9.4. *Here we illustrate the process of down- and up-sampling of an image by using Gaussians as interpolation functions.*

- *In Fig. 9.7 we show a graph representing an image where red x-marks represent pixels. The image will be reduced by a factor of 2. At every second row and column, a blue ring shows where the interpolation functions are placed and a scalar is computed by the scalar product of the local image and the interpolation image. These scalars represent the pixel values of the reduced image. The concentric magenta rings show the isovalues of the interpolation function. The rings show the loci where the (continuous) interpolation function, with $\sigma = 0.8$, reaches 0.5 (dashed) and 0.01 (solid) levels, respectively. The interpolation function can be sampled at points marked with $\times$, and it can be truncated at the level of the solid magenta ring.*

- *In Fig. 9.8 we show another drawing representing an image that has been enlarged by a factor 2. The red x-marks represent the pixels copied from the original (smaller) image, whereas at every second row and column, the blue dots mark the pixels that are initially put to zero. The interpolation function convolves this image to deliver the final enlarged image, via local scalar products. The concentric magenta rings show the isovalues of the interpolation function. The rings show*

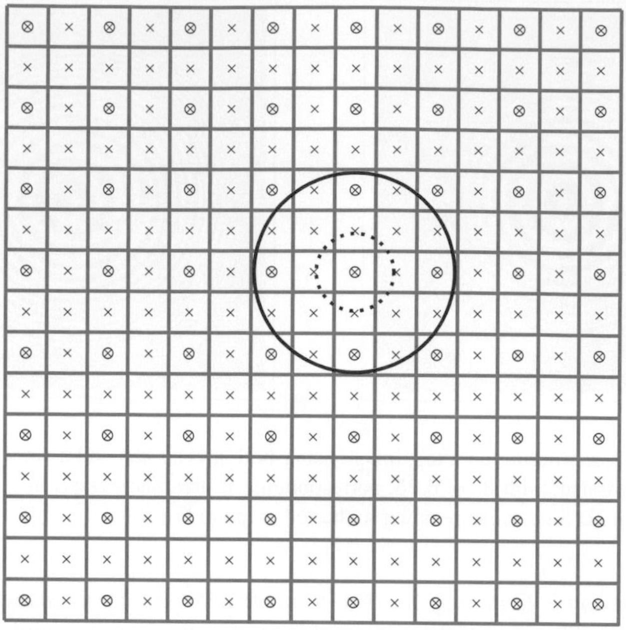

Fig. 9.7. The graph represents the down-sampling of an image with the factor 2. The crosses with rings show the placement of the interpolation function where the scalar products between the filter and the image data are computed. Two isocurves of such an interpolation function are shown by two concentric circles (magenta)

> the loci where the (continuous) interpolation function with $\sigma = 0.8$ reaches 0.5 (dashed) and 0.01 (solid) levels, respectively.

9.5 Gaussian and Laplacian Pyramids

Useful shape properties describing size and granularity can be extracted for numerous applications by *scale space* [128, 141, 152, 221, 230] techniques, of which image pyramids [43, 160] are among the most efficient to compute. The *Gaussian pyramid* of an image is a hierarchy of lowpass filtered versions of the original image, such that successive levels correspond to lower frequencies. Because of the effectuated lowpass filtering one can reduce the image size without loss of information. The lowpass filtering is done using convolution with either a Gaussian filter or similar. Where the filter overhangs the image boundaries one can reflect the image about its boundary. There are, however, other possibilities to avoid the artificial irregularities the boundaries cause, e.g., extension with zeros. The cut-off frequency of the Gaussian filters can be controlled using the parameter σ, offering different size reduction rates, κ, which refers to the reduction rate an image boundary, e.g., the image width in 2D. If the reduction rate is $\kappa = 2$ then the pyramid is called an octave pyramid. One may

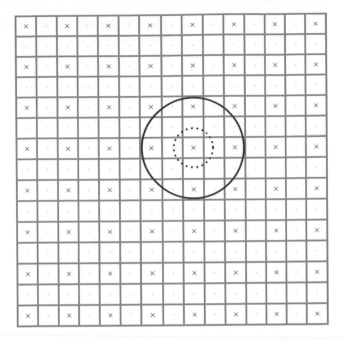

Fig. 9.8. The up-sampling of an image with the factor 2. *Dots* and *crosses* show where the interpolation functions are placed. The *dots* represent zero values, whereas crosses represent the data of the original grid, which has fewer samples than the new grid consisting of crosses and dots. Two isocurves of the interpolation function are shown by *two concentric circles (dotted and solid)*

define the resizing operator that effectuates the down-sampling as $\mathbf{R}_{\downarrow}$, which is a filtering followed by dropping appropriate number of pixels in each of the dimensions, e.g., both rowwise and columnwise in 2D. For a Gaussian kernel g and an image f, $\mathbf{R}_{\downarrow}$ is defined as:

$$(\mathbf{R}_{\downarrow}f)(\mathbf{r}'_i) = \sum_m g(\mathbf{r}_m)f(\mathbf{r}'_i + \mathbf{r}_m), \qquad \text{with} \qquad \mathbf{r}'_i \in \mathcal{N}_{R_{\downarrow}} \subset \mathcal{N}. \qquad (9.28)$$

Here $\mathcal{N}$ is the regular discrete grid on which the original image f is sampled and from which $\mathbf{r}_m$ is drawn. The points defining $\mathcal{N}_{R_{\downarrow}}$ are a subset of $\mathcal{N}$ obtained by retaining every κ row and column of it. Consequently, $\mathbf{r}_m$ tessellates a finer grid compared to $\mathbf{r}'_i$. In Fig. 9.7, which shows the case with the down-sampling rate of $\kappa = 2$, and in 2D, $\mathcal{N}_{R_{\downarrow}}$ is illustrated by the encircled grid points, whereas the points of $\mathcal{N}$ are marked with crosses. Applying the operator $R_{\downarrow}$ to f k times:

$$R_{\downarrow}^k f = R_{\downarrow} \cdots R_{\downarrow}R_{\downarrow}f \qquad (9.29)$$

yields the kth level in the pyramid, with $k = 0$ defining the original image f.

The *Laplacian pyramid* of an image is obtained from the Gaussian pyramid. It consists of a hierarchy of images such that successive levels of the Laplacian pyra mid correspond to the differences of two successive layers of the Gaussian pyramid. Because the layers of the Gaussian pyramid have different sizes, one may define the resizing operator that yields the up-sampling as $\mathbf{R}_\uparrow$. This is a lowpass filtering applied to an image after having injected an appropriate number of points in each of the dimensions having zero image values. The kernel is the same kernel g used when building the Gaussian pyramid. Assuming that the image to be up-sampled is f, then the $R_\uparrow$ is defined as:

$$(\mathbf{R}_\uparrow f)(\mathbf{r}'_i) = \sum_m g(\mathbf{r}_m) f(\mathbf{r}'_i + \mathbf{r}_m), \qquad \text{with} \qquad \mathbf{r}'_i \in \mathcal{N}_{R_\uparrow} \supset \mathcal{N}. \qquad (9.30)$$

As before, $\mathcal{N}$ is the regular discrete grid on which the image f is originally sampled and from which $\mathbf{r}_m$ is drawn. The $\mathcal{N}_{R_\uparrow}$ is now a superset of $\mathcal{N}$ and is determined in such a way that when every κ row and column of it is retained then $\mathcal{N}$ is obtained. In other words, $\mathbf{r}_m$ tessellates a sparser grid compared to $\mathbf{r}'_i$. In Fig. 9.8, which illustrates the up-sampling rate of $\kappa = 2$ in 2D, $\mathcal{N}_{R_\downarrow}$ is illustrated by all of the grid points (blue dots as well as red crosses), whereas the points of $\mathcal{N}$ stand out as red crosses. Applying the operator $R_\uparrow$ to f k times:

$$R_\uparrow^k f = R_\uparrow \cdots R_\uparrow R_\uparrow f \qquad (9.31)$$

yields the kth level in the pyramid, with $k = 0$ defining the lowest level corresponding to the highest resolution.

Example 9.5. *In Fig. 9.9, we illustrate the Gaussian pyramid in 2D for $\kappa = 2$ as layers of images, stacked on top of another.*

- *The Gaussian pyramid is shown on the right of Fig. 9.9. The largest image of the Gaussian pyramid containing all frequencies is marked as level 0 of the pyramid. Each level is obtained by lowpass filtering the level below and retaining every second row and column of the result. The filter used was a Gaussian with $\sigma = 0.75$.*
- *The Laplacian pyramid is shown on the left of Fig. 9.9 . The largest image of the Laplacian pyramid containing the highest frequencies is level 0. Each level is obtained by the difference of two levels of the Gaussian pyramid. Before subtraction, the smaller of the two images is up-sampled by injecting zeros and lowpass filtering using a Gaussian filter with $\sigma = 0.75$.*
- *In Fig. 9.10 the concentric annuli show the effective frequency bands captured by each level of the Laplacian pyramid. The color code matches the corresponding level of the pyramid. The figure also shows which bands the Gaussian pyramid retains. Level 0 of the Gaussian pyramid corresponds to the full circle precisely containing the annulus marked with 0. Level 1 corresponds to the full circle precisely containing the annulus marked with 1, and so on.*

9.6 Discrete Local Spectrum, Gabor Filters

For convenience, we start this section by discussing 1D time–frequency sampling. Music scores, which are a way to encode music, illustrate time–frequency sampling.

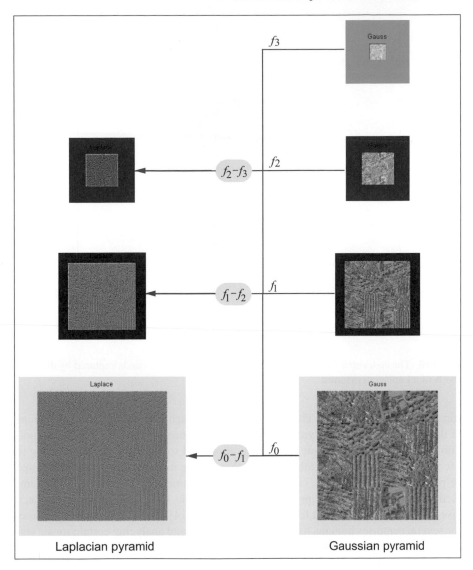

Fig. 9.9. Gaussian and Laplacian pyramids applied to a texture patch image, shown in *bottom, right*

Although forms of notational aids to remember music existed even in ancient Greek, Judeo-Aramaic as well as Oriental traditions, western music notation offering elaborate tone and duration representation came to dominate. This notation is traced back to pre-medieval ages. In its modern version, a sequence of tones with duration on a time axis represents the music, as illustrated by Fig. 9.11. The printed *music* score is brought to life through musical instruments producing an air pressure variation that

Fig. 9.10. The *concentric annuli* illustrate the effective frequency bands captured by different levels of a Laplacian pyramid

can be represented by $f(t)$. There are symbols for each tone, and the time durations of the tones are part of the symbols. Hence a given tone is played for a certain duration, followed by another tone with its duration, and so on. This way of bringing to life a 1D function f is radically different than telling how much air pressure should be produced at a given time, i.e., a straightforward time sampling of $f(t)$. The sampled joint time–frequency representations of functions have relatively recently been given a formal mathematical frame [10,53,77,203,214]. This is remarkable because, for several centuries humans have been synthesizing and analyzing certain music signals by using sequences of tones chosen from a limited set, differing from each other either in their (basic) frequencies or durations. Below, we discuss time–frequency sampling concept in further detail. We subsequently extend these results to 2D and higher dimensional images.

Let $f(t)$ be a 1D function and $w(t)$ be a window function with limited support, where the coordinate t varies continuously in $]-\infty, \infty[$.

Definition 9.2. *The local function around t_0 is $f'(t_0, t) = f(t - t_0)w(t)$, and the local spectrum is defined as the Fourier transform of the local function f':*

$$F'(t_0, \omega_0) = \int f(t - t_0)w(t)\exp(-i\omega_0 t)dt$$
$$= \langle w(t)\exp(i\omega_0 t), f(t - t_0)\rangle \qquad (9.32)$$

Fig. 9.11. Music scores are represented and interpreted as a sequence of time-limited musical tones chosen from a finite set

As originally proven by Gabor [77], for a Gaussian window function w, it is possible to discretize $F'(t_0, \omega_0)$ w.r.t. t_0 and ω_0 such that f as well as F can be recovered, up to a lowpass filtering, from the samples. The samples corresponding to the same t_0 constitute the discrete local spectrum. It is also possible to come to this conclusion by using our results in the previous sections. From the definition, the local spectrum around t_0 equals:

$$F'(t_0, \omega_0) = \langle w(t) \exp(it\omega_0), f(t - t_0) \rangle$$
$$= 2\pi \langle W(\omega - \omega_0), F(\omega) \exp(-it_0\omega) \rangle$$

$$(9.33)$$

where we have used theorem 7.2 (Parseval–Plancherel). In consequence, we can write

$$F'(t_0, \omega_0) = \int f(t - t_0) w(t) \exp(-i\omega_0 t) dt \qquad (9.34)$$

$$= 2\pi \int W(\omega - \omega_0) F(\omega) \exp(-it_0\omega) d\omega \qquad (9.35)$$

For $t_0 = 0$, Eq. (9.35) represents a lowpass filtering of the spectrum, F, with the filter W. Accordingly, the smoothed spectrum can be sampled with the discretization step $\Omega = \frac{2\pi}{T}$ where T is the effective width of the window, $w(t)$, in the time domain. However, the smoothed spectrum, $F'(0, \omega)$, continuous in ω, represents the Fourier transform of the local function around the origin, i.e., $f(t)w(t)$ in Eq. (9.34), and it can be recovered from the samples, $F'(0, n\frac{2\pi}{T})$ precisely because the function $f(t)w(t)$ has limited extension due to windowing. The origin $t = 0$ is in no way unique, because the function $F(\omega) \exp(-it_0\omega)$ appearing in Eq. (9.35) is the Fourier transform of $f(t - t_0)$, which is a shifted version of the function such that t_0 is the origin. In consequence, what we have in Eq. (9.35) is a smoothing of the Fourier transform of the shifted function $f(t - t_0)$ with the filter $W(\omega)$ in the Fourier domain. Because of this, $F'(t_0, \omega)$ can be sampled w.r.t. ω, i.e., $F'(t_0, n\frac{2\pi}{T})$, where T corresponds to the effective extension of $w(t)$ and recovered from the samples. Using the same arguments, and using the symmetry of the Fourier transform, we conclude that $F'(t, \omega_0)$ can be sampled w.r.t. t, i e., $F'(mT, \omega_0)$, which in turn can be recovered from the samples of $F'(mT, n\frac{2\pi}{T})$. The discrete local spectrum, also called the *Gabor spectrum*, or *Gabor decomposition*, is computed as a projection onto a filter bank either in the spatial domain, as in Eq. (9.36), or in the Fourier domain, as in Eq. (9.37).

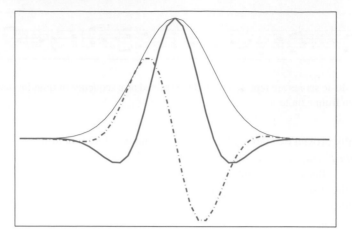

Fig. 9.12. The graphs of the real (*green, solid*) and the imaginary (*green, dashed*) parts of a *Gabor filter*, $g(t)$, in the time domain. The window function $w(t) = |g(t)|$ is drawn in *black*

$$F'(mT, n\frac{2\pi}{T}) = \int f(t - mT)w(t)\exp(-in\frac{2\pi}{T}t)dt \qquad (9.36)$$

$$= 2\pi \int W(\omega - n\frac{2\pi}{T})F(\omega)\exp(-imT\omega)d\omega \qquad (9.37)$$

The filters of the filter bank are called the Gabor filters and they are given as follows in the spatial domain:

$$g_n(t) = w(t)\exp(-in\frac{2\pi}{T}t) = \frac{1}{(2\pi\sigma^2)^{\frac{1}{2}}}\exp(-\frac{t^2}{2\sigma^2})\exp(-in\frac{2\pi}{T}t) \qquad (9.38)$$

with $2\sigma > T$. In Fig. 9.12, the real and the imaginary part of one such Gabor filter, $g_n(t)$ with $n = 2$, is shown. The black curve shows $|g_n(t)|$, which is the Gaussian window. The Fourier transform of the filter is shown in magenta in Fig. 9.13. The higher the value of n, the more the filter function oscillates in the Gaussian window. In the Fourier domain, Gabor filters are translated Gaussians:

$$G_n(\omega) = W\left(\omega - n\frac{2\pi}{T}\right) = \exp\left(-\frac{(\omega - n\frac{2\pi}{T})^2}{\frac{2}{\sigma^2}}\right) \qquad (9.39)$$

Figure 9.13 illustrates $G_n(\omega)$ for $n = 1 \cdots 7$ for a set of Gabor filters uniformly distributed in the angular frequency range $[0.05\pi, 0.95\pi]$. The filters are centered in the seven equally sized frequency cells and attenuate 50% at the cell boundaries. Note that this amount of overlap is motivated by our finding in Sect. 9.3, according to which a Gaussian approximates the characteristic function of its frequency cell best in the $\mathcal{L}_\infty$ norm when it attenuates at the characteristic function boundaries with 50%. The characteristic function of one frequency cell that is to be approximated by its underlying Gaussian (magenta) is drawn in black in Fig. 9.13. Note that we only

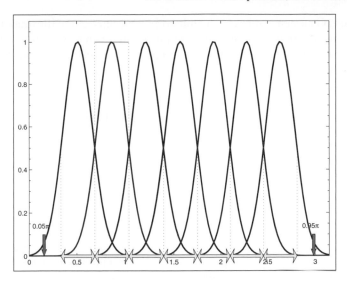

Fig. 9.13. The graphs of a Gabor filter bank with minimum and maximum angular frequencies 0.1π and 0.9π, respectively. The filter corresponding to Fig. 9.12 is shown in *magenta*, where the *black* constant illustrates the characteristic function of the corresponding frequency cell

need to have filters that cover the positive ω-axis in the Gabor filter bank because we assumed that the function f is real-valued so that there is no additional information in the negative axis, compared to the positive axis due to $F(\omega) = F^*(-\omega)$. Given the attenuation at the cell boundaries, for a 1D Gabor filter bank we thus need the minimum frequency, the maximum frequency, shown as vertical arrows, as well as the total number of filters to determine the filter bank. The tip of a green arrow and the closest yellow arrow is half the width of the angular frequency cells.

One can extend these results into 2D and higher dimension in a straightforward manner. The window function is an ND Gaussian, which has certain advantages, e.g., direction isotropy in image analysis applications. The resulting Gabor filters are given in the spatial domain by

$$g_n(\mathbf{r}) = w(t)\exp(-i\boldsymbol{\omega}_n^T\mathbf{r}) = \frac{1}{(2\pi\sigma^2)^{\frac{N}{2}}}\exp(-\frac{\|\mathbf{r}\|^2}{2\sigma^2})\exp(-i\boldsymbol{\omega}_n^T\mathbf{r}) \qquad (9.40)$$

and in the frequency domain, by

$$G_n(\boldsymbol{\omega}) = W(\boldsymbol{\omega} - \boldsymbol{\omega}_n) = \exp(-\frac{\|\boldsymbol{\omega} - \boldsymbol{\omega}_n\|^2}{2\sigma_\omega^2}) \qquad (9.41)$$

where $\sigma_\omega = 1/\sigma$. Without loss of generality, we assumed that the frequency cells are axis-parallel hypercubes with equal edge sizes in all directions and that the cell centers are at $\boldsymbol{\omega}_n$. We illustrate the frequency cells of one such Gabor filter bank in Fig. 10.28 along with the isocurves of the filters (green). The DC extracting Gabor filter is a pure Gaussian and has been omitted in the center of the illustration but can be included in the filter bank. In certain applications, such as direction estimation, the DC-filter is not needed. As before, the filters need tessellate only one half of the frequency plane because the image f is assumed to be real.

9.7 Design of Gabor Filters on Nonregular Grids

It has been shown that both in audio and in visual signal processing pathways of primates the sensitivity is not uniformly distributed across the perceivable frequencies. We start to hear low-frequency audio at lower amplitudes than we do for high-frequency audio. In images, we start to perceive low spatial-frequencies at lower contrasts (amplitudes) than high spatial-frequencies. Therefore, noise in high frequencies does not influence our audio or image quality assessment as much as it does at the low frequencies of the respective spectra. The nonuniformity in sensitivity is attributable to the limited signal processing and communication resources, such as the number of specialized cells, the connections between the cells and the statistics of real-world signals that matter for the organism. Images that humans and other primates encounter have a decreasing spectral power with increased frequencies. It is therefore plausible that an organism devotes its limited processing resources in proportion to their actual use.

As to human-made systems, these observations help to design more efficient systems, such as in compression of audio and video, where the largest part of the limited vocabulary represents the frequencies to which humans are most sensitive. Pattern recognition in audio and video increasingly uses filters with bandwidths that increase with *tune-on frequencies*.[6] Applications such as biometric person authentication also increasingly use such filter banks. Here, we will discuss *Gabor filter design* on multidimensional and nonuniform grids, [21], to yield higher sensitivies at certain bands than others.

We illustrate the approach first in 1D for convenience, and then we generalize the approach by extending it to 2D. To reduce the amount of manual labor, we suggest using an analytic coordinate transformation:

$$\xi = u(\omega), \qquad \text{or} \qquad \omega = u^{-1}(\xi), \qquad (9.42)$$

where u is yet to be specified. The transformation u is an injective reparametrization of ω such that u is invertible and establishes a differentiable (continuous) map between $\omega \in [\omega_{\min}, \omega_{\max}]$ and $\xi[0, 1]$. In consequence, we require that u satisfies the boundary conditions:

$$\omega(0) = u^{-1}(0) = \omega_{min} \qquad \text{and} \qquad \omega(1) = u^{-1}(1) = \omega_{max} \qquad (9.43)$$

Additionally, we demand that there will be higher sensitivity at low frequencies. We must translate sensitivity to a mathematical concept by defining it as a property of

---

[6] The tune-on frequency of a bandpass filter is the frequency where the Fourier transform of the filter has the maximum amplitude.

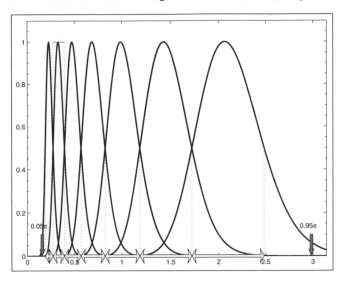

Fig. 9.14. The Gabor filters in the Fourier domain plotted against the linear frequency scale. The green arrows show the used minimum and the maximum frequency parameters

a coordinate transformation, such that a small interval is mapped to a larger interval through the transformation. To be precise, we want a mapping such that when we divide ξ equally, where each interval has the width $\delta\xi$, the widths of the corresponding intervals in ω should be larger and larger in such a way that $\delta\omega$ grows proportionally with ω, i.e.,

$$\delta\omega = C\omega\delta\xi \qquad (9.44)$$

where C is an unknown constant. Effectively, this leads to a differential equation when we let $\delta\xi$, and thereby $\delta\omega$ (which depends on it), be ever smaller.

$$\delta\omega = C\omega\delta\xi \qquad \rightarrow \qquad \frac{d\omega}{d\xi} = C\omega \qquad (9.45)$$

The differential equation with the boundary condition yields the unique solution

$$\omega = u(\xi) = \omega_{\min} \exp\left[\left(\log\frac{\omega_{\max}}{\omega_{\min}}\right)\xi\right] \qquad (9.46)$$

or

$$\omega = \omega_{\min} \cdot \left(\frac{\omega_{\max}}{\omega_{\min}}\right)^{\xi} \qquad (9.47)$$

with the inverse

$$\xi = \left[\log\left(\frac{\omega_{\max}}{\omega_{\min}}\right)\right]^{-1} \cdot \log\left(\frac{\omega}{\omega_{\min}}\right) \qquad (9.48)$$

We divide ξ into N equal cells and place the same translated Gaussian in each ξ cell in analogy with Fig. 9.13 and Eq. (9.41):

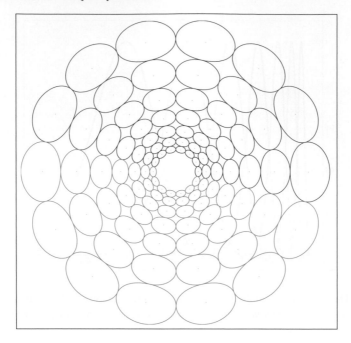

Fig. 9.15. The isocurves of Gabor filters at cell boundaries. The orientation and frequency resolution of the decomposition are explicitly controllable because the frequency cells are designed in the $\log(z)$ coordinates

$$G_n(\xi) = W(\xi - \xi_n) = \exp\left(-\frac{\|\xi - \xi_n\|^2}{2\sigma_\xi^2}\right) \tag{9.49}$$

where ξ_n and σ_ξ^2 are tune-on and bandwidth parameters as before, but in the mapped frequency, i.e., the ξ-domain. These filters can be remapped to the ω-domain via

$$G_n[\xi(\omega)] = W[\xi(\omega) - \xi_n] = \exp\left[-\frac{\|\xi(\omega) - \xi_n\|^2}{2\sigma_\xi^2}\right] \tag{9.50}$$

where $\xi(\omega)$ is the log function given by Eq. (9.48). As illustrated by Fig. 9.14, the only difference in this approach is the transformation Eq. (9.48) applied to the ω-coordinate, which enables the filters to be designed in the ξ-domain such that they all become identical and ordinary Gaussians translated to the centers of equally sized cells. The function $G_n(\xi(\omega))$ vanishes at $\omega = 0$, which guarantees that these filters have no DC bias, as opposed to $G_n(\omega) > 0$ at $\omega = 0$, discussed in the previous section. The filter bank uses the same parameters as the one illustrated by Fig. 9.13. Note that there are more resources (filters) devoted to the low-frequency range where humans are most sensitive. More precisely, the tune-on frequencies, the cell-boundaries, as well as the bandwidths, defined as the size of cells in ω-domain, progress geometrically with the factor $(\omega_{\max}/\omega_{\min})^{\Omega_\xi}$, with Ω_ξ being the cell size.

This should be contrasted to progressing arithmetically, with the increment of the cell size Ω, see Fig. 9.13. The explanation lies in the coordinate transformation, which does not introduce new filter values (heights), but rather stretches the ω-axis. The t-domain versions of these filters are found by inverse Fourier transforming $G_n(\xi(\omega))$ w.r.t. ω. It is possible to show that functions sampled on nonregular lattices can be reconstructed from their samples [5]. In consequence, the discretization discussed here is a representation of the local spectrum.

In 2D, we wish to have a geometric progression of the tune-on frequencies in the radial direction, whereas we wish to have an arithmetic progression of them in the angular direction. The latter is easily justifiable because, a priori, all directions appear to have equal importance in human vision as well as in numerous applications in machine vision. Assuming that ω is the 2D Cartesian coordinate vector in the Fourier domain, the transformation that has the desired properties is given by the log–polar coordinates:

$$\xi = \left[\log \left(\frac{\omega_{\max}}{\omega_{\min}} \right) \right]^{-1} \cdot \log \left(\frac{\|\omega\|}{\omega_{\min}} \right) \tag{9.51}$$

$$\eta = \frac{1}{\pi} \tan^{-1}(\omega) \tag{9.52}$$

where ξ varies between $[0, 1]$ to the effect that $\|\omega\|$ varies between $[\omega_{\min}, \omega_{\max}]$, and $\tan^{-1}(\omega)$ varies between $[-\pi, \pi]$, yielding a variation between $[-1, 1]$ for η.

We divide ξ, η uniformly as in Fig. 10.28 and place Gaussians in the marked cell centers (ξ_n, η_n):

$$G_n(\xi, \eta) = \exp \left(-\frac{|\xi - \xi_n|^2}{2\sigma_\xi^2} \right) \cdot \exp \left(-\frac{|\eta - \eta_n|^2}{2\sigma_\eta^2} \right) \tag{9.53}$$

The radial part of this filter function was previously suggested in [137] when designing quadrature mirror filters. Note that only one half of the plane needs to be covered because we assume that the filter bank will be used to analyze real-valued signals having $F(\omega) = F^*(-\omega)$ only. The filters, $G_n(\xi(\omega), \eta(\omega))$ can be readily sampled on the original *Cartesian grid* ω_l because $\xi(\omega)$ and $\eta(\omega)$ are available via Eqs. (9.51)–(9.52). We show the isocurves of the resulting filters at the cell-boundaries for seven frequencies and seven directions in Fig. 9.15 and mark the *tune-on frequencies* by dots. The radial cross-section of the filters through the tune-on frequencies is shown in Fig. 9.14.

The extension of the approach to ND Gabor filter banks is possible, provided that the type of resource allocation scheme is at least roughly known. The latter is crucial because it determines the coordinate transformations, which are not always intuitively derivable. For an example, we refer to [21], where Gabor filters on an irregular grid in 3D are suggested. The purpose is to mimic the speed sensitivity of humans to apparent motion in image sequences. For example, using coordinate transformations one can obtain a decomposition that is specially sensitive to low spatial frequencies moving quickly and high frequencies moving slowly as compared to other speed, and spatial frequency combinations.

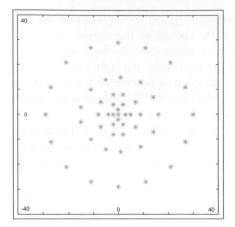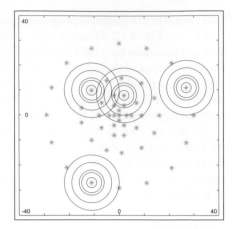

Fig. 9.16. *Left*: the retinotopic sampling grid with axes graded in pixels. *Right*: a few receptive fields are represented as sets of concentric circles. Adapted from [205].

9.8 Face Recognition by Gabor Filters, an Application

We discuss here a *saccadic search* strategy, a general-purpose attentional mechanism that identifies semantically meaningful structures in images by performing "jumps" (*saccades*) between relevant locations [205]. The saccade paths are chosen automatically and rely on apriori knowledge of facial features that are modeled by means of Gabor filters discussed in the previous sections. Additionally, here they are applied to a *log–polar grid*, that is, together with Gabor filter responses, called the *retinotopic sensor*, because each sampling point is not a gray value but is instead an array of responses coming from Gabor filters designed in the *log–polar frequency plane*. The usefulness of the concept to complex cognitive tasks is demonstrated by facial landmark detection and identity authentication experiments over the M2VTS and Extended M2VTS (XM2VTS) databases.

The saccadic search strategy and *face recognition* are based on a sparse retinotopic grid obtained by *log–polar mapping* [196], also called *log-z mapping* :

$$\begin{cases} \xi = \log \sqrt{x^2 + y^2} \\ \eta = \tan^{-1}(y, x) \end{cases} \tag{9.54}$$

that we will return to in Chap. 11 in the context of other image processing tasks. The log–polar grid, which is in the spatial domain, is used to sample the original image but not for extracting the grayvalues. At each log–polar grid point in the image, a local Gabor decomposition of the image is performed to the effect that they mimic the *simple cells* of the primary visual cortex having the same *receptive field* but different spatial directions and frequencies [113].

The log–polar grid used in the example results discussed here is shown in Fig. 9.16. The simple cells are modeled by computing a vector of 30 Gabor filter re-

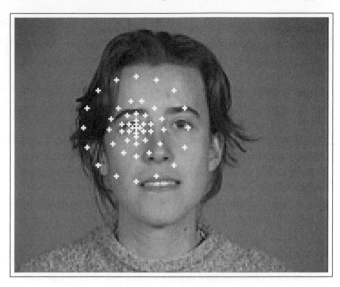

Fig. 9.17. The local and the extended models of a facial landmark are obtained by centering the retina on that landmark in the images of the training set. Adapted from [205]

sponses at each point of the retina. The filters are organised in five frequency channels and six equally spaced orientation channels as discussed in Sect. 9.7. Filter wavelengths span the range from 4 to 16 pixels in half-octave intervals.

The same sensor is used both for facial landmark localization and for person authentication. The sparse nature of the sensor and the relatively low number of fixations, thanks to the quick convergence of the saccades to the facial landmarks, make the computation of Gabor responses by direct filtering in the image domain more advantageous [205] and feasible even in real time [27].

The *saccadic search* strategy requires the use of two levels of modeling for each of the *facial landmarks*: left eye, right eye, mouth+nose. The latter is a metafeature that consists in joining the mouth and the nose. The *local model* is obtained for each facial landmark as the vector of Gabor responses extracted at the central grid point at the location of that landmark in the images of a training set. The *extended model* is obtained by placing the whole retina at the location of the facial landmark on the training images (Fig. 9.17) and collecting the set of Gabor filter responses from all of the retinal points. The local model and the global model are given to supervised classifiers[7] that learn a compact representation of the facial landmarks by examples, to identify them later in other images.

Facial landmark detection is achieved by a saccadic search, starting with the retinotopic sensor placed at random on the image. The search initially aims to find any of the three facial landmarks. Gabor response vectors are computed at all reti-

---

[7] Examples of supervised classifers are neural networks [29, 49], and Bayes classifiers [125, 153]. and support vector machines.

nal points at a random starting point. The response vectors at each log–polar grid point on a circle are rated by the local model of the selected landmark according to the output of the corresponding classifier, $f_{loc}(\mathbf{v}_{c\gamma})$. The retina is subsequently centered at the grid point in the image that maximizes the similarity provided by the classifier. This procedure is iterated until the retinotopic sensor is centered on a local maximum. One advantage of this search strategy is that the search automatically becomes finer as a local maximum is approached, since the artificial retina is denser at the center (*fovea*) than at the periphery. As this application demonstrates, the acuity gradient existing between the peripheral and the foveal vision in the topology of the human retina plays a plausible role in achieving fast convergence to targets by use of the saccades, *homing*. After saccades have converged to a local optimum, the retinotopic grid is displaced in a pixel-by-pixel fashion to maximize the output of the more accurate, but computationally heavier, extended model for the detected facial landmark. Matches that score less in classifier-provided comparisons are discarded at this stage.

Once a match for a facial landmark has been found, a saccade to the average assumed location of one of the others is performed. An attempt at detection is made directly with the corresponding extended model. If this fails, the search is restarted at random to look for this feature.

A global *configuration score* is computed based exclusively on the quality of the matches detected. Saccadic search is continued until a complete set of facial landmarks that has a very high configuration score is found. In Fig. 9.18, the performance of this technique is illustrated by way of examples. These images belong to the XM2VTS database [162]. In 99.5% of the database images, at least two facial features were correctly positioned. Following the facial feature detection, an authentication of the found face against a reference or client person can be performed by comparing the measured features of the global models for all three facial features with those of the client. This can be done by the same classifier that was used to locate the eyes and mouth+nose, but adapted to the identity of the person, yielding 0.5% false acceptance at 0.5% false rejection threshold. The details of the system and its performance are presented in [205].

Fig. 9.18. Eye and mouth localization by retinotopic sensor and saccades. Performance in the presence of eyeglasses (*1st row*), partial occlusion (*2nd row*), pose changes (*3rd row*), facial hair (*4th row*), and ethnic diversity (*5th row*) are illustrated. Adapted from [205]

Fig. 3.18.

Part III

Vision of Single Direction

Humility is the barrier against all evils.

St. Isaac (died 7th century)

10

Direction in 2D

Directional processing of visual signals is the largest single analysis toolbox of *mammalian visual* system: it feeds other specialized visual processing areas [114, 173, 235], e.g., face recognition. Directional analysis is gaining increased traction even in computer vision, as it moves from single-problem-solving systems towards multi-problem-solving platforms. Nearly all applications of image analysis now have alternatives using direction tensor fields. The necessary tools are more modern and offer advanced low-level signal processing that was hitherto reserved to processing of high-level tokens, such as binarized or skeletonized edge maps. In 2D, the earliest solutions to the problem of finding the direction of an image patch, e.g., [51, 116], consisted in projecting the image onto a number of fixed orthogonal functions. The projection coefficients were then used to evaluate the orientation parameter of the model. When the number of filters used is increased, the local image is described better and better, but the inverse function, mapping the coefficients to the optimal orientation, increase greatly in complexity. A generalization of the inverse projection approach to higher dimensions becomes therefore computationally prohibitive. Here we will follow a different approach by modeling the shapes of isocurves via tensors.

10.1 Linearly Symmetric Images

We will refer to a small 2D image patch around a point as an *image*, to the effect that we will treat the local image patches in the same way as the global image. Let the scalar function f, taking a two-dimensional vector $\mathbf{r} = (x, y)^T$ as argument, represent an image. Assume that $\mathbf{fr} = (x, y)^T$ is a two-dimensional real vector that represents the coordinates of a point in a plane on which an image $f(x, y)$ is defined. Furthermore, assume that $\mathbf{k} = (k_x, k_y)^T$ is a two-dimensional unit vector representing a constant direction in the plane of the image.

Definition 10.1. *The function f is called a* linearly symmetric image *if its isocurves have a common direction, i.e., there exists a scalar function of one variable g such that*

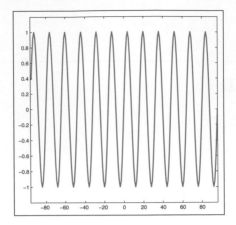

Fig. 10.1. The graph represents the 1D function $g(t) = \sin(\omega t)$ that will be used to construct a linearly symmetric 2D function

$$f(x, y) = g(\mathbf{k}^T \mathbf{r}) = g(k_x x + k_y y) \tag{10.1}$$

The direction of the linear symmetry is $\pm \mathbf{k}$.

The term is justified in that F, the 2D Fourier transform of f, is concentrated to a line, as will be shown below. In addition, all isocurves of linearly symmetric functions are lines that have a common direction $\mathbf{k}$, i.e., they are parallel to each other. Note that the term *isocurves* refers to the fact that the values of g and thereby f are invariant when one moves along certain curves in the argument domain. For linearly symmetric images, these curves are lines.

It should be noted that while $g(t)$ is a function of one free variable, $g(\mathbf{k}^t\mathbf{r})$ is a function of two free variables, $(x, y)^T$ since $\mathbf{k}$ is constant. In the rest of this section we will assume that the argument domain of f is two-dimensional, whereas that of g is one-dimensional and g is a "constructor" of f via Eq. (10.1) and $\mathbf{k}$ whenever f is linearly symmetric. Therefore $g(\mathbf{k}^T\mathbf{r})$ generates an image despite that g by itself is a one-dimensional function. By definition, images with the linear symmetry property have the same gray value at all points $\mathbf{r}$ satisfying $\mathbf{k}^T\mathbf{r} = C$ for a given value C. Because $\mathbf{k}^T\mathbf{r} = C$ describes a line in the (x, y)-plane, it follows that along this line the gray values of the image do not change and this gray value equals to $g(C)$. In such images, the only occasion when g can change is when the argument of g changes, i.e., when the constant C assumes another value. However, the curves $\mathbf{k}^T\mathbf{r} = C_1$, $\mathbf{k}^T\mathbf{r} = C_2$, ... $\mathbf{k}^T\mathbf{r} = C_n$, with C_i being different constants, represent lines that are shifted versions of each other, all having the same direction, $\mathbf{k}$. Consequently, the one-dimensional function g is a profile of the two-dimensional function $g(\mathbf{k}^T\mathbf{r})$ along any line perpendicular to the line $\mathbf{k}^T\mathbf{r} = C$.

Local images, which can be extracted by multiplying the original image with an appropriate window function, are, from mathematical viewpoint, no different than the larger original image from which these local images are "cut". For notational

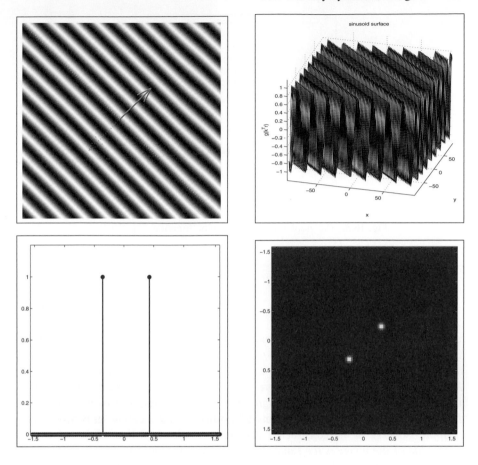

Fig. 10.2. In *top, left*, the image generated by $g(t) = \sin(\omega t)$ with the argument $t = \mathbf{k}^T \mathbf{r}$ is given. The *solid* and *dashed* vectors represent $\mathbf{k}$ and $-\mathbf{k}$ respectively. The 3D graph in *top, right* represents $g(\mathbf{k}^T \mathbf{r})$ as a surface. The FT magnitudes of $g(t)$ and $g(\mathbf{k}^T \mathbf{r})$ are shown in *bottom, left* and *bottom, right*. The FT coordinates are in the angular frequencies ω and $(\omega_x, \omega_y)^T$

simplicity, we will therefore not make a distinction between an image and a local neighborhood of it. Both variants will be referred to as an image here, unless an ambiguity calls for further precision.

Below we detail the process of constructing linearly symmetric images from 1D functions first by three examples of 1D functions g, that are continuous. The last one of these will model an ideal line. Then we will study a discontinuous g which will be a model of ideal edges. Both ideal lines and ideal edges have been used to model and to detect discontinuities in image processing.

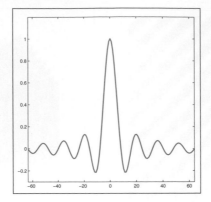

Fig. 10.3. The graph represents the sinc function, Eq. (10.4)

Example 10.1. *The function*

$$g(t) = \sin(\omega t), \qquad with \qquad \omega = \frac{2\pi}{12}, \tag{10.2}$$

rendered in Fig. 10.1, oscillates around zero. This sinusoid is used to construct the linearly symmetric 2D function, shown in Fig. 10.2 (top, left). To render the negative values of g we added the constant 0.5 and rescaled the gray values to the effect that white represents 1 and black represents -1 by using 256 gray tones and linear mapping. The dashed green line through origin shows an example isocurve. We recall that an isocurve is the curve that joins all points having a certain gray value. The example isocurve in the constructed 2D image is clearly a line, along which the gray value of the image is the same. In fact, every other isocurve is also a line and all isocurves are parallel, just as they should be, because of the way the image is constructed.

By construction, the gray value of the image cannot change unless the argument of g changes. For any given image location $\mathbf{r}$, the argument of g will equal a constant value C

$$\mathbf{k}^T \mathbf{r} = C \tag{10.3}$$

to the effect that it is possible to restrict the changes of $\mathbf{r}$ to a curve so that C is invariant. By virtue of Eq. (10.3), this path is a line, and the equation represents parallel lines when $\mathbf{k}$ is fixed. We can move in the image, i.e., change $\mathbf{r}$, along a line which is perpendicular to the line shown as the dashed diagonal, i.e along $\mathbf{k}$, and obtain changes in the value of C, which in turn changes the values of g. In this path, the obtained function or gray values are identical to the values of the original 1D function, $g(t) = \cos(\omega t)$. The solid green arrow illustrates the vector $\mathbf{k} = (\cos(\frac{\pi}{4}), \sin(\frac{\pi}{4}))^T$ used to build the image. The representation is not unique because $-\mathbf{k}$ (dashed) would have generated the same image. In other words, the $\mathbf{k}$ used to construct the linearly symmetric image is unique only up to a sign factor ± 1. The color surface in 3D shows *the same image as a landscape, illustrating that the gray variations are identical to those in Fig. 10.1 across the isocurves.*

The magnitudes of the Fourier transforms of $g(t)$ and $g(\mathbf{k}^T\mathbf{x})$ are also given in Fig. 10.2 (bottom). The red color represents the value zero in the image. The bright (yellowish) spots represent the largest values. In the image, the 2D Fourier transform magnitudes clearly equal zero outside of the two bright points.

Example 10.2. *The* 1D *sinc function*

$$g(t) = \frac{\sin(\omega t)}{\omega t}, \qquad \text{with} \qquad \omega = \frac{2\pi}{12}, \tag{10.4}$$

is plotted in Fig. 10.3. The synthetic image represented by the function $g(\mathbf{k}^T\mathbf{r})$ is linearly symmetric. Its image is illustrated by the gray image in Fig. 10.4 (top) which differs from the one in Fig. 10.2 only by the choice of g. The function values are scaled and shifted to be rendered by the available 256 gray tones. The solid and the dashed vectors represent $\mathbf{k}$ and $-\mathbf{k}$ respectively. Both this image and the gray image in Fig. 10.2 have the same vector $\mathbf{k} = (\cos(\frac{\pi}{4}), \sin(\frac{\pi}{4}))^T$, which represents the direction orthogonal to the isocurve direction. In the direction of $\mathbf{k}$, any cross section of the image is identical to the sinc function of Fig. 10.3, as illustrated by the 3D graph in Fig. 10.4, which shows $g(\mathbf{k}^T\mathbf{r})$ as a surface.

In the (2D) color image, we note that the magnitudes of the Fourier transformed function equal zero (red color) outside of a line passing through the origin, indicated by bright yellow in Fig. 10.4 (bottom, right). The line has the direction $\mathbf{k}$. Along the line, the Fourier transform magnitude has the same shape as the (1D) Fourier transform magnitude (bottom, left).

We will bring further precision to the relationship between the 1D and 2D Fourier transforms of the linearly symmetric functions below. For now we note the result of this example, as illustrated by Fig. 10.5 bottom, right, is consistent with that of the previous example, Fig. 10.2 bottom, right. Both Fourier transform magnitudes vanish outside a central line having the direction $\mathbf{k}$, whereas on the line itself both magnitudes have at least the same magnitude variations as their 1D counterparts shown on the respective left. From the magnitudes of Fourier transformed functions, we can in general not deduce the underlying complex values. However, there is one exception to that which is a result of the null property of norms, i.e., the magnitudes of complex numbers are zero if and only if the complex numbers are zero. The Fourier transforms of the two illustrated example images possessing linear symmetry must consequently have not only magnitudes but also complex values that equal zero outside the referenced line.

The second example actually showed the same sinusoid as in the first one, with the difference that in the second example, the sinusoid attenuates gradually as $1/t$. The Fourier transform magnitude of the Sinc example is therefore more spread as compared to that of the pure sinusoid, which consists of a pair of Dirac pulses.

The sinusoid is neither a pure line nor a pure edge, but yet it has a direction. The classical edge and line detection techniques in image processing model and detect

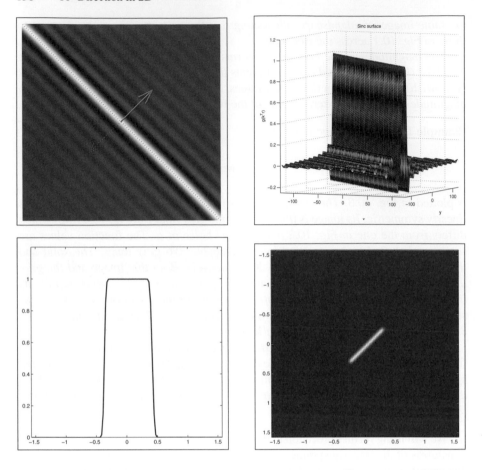

Fig. 10.4. (*Top*) The gray image is generated by substituting $t = \mathbf{k}^T\mathbf{r}$ in Eq. (10.4). The 3D graph shows $g(\mathbf{k}^T\mathbf{r})$. (*Bottom*) The 1D and 2D FT magnitudes of $g(t)$ and $g(\mathbf{k}^T\mathbf{r})$, respectively

pure lines and edges, without a provision for other types of patterns that have well-defined directions. In the next example, we show that pure lines can be modeled as a linearly symmetric function generated by means of an analytic function, a Gaussian.

Example 10.3. *The 1D Gaussian*

$$g(t) = \exp\left(-\frac{t^2}{2\sigma^2}\right), \qquad with \qquad \sigma = 3, \qquad (10.5)$$

is plotted as the green curve in Fig. 10.5. The synthetic image represented by the function $g(\mathbf{k}^T\mathbf{r})$ is linearly symmetric and is illustrated by the gray image in Fig. 10.5. The function values are scaled and linearly mapped to 256 gray tones, with 0 corresponding to black, and 1 corresponding to white. In the direction of $\mathbf{k}$, any cross section of the image is identical to the 1D Gaussian of Fig. 10.5.

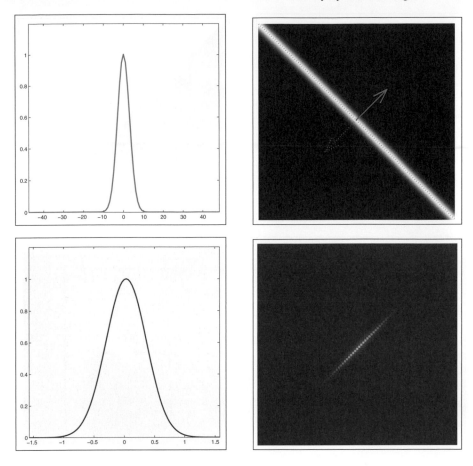

Fig. 10.5. (*Top*) The graph represents the 1D Gaussian $g(t)$ in Eq. (10.5) and the 2D function generated by the substitution $t = \mathbf{k}^T\mathbf{r}$. The *solid* and *dashed* vectors represent $\mathbf{k}$ and $-\mathbf{k}$ respectively. (*Bottom*) The 1D FT magnitude of $g(t)$ and the 2D FT magnitude of $g(\mathbf{k}^T\mathbf{r})$ are illustrated by the (*left*) graphics and the (*right*) image, respectively

When we study the 2D Fourier transform magnitudes of this linearly symmetric image, Fig. 10.5 bottom, right, we note that it too equals to zero (red) outside of the same central line (bright yellow) as in the previous two examples. The line has a profile matching the 1D version of the Fourier transform magnitude, shown in the bottom, left graph.

Example 10.4. *The 1D step function*

$$g(t) = \begin{cases} 1, & \text{if } t \geq 0, \\ 0, & \text{otherwise,} \end{cases} \qquad (10.6)$$

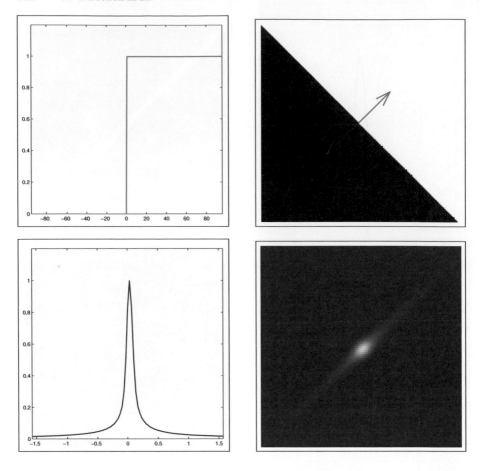

Fig. 10.6. (*Top*) The graph represents the 1D step function $g(t)$ in Eq. (10.6) and the 2D function generated by the substitution $t = \mathbf{k}^T \mathbf{r}$. The *solid* and *dashed* vectors represent $\mathbf{k}$ and $-\mathbf{k}$, respectively. (*Bottom*) The 1D FT magnitude of $g(t)$ and the 2D FT magnitude of $g(\mathbf{k}^T \mathbf{r})$ are illustrated by the (*left*) graphics and the (*right*) image, respectively

is discontinuous. It is shown as the green curve in Fig. 10.6 with its corresponding linearly symmetric gray image, which is obtained by sampling

$$g(\mathbf{k}^T \mathbf{r}) = \begin{cases} 1, & if k_x x + k_y y \geq 0, \\ 0, & otherwise. \end{cases} \tag{10.7}$$

The gray image models what came to be known as the ideal edge in image processing. Despite the fact that it is discontinuous, it too is a linearly symmetric function, with Fourier transform magnitudes concentrated to the same central line as in the previous examples.

Fig. 10.7. In real images, linear symmetries often resemble "lines" or "edges" but they appear even as object or texture boundaries. Some of these are illustrated by colored lines in this photograph

Example 10.5. *The linearly symmetric images appear frequently as local images, both in images of human-made environments and in images of the nature. They are often perceived as lines or edges (Fig. 10.7), or as repetitive patterns called textures (Fig. 10.8). Although the cross sections of their isocurves are seldom like the ideal lines and edges, they usually have a well-distinguishable direction, with Fourier transform magnitudes concentrated to a line, as Fig. 10.8 shows.*

We give precision to the example indications regarding the Fourier transform of the linearly symmetric images in the following lemma. We recall that the argument domain of g is one-dimensional, whereas that of f is two-dimensional by the adopted convention.

Lemma 10.1. *A linearly symmetric image $f(\mathbf{r}) = g(\mathbf{k}^T \mathbf{r})$ has a 2D Fourier transform concentrated to a line through the origin:*

$$F(\omega_x, \omega_y) = G(\mathbf{k}^T \boldsymbol{\omega})\delta(\mathbf{u}^T \boldsymbol{\omega}) \qquad (10.8)$$

where $\mathbf{k}$, $\mathbf{u}$ are orthogonal vectors and δ is the Dirac distribution. The vector $\boldsymbol{\omega}$ is the angular frequency vector $\boldsymbol{\omega} = (\omega_x, \omega_y)^T$. The function G is the one-dimensional Fourier transform of g.

♦

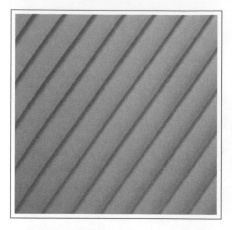 

Fig. 10.8. *Left*: a real image that is linearly symmetric. It shows a close-up view of the blinds. *Right*: The Fourier transform magnitude of a neighborhood in the central part of the original (brightness) image. Notice that the power is concentrated to a line orthogonal to isocurves in the original

Lemma 10.1 is proven in the Appendix of this chapter, Sect 10.17. It states that the function $g(\mathbf{k}^T \mathbf{r})$, which is in general a "spread" function such as a sinusoid or an edge, is compressed to a line having the direction $\mathbf{k}$ in the Fourier domain. Even more important, it says that as far as $\mathbf{k}$ is concerned the choice of G, and thereby g, has no significance. This is because, $\mathbf{k}$, the angle at which all nonzero F reside, remains the same no matter what G is. This is achieved by the Dirac pulse δ, which becomes a line pulse along the infinite line $\mathbf{u}^T \boldsymbol{\omega} = 0$ by the expression $\delta(\mathbf{u}^T \boldsymbol{\omega})$. Because $\mathbf{u}$ is the normal direction of this line and $\mathbf{u}$ is orthogonal to $\mathbf{k}$, the direction of the spectral line $\mathbf{u}^T \boldsymbol{\omega} = 0$ coincides with the vector $\mathbf{k}$. We have already observed this line in red-colored images of Examples 10.1–10.5, as a concentration of the magnitudes to the central line, in the same direction as the same (green) $\mathbf{k}$ vector shown in the gray images, regardless g.

Along this central spectral line, not only the magnitude but also the complex values conform to that of the 1D Fourier transform. This is because G is the 1D Fourier transform of g, and Eq. (10.8) is a formula for how to produce the 2D Fourier transform of the linearly symmetric functions only from the 1D Fourier transform G and the isocurve normal $\mathbf{k}$. According to the lemma, the vector $\mathbf{u}$ can always be deduced from $\mathbf{k}$ up to a sign factor, because it is orthogonal to $\mathbf{k}$. Due to limitations of the illustration methods, Examples 10.1–10.5 could only be indicative about this more powerful result, and this only as far as the Fourier transform magnitudes are concerned.

Consequently, to determine whether or not an image is linearly symmetric is the same thing as to quantitate to what extent the Fourier transform vanishes outside of a line, a property which will be exploited to construct computer algorithms below. Such algorithms can be constructed conveniently to describe textures possessing a direction, or by detecting the lack of linear symmetry, to describe textures lacking direction. Measuring the lack of linear symmetries has been frequently used as a way of detecting corners in image processing.

10.2 Real and Complex Moments in 2D

In image analysis there are a variety of occasions when we need to quantitate functions by comparing them to other functions. Assuming that the integral exists, the quantity

$$m_{pq}(\kappa) = \langle x^p y^q, \kappa \rangle = \int\int x^p y^q \kappa(x,y) dx dy \qquad (10.9)$$

with p and q being nonnegative integers, is the *real moment* p, q of the function κ. If κ has a finite extension, then the real moments defined as above are projections of an integrable function onto the vector space of polynomials. It follows from the Weierstrass theorem [193, 209], that the vector space of the polynomials is powerful enough to approximate a finite extension function κ to a desired degree of accuracy. In that, the approximation property of moments is comparable to the FCs, although the polynomial basis of moments is not orthogonal, whereas the Fourier basis is. Nonetheless, moments are widely used in applications. If κ is a positive function then it is possible to view it as a probability distribution, after a normalization with m_{00}. Accordingly,

$$\bar{c} = (\bar{x}, \bar{y})^T = \frac{1}{m_{00}}(m_{10}, m_{01})^T \qquad (10.10)$$

represents the centroid or the mean vector of the function κ. The quantity

$$\tilde{m}_{pq}(\kappa) = \left\langle \left(x - \frac{m_{10}}{m_{00}} \right)^p \left(y - \frac{m_{01}}{m_{00}} \right)^q, \kappa \right\rangle$$

$$= \int\int \left(x - \frac{m_{10}}{m_{00}} \right)^p \left(y - \frac{m_{01}}{m_{00}} \right)^q \kappa(x,y) dx dy \qquad (10.11)$$

related to real moments, is called the central moment p, q of the function κ. Both real moments and central moments have been utilized as tools to quantitate the shape of a finite extension image region. We will discuss this further in Sect. 17.4.

Another type of moment, which we will favor over real moments in what follows, is

$$I_{pq}(\kappa) = \langle (x - iy)^p (x + iy)^q, \kappa \rangle = \int\int (x + iy)^p (x - iy)^q \kappa(x,y) dx dy \quad (10.12)$$

with p and q being nonnegative integers. This is the *complex moment* p, q of the function κ. Notice that complex moments are linear combinations of the real moments, e.g.,

$$I_{20} = m_{20} - m_{02} + i2m_{11}$$
$$I_{11} = m_{20} + m_{02}$$

The order number and the symmetry number of a complex moment refer to $p+q$ and $p-q$, respectively. The integrals above should be interpreted as summations when the complex moments of discrete functions are to be computed. In the following sections we will make use of (real and complex) moments as a spectral regression tool, i.e., to fit a line to the FT of a function.

10.3 The Structure Tensor in 2D

The *structure tensor*[1] or the *direction tensor* models linearly symmetric structures that are frequently found in images. To represent certain geometric properties of images, it associates 2×2 symmetric matrices that are tensors to them. This is not different from the fact that in a color image there are several color components per image point, e.g., HSB, to every point. Typically, however, the structure tensor is used to quantify shape properties of local images. As such, structure tensors are assigned to every image point to represent properties of neighborhoods.

Let the scalar function $f(\mathbf{r})$, taking the two-dimensional vector $\mathbf{r} = (x, y)^T$ as argument, represent an image, which is usually a neighborhood around an image point. As before, the (capitalized) letter F is the Fourier transform of f. We denote with $|F(\boldsymbol{\omega})|$ the magnitude spectrum of f, where $\boldsymbol{\omega} = (\omega_x, \omega_y)^T$ is the Fourier transform coordinates in angular frequencies, and with $|F(\boldsymbol{\omega})|^2$ we denote the power spectrum of f. We will use the power spectrum rather than $|F|$ to measure the significance of a given frequency in the signal because it will turn out that the average values of $|F|^2$ are easier to measure in practice than $|F|$.

The direction of a linearly symmetric function $f(\mathbf{r}) = g(\mathbf{k}^T\mathbf{r})$ is well-defined by the vector $\mathbf{k}$, but only up to a sign factor. According to lemma 10.1, if and only if f is linearly symmetric is its power spectrum, $|F|^2$ concentrated to a central line with the direction $\mathbf{k}$. The direction of this line represents the direction of the linear symmetry. We will approach estimating $\mathbf{k}$ by fitting the image power spectrum, $|F|^2$, a line in the total least square TLS sense. Consequently, it will be possible to "measure" if f is linearly symmetric by studying the error of the fit. If the error is "small" in the sense that has been defined, then our method will take this as a provision that the fit was successful and that the image approximates a linearly symmetric image $g(\mathbf{k}^T\mathbf{r})$ well. It turns out that, in this procedure, g need not be known beforehand. It will be automatically determined when the error of the fit is near zero, because we will then obtain a reliable direction along which to "cut" the image. In turn the 1D function obtained by cutting the image is g only if f is linearly symmetric. Owing to the continuity of the TLS error function, the decrease or the increase of the error will be graceful when f approaches to a linearly symmetric function or departs from one. We discuss the details of the line-fitting next.

We wish to fit an axis through the origin of the Fourier transform of an image, f, which *may or may not* be linearly symmetric. Fitting an axis to a finite set of points is classically performed by minimizing the error function:

---

[1] Other names of this tensor include the "second order moment tensor", "inertia tensor", "outer product tensor", and "covariance matrix".

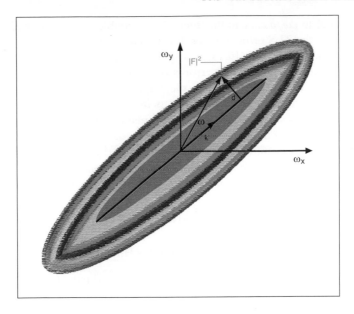

Fig. 10.9. The line-fitting process is illustrated by the linear symmetry direction vector $\mathbf{k}$, the angular frequency vector $\boldsymbol{\omega}$, and the distance vector $\mathbf{d}$. The function values $|F(\boldsymbol{\omega})|^2$ are represented by color. The frequency coordinate vector is $\boldsymbol{\omega}$

$$e(\mathbf{k}) = \sum_{\boldsymbol{\omega}} d^2(\boldsymbol{\omega}, \mathbf{k}) \tag{10.13}$$

where $d(\boldsymbol{\omega}, \mathbf{k})$ is the shortest distance between a point $\boldsymbol{\omega}$ and a candidate axis $\mathbf{k}$. This is the TLS error function for a discrete data set. Noting that $\|\mathbf{k}\| = 1$, then the projection of $\boldsymbol{\omega}$ on the vector $\mathbf{k}$ is $(\boldsymbol{\omega}^t \mathbf{k})\mathbf{k}$. As illustrated by Fig. 10.9, the vector $\mathbf{d}$ represents the difference between $\boldsymbol{\omega}$ and the projection of $\boldsymbol{\omega}$. This difference vector is orthogonal to $\mathbf{k}$

$$\mathbf{d} = \boldsymbol{\omega} - (\boldsymbol{\omega}^t \mathbf{k})\mathbf{k} \tag{10.14}$$

with its norm being equal to the shortest distance, i.e., $\|\mathbf{d}\| = d(\boldsymbol{\omega}, \mathbf{k})$. Consequently, the square of the norm of $\mathbf{d}$ provides:

$$d^2(\boldsymbol{\omega}, \mathbf{k}) = \|\boldsymbol{\omega} - (\boldsymbol{\omega}^t \mathbf{k})\mathbf{k}\|^2 \tag{10.15}$$

$$= (\boldsymbol{\omega} - (\boldsymbol{\omega}^t \mathbf{k})\mathbf{k})^T (\boldsymbol{\omega} - (\boldsymbol{\omega}^t \mathbf{k})\mathbf{k}) \tag{10.16}$$

Since we have a Fourier transform function, F, defined on dense angular frequency coordinates in E_2, instead of a sparse point set, Eq. (10.13) needs to be modified. The following error function is a generalization of Eq. (10.13) to dense point sets. It weights the squared distance contribution at an angular frequency point $\boldsymbol{\omega}$ with the energy $|F(\boldsymbol{\omega})|^2$ and integrates all error contributions

$$c(\mathbf{k}) = \int d^2(\boldsymbol{\omega}, \mathbf{k})|F(\boldsymbol{\omega})|^2 d\boldsymbol{\omega} \tag{10.17}$$

where $d\boldsymbol{\omega}$ equal to $d\omega_x d\omega_y$, and the integral is a double integral over E_2. The expression defines the TLS error function for a continuous data set. Interpreting the integral as a summation, Eq. (10.13) is a special case of Eq. (10.17). By construction, the contribution of $F(\boldsymbol{\omega})$ to the error is zero for points $\boldsymbol{\omega}$ along the line $t\mathbf{k}$. With increased distance of $\boldsymbol{\omega}$ to the latter axis, the contribution of $F(\boldsymbol{\omega})$ will be amplified.

If $|F(\boldsymbol{\omega})|^2$, which is the spectral energy, is interpreted as the mass density, then the error $e(\mathbf{k})$ corresponds to the inertia of a mass with respect to the axis $\mathbf{k}$ in mechanics [83]. Besides this function being continuous in $\mathbf{k}$, it has also a screening effect; that is, F is concentrated to a line if and only if $e(\mathbf{k})$ vanishes for some $\mathbf{k}$. This is to be expected because the resulting quantity when substituting Eq. (10.16) in Eq. (10.17) is the square norm of a vector-valued function, i.e.,

$$e(\mathbf{k}) = \|\left(\boldsymbol{\omega} - (\boldsymbol{\omega}^T\mathbf{k})\mathbf{k}\right)F\|^2 = \int \|\left(\boldsymbol{\omega} - (\boldsymbol{\omega}^T\mathbf{k})\mathbf{k}\right)\|^2 |F(\boldsymbol{\omega})|^2 d\boldsymbol{\omega} \qquad (10.18)$$

Because $e(\mathbf{k})$ is the square norm of a function in a Hilbert space,

$$e(\mathbf{k}) = \|\left(\boldsymbol{\omega} - (\boldsymbol{\omega}^T\mathbf{k})\mathbf{k}\right)F\|^2 = 0 \quad \Rightarrow \quad \left(\boldsymbol{\omega} - (\boldsymbol{\omega}^T\mathbf{k})\mathbf{k}\right)F = \mathbf{0} \qquad (10.19)$$

Consequently, when $e(\mathbf{k}) = 0$, the expression $\left(\boldsymbol{\omega} - (\boldsymbol{\omega}^T\mathbf{k})\mathbf{k}\right)F$ equals (the vector-valued function) zero. Assuming $F \neq 0$, this will happen if and only if F equals zero outside the line $t\mathbf{k}$. A nontrivial F can thus be nonzero only on the line $t\mathbf{k}$, which, according to lemma 10.1, means that f must be linearly symmetric. Conversely, if F is zero except on the line $t\mathbf{k}$, the corresponding $e(\mathbf{k})$ vanishes. To summarize, if and only if $e(\mathbf{k}) = 0$, all spectral energy, $|F|^2$, is concentrated to a line and f is linearly symmetric with the direction $\pm\mathbf{k}$. Leaving the question whether or not $\pm\mathbf{k}$ always translates to an unambiguous direction aside (we will discuss this further below), we proceed to the more urgent question of how to calculate $\mathbf{k}$.

Noting that $\boldsymbol{\omega}^T\mathbf{k}$ is a scalar that is indistinguishable from the scalar $\mathbf{k}^T\boldsymbol{\omega}$, and $\|\mathbf{k}\|^2 = \mathbf{k}^T\mathbf{k} = 1$, the square magnitude distance can be written in quadratic form:

$$d^2(\boldsymbol{\omega}, \mathbf{k}) = \mathbf{k}^T\left(\mathbf{I}\boldsymbol{\omega}^T\boldsymbol{\omega} - \boldsymbol{\omega}\boldsymbol{\omega}^T\right)\mathbf{k} \qquad (10.20)$$

$$= \mathbf{k}^T\left[\begin{pmatrix} \omega_x^2 + \omega_y^2 & 0 \\ 0 & \omega_x^2 + \omega_y^2 \end{pmatrix} - \begin{pmatrix} \omega_x^2 & \omega_x\omega_y \\ \omega_x\omega_y & \omega_y^2 \end{pmatrix}\right]\mathbf{k} \qquad (10.21)$$

Defining the components of $\boldsymbol{\omega} = (\omega_x, \omega_y)^T$ as

$$\omega_x = \omega_1, \quad \text{and} \quad \omega_y = \omega_2, \qquad (10.22)$$

for notational convenience, Eq. (10.17) is expressed as

$$e(\mathbf{k}) = \mathbf{k}^T\mathbf{J}\mathbf{k} = \mathbf{k}^T\left(\mathbf{I} \cdot \text{Trace}(\mathbf{S}) - \mathbf{S}\right)\mathbf{k} \qquad (10.23)$$

where $\mathbf{I}$ is the identity matrix,

$$\mathbf{J} = \mathbf{I} \cdot \text{Trace}(\mathbf{S}) - \mathbf{S} \qquad (10.24)$$

and

$$\mathbf{S} = \int \boldsymbol{\omega}\boldsymbol{\omega}^T |F|^2 d\boldsymbol{\omega} = \begin{pmatrix} \mathbf{S}(1,1) & \mathbf{S}(1,2) \\ \mathbf{S}(2,1) & \mathbf{S}(2,2) \end{pmatrix}, \quad \text{with} \quad \mathbf{S}(i,j) = \int \omega_i \omega_j |F(\boldsymbol{\omega})|^2 d\boldsymbol{\omega}$$
$$(10.25)$$

Definition 10.2. *The matrix* $\mathbf{S}$ *in Eq. (10.25), which consists of the second-order moments of the power spectrum,* $|F|^2$, *is called the* structure tensor *of the image* f.

The matrix $\mathbf{J}$ is the *inertia tensor* of the power spectrum using a term of mechanics. The matrix $\mathbf{S}$ is also called the *scatter tensor* of the power spectrum in statistics. The structure tensor can be readily obtained from $\mathbf{J}$, and vice versa via Eq. (10.24). There is another related tensor called the *covariance tensor* or the covariance matrix in statistics, $\mathbf{C} = \mathbf{S} - \mathbf{m} \cdot \mathbf{m}^T$, where $\mathbf{m} = \int \boldsymbol{\omega} |F|^2 d\boldsymbol{\omega}$. However, for real images $\mathbf{m} = 0$, since $|F|$ is even when f is real. Because of the tight relationship between the notions inertia, scatter, and covariance, they are used in an interchangeable manner in many contexts. Since different notions of the structure tensor coexist, the following lemma, which establishes the equivalence of $\mathbf{J}$ and $\mathbf{S}$ (and of $\mathbf{C}$), is useful to remember.

Lemma 10.2. *With eigenvalue, eigenvector pairs of* $\mathbf{J}$ *being* $\{\lambda_1, \mathbf{u}_1\}$ *and* $\{\lambda_2, \mathbf{u}_2\}$, *and those of* $\mathbf{S}$ *being* $\{\lambda_1', \mathbf{u}_1'\}$ *and* $\{\lambda_2', \mathbf{u}_2'\}$, *we have*

$$\{\lambda_1', \mathbf{u}_1'\} = \{\lambda_1, \mathbf{u}_2\}, \quad \text{and} \quad \{\lambda_2', \mathbf{u}_2'\} = \{\lambda_2, \mathbf{u}_1\}. \quad (10.26)$$

$\blacklozenge$

The eigenvector with a certain eigenvalue in the first matrix is an eigenvector with the *other* eigenvalue in the second matrix. The lemma can be proven by utilizing Eq. (10.24) and operating with $\mathbf{J}$ on $\mathbf{u}_i'$, which is assumed to be an eigenvector of $\mathbf{S}$:

$$\mathbf{J}\mathbf{u}_i' = (\mathbf{I} \cdot \text{Trace}(\mathbf{S}) - \mathbf{S})\mathbf{u}_i' = (\lambda_1 + \lambda_2)\mathbf{u}_i' - \lambda_i \mathbf{u}_i' \quad i = 1, 2 \quad (10.27)$$

The error minimization problem formulated in Eq. (10.17) is reduced to a minimization of a quadratic form, $\mathbf{k}^T \mathbf{J}\mathbf{k}$ with the matrix $\mathbf{J}$ given by Eq. (10.24). This is in turn minimized by choosing $\mathbf{k}$ as the least eigenvector of the inertia matrix, $\mathbf{J}$ [231]. All eigenvalues of $\mathbf{J}$ are real and nonnegative because the error expression Eq. (10.17) is real and nonnegative. Calling the eigenvalue and eigenvector pairs of $\mathbf{J}$ $\{\lambda_{min}, \mathbf{k}_{min}\}$ and $\{\lambda_{max}, \mathbf{k}_{max}\}$, the minimum of $e(\mathbf{k})$ will occur at $e(\mathbf{k}_{min}) = \lambda_{min}$. In other words, the matrix $\mathbf{J}$, or equivalently $\mathbf{S}$, contains sufficient information to allow the computation of the optimal $\mathbf{k}$ in the TLS error sense. We will discuss the motivation behind this choice of error in some detail in Sect. 10.10.

The matrix $\mathbf{S}$ is defined in the frequency domain, which is inconvenient, particularly if $\mathbf{S}$ must be estimated numerous times. For example, when computing the direction for all local patches of an image, we would need to perform numerous Fourier transformations if we attempt to directly estimate the structure tensor from its definition. We can, however, eliminate the need for a Fourier transformation by utilizing (Parseval–Plancherel) theorem 7.2 which states that the scalar products are conserved under the Fourier transform. Applying it to Eq. (10.28), the computation of the matrix elements will be lifted from the Fourier domain to the spatial domain:

$$\mathbf{S}(i,j) = \int \omega_i \omega_j |F(\boldsymbol{\omega})|^2 d\boldsymbol{\omega} = \frac{1}{4\pi^2} \int \frac{\partial f}{\partial x_i} \frac{\partial f}{\partial x_j} d\mathbf{x} \qquad i,j : 1,2 \qquad (10.28)$$

where $x_1 = x$, $x_2 = y$, $d\mathbf{x} = dxdy$, and the integral is a double integral over the entire 2D plane. This is rephrased in matrix form,

$$\mathbf{S} = \int \boldsymbol{\omega}\boldsymbol{\omega}^T |F|^2 d\boldsymbol{\omega} = \frac{1}{4\pi^2} \int (\nabla f)(\nabla^T f) d\mathbf{x} \qquad (10.29)$$

where

$$\nabla f = \left(D_x f, D_y f\right)^T = \left(\frac{\partial f(\mathbf{r})}{\partial x}, \frac{\partial f(\mathbf{r})}{\partial y}\right)^T. \qquad (10.30)$$

We summarize our finding on 2D direction estimation via the following theorem, where the integrals are double integrals taken over the 2D spatial domain.

Theorem 10.1 (Structure tensor I). *The extremal inertia axes of the power spectrum, $|F|^2$ are determined by the eigenvectors of the* structure tensor:

$$\mathbf{S} = \frac{1}{4\pi^2} \int (\nabla f)(\nabla^T f) d\mathbf{x} \qquad (10.31)$$

$$= \frac{1}{4\pi^2} \begin{pmatrix} \int (D_x f)^2 d\mathbf{x} & \int (D_x f)(D_y f) d\mathbf{x} \\ \int (D_x f)(D_y f) d\mathbf{x} & \int (D_y f)^2 d\mathbf{x} \end{pmatrix} \qquad (10.32)$$

The eigenvalues $\lambda_{\min}$, $\lambda_{\max}$ and the corresponding eigenvectors $\mathbf{k}_{\min}$, $\mathbf{k}_{\max}$ of the tensor represent the minimum inertia, the maximum inertia, the axis of the maximum inertia, and the axis of the minimum inertia of the power spectrum, respectively.

♦

We note that $\mathbf{k}_{\min}$ is the least eigenvector, but it represents the axis of the maximum inertia. This is because the inertia tensor $\mathbf{J}$ is tightly related to the scatter tensor $\mathbf{S}$ according to lemma 10.2. The two tensors share eigenvalues in 2D, although the correspondence between the eigenvalues and the eigenvectors is reversed.

While the major eigenvector of $\mathbf{S}$ fits the minimum inertia axis to the power spectrum, the image itself does not need to be Fourier transformed according to the theorem. The eigenvalue $\lambda_{\max}$ represents the largest inertia or error, which is achieved with the inertia axis having the direction $\mathbf{k}_{\min}$. The worst error is useful too, because it indicates the scale of the error when judging the size of the smallest error, $\lambda_{\min}$ (the *range problem*). By contrast, the axis of the maximum inertia provides no additional information, because it is always orthogonal to the minimum inertia axis as a consequence of the matrix $\mathbf{S}$ being symmetric and positive semidefinite.

10.4 The Complex Representation of the Structure Tensor

Estimating the structure tensor, and thereby the direction of an image, can be simplified further by utilizing the algebraic properties of the complex z-plane. Multiplication and division are well-defined in complex numbers, whereas conventional vectors cannot be multiplied or divided with each other to yield new vectors. As a result, an explicit computation of the matrix eigenvalues will become superfluous, and the structure tensor will be automatically decomposed into a directional and a nondirectional part.

Central to the structure tensor theory is the maximization of the scatter $\mathbf{k}^T\mathbf{Sk}$. Because $\mathbf{S}$ is a positive semidefinite matrix with real elements, i.e., $\mathbf{S} = \mathbf{A}^T\mathbf{A}$ for some $\mathbf{A}$ having real coefficients, the scatter is $(\mathbf{Ak})^T(\mathbf{Ak})$ and is either zero or positive for *any* real vector $\mathbf{k}$. By incorporating the complex conjugation into transposition, i.e., $\mathbf{B}^H = (\mathbf{B}^*)^T$, the Hermitian transposition, the expression $(\mathbf{Ak})^H(\mathbf{Ak}) = \mathbf{k}^H\mathbf{Sk}$ will be either zero or positive even if the vector $\mathbf{k}$ has been expressed in a basis that has complex elements. Here, we will maximize $\mathbf{k}^H\mathbf{Sk}$, assuming $\mathbf{k}$ may have complex elements. First, we introduce a new basis that has complex vector elements, using the unitary matrix: $\mathbf{U}^H$

$$\mathbf{k'} = \mathbf{U}^H\mathbf{k}, \quad \text{where} \quad \mathbf{U}^H = \frac{1}{\sqrt{2}}\begin{pmatrix} 1 & i \\ i & 1 \end{pmatrix}, \quad \text{and} \quad U = \frac{1}{\sqrt{2}}\begin{pmatrix} 1 & -i \\ -i & 1 \end{pmatrix}. \quad (10.33)$$

Unitary matrices generalize orthogonal matrices in that a unitary matrix has in general complex elements, and it obeys the relationship $\mathbf{U}^H\mathbf{U} = \mathbf{UU}^H = \mathbf{I}$. Consequently,

$$\mathbf{k}^H\mathbf{Sk} = (\mathbf{Uk'})^H\mathbf{SUk'} = \mathbf{k'}^H\mathbf{Zk'} \quad (10.34)$$

where

$$\mathbf{Z} = \mathbf{U}^H\mathbf{SU} \quad (10.35)$$
$$= \frac{1}{2}\begin{pmatrix} \mathbf{S}(1,1) + \mathbf{S}(2,2) & -i(\mathbf{S}(1,1) - \mathbf{S}(2,2) + i2\mathbf{S}(1,2)) \\ i(\mathbf{S}(1,1) - \mathbf{S}(2,2) + i2\mathbf{S}(1,2))^* & \mathbf{S}(1,1) + \mathbf{S}(2,2) \end{pmatrix}$$

We call $\mathbf{Z}$ the *complex structure tensor*, and we can conclude that it represents the same tensor as $\mathbf{S}$, except for a basis change, and that both matrices have common eigenvalues. They share also eigenvectors, but only up to the unitary transformation, so that $\mathbf{k_Z}$ representing an eigenvector of $\mathbf{Z}$ is given by $\mathbf{k_Z} = \mathbf{U}^H\mathbf{k_S}$, with $\mathbf{k_S}$ being an eigenvector of $\mathbf{S}$. We define the elements of $\mathbf{Z}$ via the complex quantities I_{20} and I_{11} as follows.

Definition 10.3. *The matrix*

$$\mathbf{Z} = \frac{1}{2}\begin{pmatrix} I_{11} & -iI_{20} \\ iI_{20}^* & I_{11} \end{pmatrix}, \quad \text{where} \quad \begin{array}{l} I_{20} = \mathbf{S}(1,1) - \mathbf{S}(2,2) + i2\mathbf{S}(1,2) \\ I_{11} = \mathbf{S}(1,1) + \mathbf{S}(2,2) \end{array} \quad (10.36)$$

is the complex representation of the structure tensor.

A matrix $\mathbf{Z}$ is called Hermitian if $\mathbf{Z}^H = \mathbf{Z}$, and if additionally $\mathbf{k}^H\mathbf{Zk} \geq 0$, which is the case by definition for the complex structure tensor, is called Hermitian positive semidefinite. With the above representation, the elements of $\mathbf{Z}$ encode the $\lambda_{\max}$, $\lambda_{\min}$ as well as $\mathbf{k}_{\max}$ more explicitly than $\mathbf{S}$. This is summarized in the following theorem [28], the proof of which is found in Sect. 10.17.

Theorem 10.2 (Structure tensor II). *The minimum and the maximum inertia as well as the axis of minimum inertia of the power spectrum are given by*

$$I_{20} = (\lambda_{\max} - \lambda_{\min})e^{i2\varphi_{\min}} = \frac{1}{4\pi^2} \int \daleth(f)(x,y)d\mathbf{x} \tag{10.37}$$

$$I_{11} = \lambda_{\max} + \lambda_{\min} = \frac{1}{4\pi^2} \int |\daleth(f)(x,y)|d\mathbf{x} \tag{10.38}$$

with the infinitesimal linear symmetry tensor *(ILST) defined as*[2]

$$\daleth_{x,y}(f)(x,y) = \left(\frac{\partial f}{\partial x} + i\frac{\partial f}{\partial y}\right)^2 \tag{10.39}$$

The quantities $\lambda_{\min}$, $\lambda_{\max}$, and $\varphi_{\min}$ are, respectively, the minimum inertia, the maximum inertia, and the axis of the minimum inertia of the power spectrum.

◆

The eigenvalues of the tensor in theorem 10.1 and the λ's appearing in this theorem are identical. Likewise, the direction of the major eigenvector $\mathbf{k}_{\max}$ and the $\varphi_{\min}$, of theorem 10.2 coincide. Accordingly, the eigenvector information is encoded explicitly in an offdiagonal element of $\mathbf{Z}$, i.e., I_{20} whereas the sum and the difference of the eigenvalues are encoded in the diagonal element as I_{11} and in the offdiagonal element as $|I_{20}|$, respectively.

For completeness, we provide the eigenvectors of $\mathbf{Z}$ as well. Because the argument angle of I_{20} is twice the direction angle of $\mathbf{k}_{\max}$, the direction of latter is obtained by the direction of the square root of I_{20}. The eigenvectors of $\mathbf{S}$ are related to the eigenvectors of $\mathbf{Z}$ via Eq. (10.33), so that we obtain

$$\mathbf{k}'_{\max} = \gamma(\sqrt{I_{20}}, i\sqrt{I_{20}^*})^T \tag{10.40}$$

$$\mathbf{k}'_{\min} = \gamma(\sqrt{-I_{20}}, i\sqrt{-I_{20}^*})^T \tag{10.41}$$

where γ is a scalar that normalizes the norms of the vectors to 1.

Summary of the complex structure tensor

- **Independence under merging.** Averaging the "square" of the complex field $(D_x f + iD_y f)^2$ and its magnitude (scalar) field $|D_x f + iD_y f|^2$, automatically fits an optimal axis to the spectrum,
- **Schwartz inequality.** The inequality $|I_{20}| \leq I_{11}$ holds with equality if and only if the image is linearly symmetric,
- **Rotation-invariance and covariance.** If the image is rotated, the absolute values of the elements of $\mathbf{Z}$, i.e., $|I_{20}|$ as well as I_{11}, will be invariant to the rotation, while the argument of I_{20} will change linearly with the rotation.

In the next two sections we discuss two simpler tensors that will be used as basis tensors for decomposing the structure tensor.

---

[2] The symbol $\daleth$ is pronounced as "doleth" or "daleth", which is intended to be a mnemonic for the fact that it is not an ordinary gradient delivering a vector consisting of derivatives but is a (complex) *scalar* comprised of derivatives.

10.5 Linear Symmetry Tensor: Directional Dominance

In this section we will mean the complex structure tensor when we refer to the structure tensor. An "ideal" linear symmetry is present in the image, when $\lambda_{\max} \gg 0$ and $\lambda_{\min} = 0$. Such images have a *directional dominance* in that there is a single and well-identified direction of isocurves. One way to quantitate this property is by measuring $\lambda_{\max} - \lambda_{\min}$ and $\mathbf{k}_{\max}$, which are jointly given by the complex scalar I_{20} [28]. When $\lambda_{\max} - \lambda_{\min}$ increases, so does the evidence for the image being linearly symmetric, and hence we have a crisp direction in the image. For Hermitian[3] positive semi definite matrices, which includes the structure tensor $\mathbf{Z}$, the dimension of the eigenvector space is equal to the multiplicity of the corresponding eigenvalue, which is the number of times the latter is repeated. The eigenvectors are orthogonal if they belong to two different eigenvalues. Because there are at most two different eigenvalues in 2D, there is no need to encode both eigenvectors. The linear symmetry quality of a 2D image has also been called the "line" property [225], and the "stick" property [161]. The *linear symmetry tensor* is a special type of structure tensor such that

$$\mathbf{Z}_L = \frac{1}{2} \begin{pmatrix} |I_{20}| & -iI_{20} \\ iI_{20}^* & |I_{20}| \end{pmatrix} \tag{10.42}$$

The tensor is fully equivalent to the scalar quantity I_{20}, which determines $\mathbf{Z}_L$ uniquely, which is, in turn, a spatial average of $\daleth(f)$:

10.6 Balanced Direction Tensor: Directional Equilibrium

Certain images lack direction, i.e., when a direction is attempted to be fit to the power spectrum there is not one optimal axis but there are an infinite (uncountable) number of them, e.g., the image of sand in Fig. 10.13 or the image in Fig. 10.10. This property is captured by the structure tensor via the condition $\lambda_{\max} = \lambda_{\min}$, i.e., the smallest (or the largest) eigenvalue is repeated twice making its multiplicity 2. The condition actually does not describe the presence of a property but the lack of it. It describes the lack of linear symmetry. An image with a structure tensor having $\lambda_{\max} = \lambda_{\min}$, has previously been called "perfectly balanced", in analogy with the terminology used in mechanics [28]. The term expresses that there is a *directional equilibrium* in that no single direction dominates over the others. Such images lack a single direction[4] that is more significant than other directions, a property that justifies the use of the term "balanced directions" or "balancedness", both referring to an equilibrium of directions. Balancedness is quantitated by $\lambda_{\min}$ because, for a fixed $\lambda_{\max}$, the larger $\lambda_{\min}$, the closer it gets to $\lambda_{\max}$. When $\lambda_{\min}$ reaches its upper bound, which is $\lambda_{\max}$, then the structure tensor has one eigenvalue which has a multiplicity 2. The least eigenvalue $\lambda_{\min}$ can be used to signal the presence of a balanced image,

---

[3] We recall that a Hermitian matrix $\mathbf{Z}$ fulfills $\mathbf{Z} = \mathbf{Z}^H$.

[4] The pattern may still have a group direction, although it may lack a single direction dominating others, e.g., see Fig. 10.10.

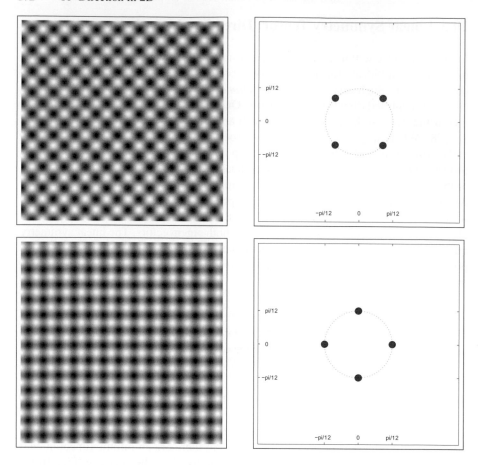

Fig. 10.10. (*Top*) A perfectly balanced image f, and its power spectrum, $|F|^2$. The *red circles* represent a concentration of the power. (*Bottom*) The same as on top, except that the image is rotated with the angle $\frac{\pi}{4}$

the state of equilibrium of directions when both eigenvalues are equal. Accordingly, the *balanced direction tensor* is given by

$$\mathbf{Z}_B = \frac{1}{2} \begin{pmatrix} I_{11} & 0 \\ 0 & I_{11} \end{pmatrix} \tag{10.43}$$

which is a special case of the complex structure tensor. The tensor is completely equivalent to the scalar quantity I_{11}, which determines $\mathbf{Z}_B$ uniquely which is in turn an average of the magnitude of $\daleth(f)$. This tensor has also been called "isotropic" [225] and being "ball-like" [161] in other studies. The term isotropic should not be interpreted as an existence of all directions in the image is a necessity. Figure 10.10 shows two images that contain only two directions and yet they are qualified for the term. Likewise the term ball tensor should not be interpreted too restrictively such as the image must look like a ball, or a junction. The images in Fig. 10.10 as well as in Fig. 10.13 represent textures in which at every point there is a "ball" tensor of approximately the same magnitude. We will discuss the use of balanced direction tensor as a corner detector in Sect. 10.9.

10.7 Decomposing the Complex Structure Tensor

In general, an image is neither perfectly linearly symmetric, e.g., Figs. 10.2-10.6, nor does it totally lack it, e.g., Fig. 10.13. Instead it has the qualities of both types. The amount of evidence for the respective type can be obtained from the structure tensor. The *structure tensor decomposition* can always be achieved into its linear symmetry and balanced direction components easily using its complex form:

$$\mathbf{Z} = \frac{1}{2} \begin{pmatrix} I_{11} & -iI_{20} \\ iI_{20}^* & I_{11} \end{pmatrix} = \mathbf{Z}_L + \mathbf{Z}_B \tag{10.44}$$

where

$$\mathbf{Z}_L = \frac{1}{2} \begin{pmatrix} |I_{20}| & -iI_{20} \\ iI_{20}^* & |I_{20}| \end{pmatrix}, \quad \text{and} \quad \mathbf{Z}_B = \frac{1}{2} \begin{pmatrix} I_{11} - |I_{20}| & 0 \\ 0 & I_{11} - |I_{20}| \end{pmatrix}. \tag{10.45}$$

Conversely, we also wish to study what happens when joining regions having different structure tensors. Without loss of generality, we consider a composition of a region consisting of two subregions, each having a different structure tensor, $\mathbf{Z}'$ and $\mathbf{Z}''$, respectively. This is a realistic scenario since two neighboring regions in an image might differ in their local structure tensors, and the local structure tensor at a border point between the two regions is needed. Because the components of the structure tensor are integrals, they can be computed as the sum of two integrals, each taken over the respective regions. Accordingly, the structure tensor, $\mathbf{Z}$, of a boundary point is obtained by the addition

$$\mathbf{Z} = p\mathbf{Z}' + q\mathbf{Z}'' \tag{10.46}$$

where p, q are two real positive scalars, with $p + q = 1$, that are proportional to the areas of the two constituent regions. Following the definition of $\mathbf{Z}$, we obtain

$$I_{20} = pI_{20}' + qI_{20}'' \tag{10.47}$$

$$I_{11} = pI_{11}' + qI_{11}'' \tag{10.48}$$

where $I'_{..}$, $I''_{..}$, and $I_{..}$ are the structure tensor parameters of the first, the second and the joint patches.

Example 10.6. *In Fig. 10.11 we have two regions labelled A, and B. There are four local images, called images here, and these are marked as $1, \cdots, 4$ with their borders shown as (color) circles. Let images 2 and 4 have the (complex) structure tensors $\mathbf{Z}'$ and $\mathbf{Z}''$, respectively. The corresponding tensor components are therefore*

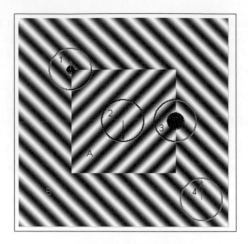

Fig. 10.11. Illustration of addition using the complex structure tensor. The linear symmetry tensor components I_{20} are shown as vectors for four local images. The balanced direction components, I_{11}, are shown as *circles filled with black*. The limits of images are marked by *color circles*

$$I'_{20} = \mathbf{k}'^2 = \exp(-i\tfrac{\pi}{2}) \quad I''_{20} = \mathbf{k}''^2 = \exp(i\tfrac{\pi}{2})$$
$$I'_{11} = 1 \qquad\qquad\qquad I''_{11} = 1$$

In these images, the linear symmetry components point at directions given by $\mathbf{k}^2$, *where,* $\mathbf{k}$ *is the direction of the respective gradient (Fig. 10.12, left). The balanced direction components are equal to zero, because both images are linearly symmetric so that* $I_{11} = |I_{20}|$. *In image 3 we have the structure tensor* $\mathbf{Z} = \tfrac{1}{2}\mathbf{Z}' + \tfrac{1}{2}\mathbf{Z}''$, *having the components*

$$I'_{20} = \tfrac{1}{2}\exp(-i\tfrac{\pi}{2}) + \tfrac{1}{2}\exp(i\tfrac{\pi}{2}) = 0$$

$$I'_{11} = \tfrac{1}{2} + \tfrac{1}{2} = 1$$

The linear symmetry component is zero, as it should be. The image is a perfectly balanced image because none of its constituent directions dominates the others. The balanced direction tensor element is, by contrast, $I_{11} - |I_{20}| = 1$, *which indicates that all spectral power is distributed in such a way that the directions balance each other perfectly. Conceptually, balanced image phenomenon is present also when the gradient directions are random (Fig. 10.12, right). Likewise in image 1 we have the structure tensor* $\mathbf{Z} = \tfrac{1}{4}\mathbf{Z}' + \tfrac{3}{4}\mathbf{Z}''$ *having the components*

$$I'_{20} = \tfrac{1}{4}\exp(-i\tfrac{\pi}{2}) + \tfrac{3}{4}\exp(i\tfrac{\pi}{2}) = \tfrac{1}{2}\exp(i\tfrac{\pi}{2})$$

$$I'_{11} = \tfrac{1}{4} + \tfrac{3}{4} = 1$$

In particular, the balanced direction component, $I_{11} - |I_{20}| = \tfrac{1}{2}$, *should be contrasted to the magnitude of the linear symmetry component* $|I_{20}| = \tfrac{1}{2}$. *The argument of* I_{20}

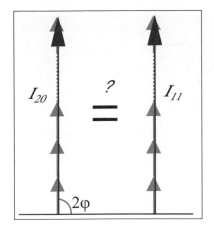 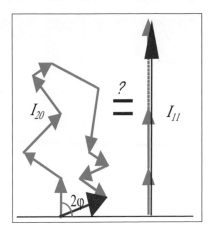

Fig. 10.12. The linear symmetry component, I_{20}, and the balanced directions component, I_{11}, of the structure tensor add as vectors and as scalars, independently. The addition of numerous structure tensors (*left*) for a merging of local images sharing a common direction, (*right*) for a merging that largely lacks such a common direction .

shows the dominant direction of the image. We paraphrase this result as follows. When the account of directions is finalized there is an excess of a single direction that is not balanced by other directions. The net excess of this dominant direction is as significant as the directions that are balanced.

10.8 Decomposing the Real-Valued Structure Tensor

Using its real form, $\mathbf{S}$, the *structure tensor decomposition* in terms of linear symmetry and balanced direction components is also possible [161, 225]. This is done by the spectral decomposition of the tensor $\mathbf{S}$

$$\mathbf{S} = \lambda_{\max}\mathbf{k}_{\max}\mathbf{k}_{\max}^T + \lambda_{\min}\mathbf{k}_{\min}\mathbf{k}_{\min}^T \tag{10.49}$$

followed by the rearrangement:

$$\mathbf{S} = (\lambda_{\max} - \lambda_{\min})\mathbf{k}_{\max}\mathbf{k}_{\max}^T + \lambda_{\min}(\mathbf{k}_{\max}\mathbf{k}_{\max}^T + \mathbf{k}_{\min}\mathbf{k}_{\min}^T)$$
$$= (\lambda_{\max} - \lambda_{\min})\mathbf{k}_{\max}\mathbf{k}_{\max}^T + \lambda_{\min}\mathbf{I} \tag{10.50}$$

where the fact that $\mathbf{k}_{\max}\mathbf{k}_{\max}^T + \mathbf{k}_{\min}\mathbf{k}_{\min}^T = \mathbf{I}$, when $\mathbf{k}_{\max}$ and $\mathbf{k}_{\min}$ are orthogonal, has been used. The first term of Eq. (10.50) is the linear symmetry tensor, and the second term is the balanced direction tensor in real matrix representation. When merging two images, each consisting of balanced directions, the result is an image that is perfectly balanced. Accordingly, adding the balanced direction components of two arbitrary images will not result in a change of the linear symmetry components. However, if two linearly symmetric images are merged the result usually has

Fig. 10.13. *Left*: a real image that is not linearly symmetric. It shows a close up view of sand. Right: The Fourier transform magnitude of a neighborhood in the central part of the original image. Notice that the power is isotropic i.e., far from being concentrated to a line

a structure tensor that has a balanced direction component unless both images have the same direction. Accordingly, the linear symmetry components do not add in a straightforward fashion in the real matrix representation, causing an "interaction" contribution to the balanced direction component. In effect, when merging two arbitrary images already decomposed into their component tensors, the linear symmetry components add (matrix addition) first according to Eq. (10.50), possibly producing a balanced direction term. This additional term should then be added to the sum of the ordinary balanced direction tensor components of the two images.

The decomposition of the complex structure tensor is conserved and closed under averaging whereas that of the real structure tensor is not. Paraphrasing, *averaging linear symmetry tensors* in their complex form yields linear symmetry tensors, whereas averaging linear symmetry tensors in their real form may produce tensors that are not linearly symmetric. This is because (i) the two components I_{20} and I_{11} add separately when merging or smoothing images, and (ii) these components are explicitly linked to eigenvalues and optimal directions.

10.9 Conventional Corners and Balanced Directions

By using other algebraic functions of $\lambda_{\max}$ and $\lambda_{\min}$, numerous measures to quantitate the amount of linear symmetry can be obtained. Likewise, we can also measure the lack of symmetry, which is the balanced directions property of an image.

Example measures include the energy invariant measure, C_{f2} [28], for linear symmetry,

$$C_{f2} = \frac{|I_{20}|}{I_{11}} = \frac{\lambda_{\max} - \lambda_{\min}}{\lambda_{\max} + \lambda_{\min}} \qquad \text{(linear symmetry)} \qquad (10.51)$$

which also immediately defines the energy-invariant measure for the *lack of linear symmetry* C_{f3}:

$$C_{f3} = 1 - \frac{|I_{20}|}{I_{11}} \qquad \text{(balanced directions)} \qquad (10.52)$$

At the heart of such measures is how well the Schwartz inequality, $|I_{20}| \leq I_{11}$, is fulfilled. The case of equality happens if and only if one has linear symmetry ($|I_{20}| = I_{11}$). The left-hand side of it vanishes and the right-hand side becomes as large as the energy permits it, if and only we have balanced directions ($0 = |I_{20}|$). Having this in mind, then other functionals that measure the distance $I_{11} - |I_{20}|$ can be used to quantitate the balancedness of the directions in the image. The popular detector of Harris and Stephen [97] (a similar measure is that of Forstner and Gulch [74]) used to measure cornerness, quantitates this distance as well

$$C_{hs} = \lambda_{\max}\lambda_{\min} - 0.04(\lambda_{\max} + \lambda_{\min})^2 \qquad (10.53)$$
$$= (I_{11} + |I_{20}|)(I_{11} - |I_{20}|)/4 - 0.04I_{11}^2$$
$$= (0.84I_{11}^2 - |I_{20}|^2)/4 \qquad (10.54)$$

albeit in the quadratic scale, which is most obvious if the empirical constant 0.84 is replaced by 1. Because of the constant, the measure C_{hs} must be combined with a threshold to reject the negative values. This will happen at (local) images that have only the linear symmetry component (e.g., on lines and edges) where $I_{11} = |I_{20}|$, yielding $C_{hs} = -0.04|I_{20}| < 0$. The measure C_{hs} responds strongly to many corner types, including a corner that consists of the junction of two orthogonal directions, or a corner that consists of the intersection of several lines. A word of caution is in place because C_{hs} will also respond strongly to other patterns, including at every point in a texture image that lacks direction. This may be a desirable property for an application at hand. However, it is also possible that the application is actually unintentionally accepting (false acceptence) many patterns as corners by using C_{f3} or C_{hs}. The texture images shown in Fig. 10.10 are perfectly balanced everywhere, meaning that every point is a *"balanced directions corner"* or *"Stephen–Harris corner"*. Likewise, all boundary points, except the boundary corners, between region A and region B in Fig. 10.11 are the strongest corners in either of the two corner senses above. All points of these four lines are, in fact, stronger "corners" than the four boundary corner points, as discussed in Sect. 10.7!

10.10 The Total Least Squares Direction and Tensors

It is in place to ask what makes the matrices $\mathbf{J}$, $\mathbf{S}$, or even $\mathbf{Z}$ (second-order) tensors. We recall that the basic difference between a second-order tensor and a matrix is subtle and lies in that a tensor represents a physical quantity on which the coordinate system has no real influence except for a numerical representation. The numerical representation of a tensor is then a matrix that corresponds to physical measurements in a specific basis. A representation of the same tensor in a different basis can only be obtained by a similarity transformation using the basis transformation matrix. For first-order tensors and vectors, a similar subtle difference exists. The first-order tensor is represented as a vector in a specific basis. Another representation of it can be obtained by a linear transformation corresponding to a basis change. The zero-order tensors represent physical quantities that are scalars. Their numerical representations do not change with basis transformations.

As illustrated by Fig. 10.9, the error function $e(\mathbf{k})$ employed by the total least square (TLS) error represents the spectral power weighted by its shortest (orthogonal) distance to the estimated axis $\mathbf{k}$. This makes e a zero-order tensor and $\mathbf{k}$ a first-order tensor. If we apply a basis change e.g., rotate the coordinates of the power spectrum, $e(\mathbf{k})$ will not change at all and only the numerical representation of $\mathbf{k}$ will change. The new direction vector, $\mathbf{k}'$, will be coupled to the old $\mathbf{k}$ linearly using the inverse of the matrix that caused the basis change.

To appreciate the TLS error in this context we compare it to the *mean square* (MS) error which is extensively used in applications where one has a black box controlled by known inputs resulting in a measurable output. In such applications there is thus a *response measurement*, $\mathbf{y}$, that may contain measurement errors and that is to be explained by means of another set of known (error-free) variables, called *explanatory variables* $\mathbf{X}$, via a linear model

$$\mathbf{y} = \mathbf{Xk} \tag{10.55}$$

Here $\mathbf{k}$ is the unknown regression parameter, which will be estimated by minimizing the following residual:

$$\min_{\mathbf{k}} \|\mathbf{y} - \mathbf{Xk}\|^2 \tag{10.56}$$

Adapted to our 2D direction estimation problem, the MS error yields:

$$\min_{\gamma} e(\gamma) = \int \|\omega_y - \gamma\omega_x\|^2 |F(\omega_x, \omega_y)|^2 d\boldsymbol{\omega} \tag{10.57}$$

This is the classical *regression problem*. Here, the direction coefficient γ is unknown and will be estimated from the data $F(\omega_x, \omega_y)$. The unknown γ is related to the direction vector $\mathbf{k} = (\cos\theta, \sin\theta)^T$ as $\gamma = -\frac{\cos\theta}{\sin\theta}$. Notice that the integrand measures the distance between the data point $\boldsymbol{\omega}$ and a point on the $\mathbf{k}$-axis to be fitted. This distance is in general not the shortest, distance as illustrated by the vector $\mathbf{d}$ in Fig. 10.14. The MS error would accordingly depend on the coordinate axis directions to the effect that after a basis change, the new error using the same data will be different. Likewise, the new direction $\mathbf{k}'$ will not be given by multiplying the inverse of the basis transformation matrix with $\mathbf{k}$, the estimated direction before the basis change. In consequence, neither the MS error nor γ are tensors. One can associate $\mathbf{k}$ and $\mathbf{k}\mathbf{k}^T$ to every $\gamma = -\frac{\cos(\theta)}{\sin(\theta)}$. These quantites are not tensors, although they are conventional vectors and matrices. A more detailed discussion of the TLS error can be found in [115] and [59].

Through Taylor expansion a spatial interpretation of $e(\mathbf{k})$, as an alternative to its original interpretation, the spectral inertia, can be obtained. In this view, the structure

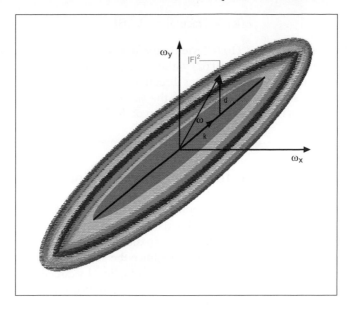

Fig. 10.14. The error, $\|\mathbf{d}\|^2$, used in the MS estimate. The error is not measured as the shortest distance between a frequency coordinate $\boldsymbol{\omega}$ and the k-axis. This should be contrasted with the TLS error, which does measure the shortest distance, as shown in Fig. 10.9

tensor, via its minor eigenvector, encodes the direction in which a small translation of the image causes it the least departure from the original. To see this, we perform the expansion, i.e., we express the image f at $\mathbf{r} + \epsilon\mathbf{k}$, where the direction $\mathbf{k} = (k_x, k_y)^T$ has the unit length, by using the partial derivatives of f at $\mathbf{r}$

$$f(\mathbf{r}+\epsilon\mathbf{k}) = f(\mathbf{r})+\epsilon\big(k_x D_x + k_y D_y\big)f(\mathbf{r})+\frac{\epsilon^2}{2}\big(k_x D_x + k_y D_y\big)^2 f(\mathbf{r})+\cdots \quad (10.58)$$

Accordingly,

$$\epsilon\big(k_x D_x + k_y D_y\big)f(\mathbf{r}) = f(\mathbf{r}+\epsilon\mathbf{k}) - f(\mathbf{r}) \qquad (10.59)$$

is the linear approximation of the difference between the function $f(\mathbf{r})$ and its translated version, $f(\mathbf{r}+\epsilon\mathbf{k})$. In consequence,

$$\big((k_x D_x + k_y D_y\big)f(\mathbf{r}))^2 \qquad (10.60)$$

is the magnitude of the rate of the change in the direction of $\mathbf{k}$, which can be viewed as the error rate or resistance rate when translating the image in the direction $\mathbf{k}$. Integrating this function and using the (Parseval–Plancherel) theorem 7.2, yields

$$\iint \big(\mathbf{k}^T \nabla f(x,y)\big)^2 dxdy = \mathbf{k}^T \left(\iint \nabla f(x,y)\nabla^T f(x,y)dxdy\right)\mathbf{k} = \mathbf{k}^T \mathbf{S}\mathbf{k}$$
$$(10.61)$$

which is the dynamic part of our original error function, $e(\mathbf{k})$, because

$$e(\mathbf{k}) = \text{Trace}(\mathbf{S}) - \mathbf{k}^T \mathbf{S} \mathbf{k} \qquad (10.62)$$

Accordingly, minimizing Eq. (10.62) yields the direction of *minimum translation error*. Evidently maximizing Eq. (10.62) yields the direction of *maximum translation error*. Both minimum and maximum error directions are given by the eigenvectors of **S**. Paraphrased, the structure tensor encodes the *minimum resistance direction* in the spatial domain, which is identical to the direction of the line fit to the power spectrum in the TLS sense.

10.11 Discrete Structure Tensor by Direct Tensor Sampling

Until this section, the theory for detection of the orientation of a scalar function in 2D space has been based on continuous signals. One such technique was summarized by theorem 10.1 which we will attempt to approximate by use of discrete functions. We call this approach *direct tensor sampling* since the suggested method examines whether or not the spectrum of an image consists of a line, by directly estimating the matrix **S**, with the elements given by Eq. (10.28) without first estimating the power spectrum by a discrete local spectrum.

We need to approximate the continuous integrand of Eq. (10.28) from a discrete image. To that end, we need the approximation of

$$\frac{\partial f(\mathbf{r})}{\partial x_i} \frac{\partial f(\mathbf{r})}{\partial x_j}, \qquad \text{with} \qquad i,j = 1,2, \quad x_1 = x, \quad \text{and} \quad x_2 = y, \qquad (10.63)$$

on a Cartesian grid i.e., $\mathbf{r} = \mathbf{r}_l$, where $\mathbf{r}_l$ is the coordinates of the grid nodes. In analogy with the theory presented in Sects. 8.2, 8.3, and 9.2, we can do this by filtering the original image linearly:

$$\frac{\partial f(\mathbf{r}_l)}{\partial x_i} = \sum_k f(\mathbf{r}_l + \mathbf{r}_k) \frac{\partial \mu(\mathbf{r}_l)}{\partial x_i} \qquad \text{with} \qquad i = 1,2 \qquad (10.64)$$

where $\frac{\partial \mu(\mathbf{r}_l)}{\partial x_i} = \frac{\partial \mu(\mathbf{r})}{\partial x_i}|_{\mathbf{r}=\mathbf{r}_l}$, $\frac{\partial f(\mathbf{r}_l)}{\partial x_i} = \frac{\partial f(\mathbf{r})}{\partial x_i}|_{\mathbf{r}=\mathbf{r}_l}$, and then applying pointwise multiplication between the two thus-obtained discrete partial derivative images:

$$\frac{\partial f(\mathbf{r}_l)}{\partial x_i} \frac{\partial f(\mathbf{r}_l)}{\partial x_j}, \qquad \text{with} \qquad i,j = 1,2, \quad x_1 = x, \quad \text{and} \quad x_2 = y. \qquad (10.65)$$

The latter is an estimate of Eq. (10.63) on a Cartesian grid. Note that the continuous form of Eq. (10.63) is not known, but we estimated nevertheless its discrete version by applying a linear discrete filtering to the discrete $f(\mathbf{r}_l)$, followed by a pointwise multiplication on the grid $\mathbf{r}_l$.

To estimate the structure tensor elements, Eq. (10.28), we first reconstruct (10.63) from its samples (10.65):

$$\frac{\partial f(\mathbf{r})}{\partial x_i} \frac{\partial f(\mathbf{r})}{\partial x_j} = \sum_l \frac{\partial f(\mathbf{r}_l)}{\partial x_i} \frac{\partial f(\mathbf{r}_l)}{\partial x_j} \mu(\mathbf{r} - \mathbf{r}_l) \qquad i,j : 1,2 \qquad (10.66)$$

where $x_1 = x$, $x_2 = y$. The vectors $\mathbf{r}_l$ represent points on a grid as before, and $\mu(\mathbf{r})$ is the continuous interpolation function, assumed to be a Gaussian with the variance σ_p^2:

$$\mu(\mathbf{r} - \mathbf{r}_l) = \exp\left(-\frac{1}{2\sigma_p^2}\|\mathbf{r} - \mathbf{r}_l\|^2\right) \tag{10.67}$$

We proceed by substituting (10.66) in Eq. (10.28)

$$\mathbf{S}(i,j) = \frac{1}{4\pi^2}\sum_l \frac{\partial f(\mathbf{r}_l)}{\partial x_i}\frac{\partial f(\mathbf{r}_l)}{\partial x_j}\int_{E_2}\mu(\mathbf{r} - \mathbf{r}_l)dxdy \tag{10.68}$$

$$= \frac{C}{4\pi^2}\sum_l \frac{\partial f(\mathbf{r}_l)}{\partial x_i}\frac{\partial f(\mathbf{r}_l)}{\partial x_j} \tag{10.69}$$

Here the integral evaluates to a constant C

$$\int_{E_2}\mu(\mathbf{r} - \mathbf{r}_l)dxdy = C$$

that is independent of $\mathbf{r}_l$ because the area under a shifted Gaussian is the same regardless of the amount of the shift. The summation in Eq. (10.69) is taken over all image points, $\mathbf{r}_l$, on the grid.

However, the structure tensor is most frequently needed for a local image, rather than the entire image. A simple way to achieve this goal is to approximate $\frac{\partial f}{\partial x_i}\frac{\partial}{\partial x_j}$ for the local image by multiplying its global version with a window function, $w(\mathbf{r})$, placed around the current point $\mathbf{r}_0$. Without loss of generality, we assume that the local image for which the structure tensor is to be estimated is the one around the origin. Using a Gaussian as a window function, this amounts to replacing μ in Eq. (10.66) with

$$\mu(\mathbf{r} - \mathbf{r}_l)w(\mathbf{r}) = \exp\left(-\frac{1}{2\sigma_p^2}\|\mathbf{r} - \mathbf{r}_l\|^2\right)\exp\left(-\frac{1}{2\sigma_w^2}\|\mathbf{r}\|^2\right) \tag{10.70}$$

where σ_w^2 is a constant that controls the effective width of the second Gaussian, the window defining the local image around the origin. Substituting this into Eq. (10.66) and then using it in Eq. (10.28) yields the local structure tensor elements:

$$\mathbf{S}(i,j) = \frac{1}{4\pi^2}\sum_l \frac{\partial f(\mathbf{r}_l)}{\partial x_i}\frac{\partial f(\mathbf{r}_l)}{\partial x_j}\int_{E_2}\mu(\mathbf{r} - \mathbf{r}_l)w(\mathbf{r})dxdy \tag{10.71}$$

$$= \frac{1}{4\pi^2}\sum_l \frac{\partial f(\mathbf{r}_l)}{\partial x_i}\frac{\partial f(\mathbf{r}_l)}{\partial x_j}\mu_l \tag{10.72}$$

where μ_l is defined as

$$\mu_l = \int_{E_2}\exp\left(-\frac{1}{2\sigma_p^2}\|\mathbf{r} - \mathbf{r}_l\|^2\right)\exp\left(-\frac{1}{2\sigma_w^2}\|\mathbf{r}\|^2\right)dxdy$$

$$= (\mu * w)(\mathbf{r}_l) = \frac{2\pi}{4}\frac{\sigma_p^2\sigma_w^2}{\sigma_p^2 + \sigma_w^2}\exp\left(-\frac{1}{2(\sigma_p^2 + \sigma_w^2)}\|\mathbf{r}_l\|^2\right). \tag{10.73}$$

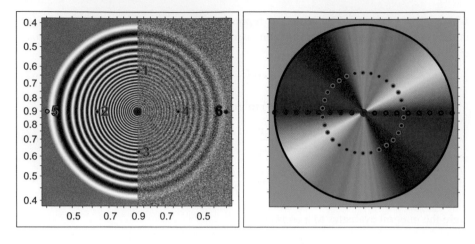

Fig. 10.15. On the *left*, the test image, the axes of which are marked with fractions of π representing the spatial frequency. On the *right* is the color code representing the directions. The marks on the axes are separated by 5 degrees when joined to the center. The *colored dots* in both images define the curves from which 1D direction measurements will be sampled

The integral represents a continuous convolution, and Eq. (10.73) is obtained by noting that both μ and w are Gaussians and that a convolution of them yields another Gaussian, with a variance that is the sum of the variances of μ and w. An easy way of seeing this is by applying the Fourier transform to $\mu * w$. Eq. (10.72), which computes the local tensor around the origin, is therefore a discrete convolution by a Gaussian if $\mathbf{S}(i, j)$ needs to be computed for local patches around *all* points of the original image grid. Since the values of μ_ls decrease rapidly outside of a circle with radius $\sqrt{\sigma_p^2 + \sigma_w^2}$, we can truncate the infinite filter when its coefficients are sufficiently small, typically when the coefficients have decreased to about 1% of the filter maximum. Thus, Eq. (10.72) implies that the local tensor (of the origin) is obtained as a window-weighted average of the gradient outer products:

$$\mathbf{S} = \frac{1}{4\pi^2} \sum_l (\nabla f_l)(\nabla f_l)^T \mu_l \tag{10.74}$$

where ∇f_l is the gradient of $f(\mathbf{r})$ at the discrete image position $\mathbf{r}_l$, and μ_l is a discrete Gaussian. Defining $D_x f_l$ and $D_y f_l$, for convenience, as the components of ∇f_l, at the mesh point $\mathbf{r}_l$:

$$\nabla f_l = (D_x f_l, D_y f_l)^T = (\frac{\partial f(\mathbf{r}_l)}{\partial x}, \frac{\partial f(\mathbf{r}_l)}{\partial y})^T, \tag{10.75}$$

We summarize our finding on tensor discretization as a theorem:

Theorem 10.3 (Discrete structure tensor I). *Assuming a Gaussian interpolator with σ_p and a Gaussian window with σ_w, the optimal discrete structure tensor approximation is given by*

$$\mathbf{S} = C \sum_l (\nabla f_l \nabla^T f_l) \mu_l \tag{10.76}$$

$$= C \sum_l \begin{pmatrix} (D_x f_l)^2 & (D_x f_l)(D_y f_l) \\ (D_x f_l)(D_y f_l) & (D_y f_l)^2 \end{pmatrix} \mu_l \tag{10.77}$$

where μ_l is a discrete Gaussian with $\sigma = \sqrt{\sigma_p^2 + \sigma_w^2}$.

◆

In analogy with Eqs. (10.63)–(10.65), the quantity

$$D_x f(\mathbf{r}_l) + iD_y f(\mathbf{r}_l) \tag{10.78}$$

can be obtained by two convolutions using real filters, one for $D_x f(\mathbf{r}_l)$ and one for $D_y f(\mathbf{r}_l)$. After that, the complex result depicted by Eq. (10.78) is squared to yield the ILST image:

$$\daleth(f)(\mathbf{r}_l) = (D_x f(\mathbf{r}_l) + iD_y f(\mathbf{r}_l))^2 \tag{10.79}$$

In consequence of theorem 10.2, the following theorem then holds true:

Theorem 10.4 (Discrete structure tensor II). *Assuming a Gaussian interpolator with σ_p and a Gaussian window with σ_w, the optimal discrete structure tensor complex elements are given by*

$$\mathbf{Z} = \frac{1}{2} \begin{pmatrix} I_{11} & -iI_{20} \\ iI_{20}^* & I_{11} \end{pmatrix} \tag{10.80}$$

where

$$I_{20} = C \sum_l (D_x f_l + iD_y)^2 \mu_l \tag{10.81}$$

$$I_{11} = C \sum_l |D_x f_l + iD_y|^2 \mu_l \tag{10.82}$$

with μ_l being a discrete Gaussian with $\sigma = \sqrt{\sigma_p^2 + \sigma_w^2}$.

◆

Figure 10.15 shows a frequency-modulated test (FM-test) image. The test image has axes marked by the spatial frequencies of the waves in the horizontal and vertical directions from the image center. The absolute frequency of the waves decreases exponentially radially, whereas the direction of the waves changes uniformly angularly. The exponential decrease occurs between the spatial frequencies 0.4π and 0.9π. The image on the right represents the color code of the ideal orientation in double-angle representation, i.e., $\exp 2\varphi$, where φ is the polar angle coordinate of a point in the image. In half of the image, spatially uncorrelated Gaussian noise X, with mean 0.5 and variance 1/36, has been added to the image signal, f, according to $pf + (1 - p)X$, where the weight coefficient $p = 0.3$, and X is the noise. On

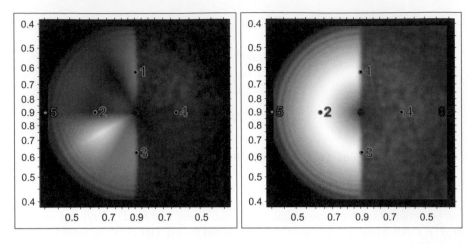

Fig. 10.16. The images illustrate the direction tensor, represented as I_{20} (*left*) and I_{11} (*right*), for Fig. 10.15 and computed by using theorem 10.4. The hue encodes the direction, whereas the brightness represents the magnitudes of complex numbers

the right the reference orientation is encoded as a color image. Using the HSB color space, the hue is modulated by 2φ, whereas the brightness and the saturation are set to the maximum at all points. The colored dots define curves along which the structure tensor measurements will be extracted and discussed in detail further below. The color coded reference image has axes marked by angles separated in $5°$ as seen from the center.

The images in Fig. 10.16 illustrate the structure tensor computed for all local images. The color image on the left represents the complex-valued I_{20}. The hue of an image point is modulated by the $\arg(I_{20})$ of its local neighborhood, whereas its brightness is given by $|I_{20}|$. The saturation of all points is set to the maximum. The computations are implemented according to theorem 10.4 where the derivative Gaussian filter had $\sigma_p = 0.8$ and the integrative Gaussian filter, $\sqrt{\sigma_p^2 + \sigma_w^2} = 2.5$. The hue of a point should be the same as the reference color given at the same point of the color image in Fig. 10.16. Visually, it appears that the colors are in good conformity with those of the reference image. The image on the right shows I_{11}, which, being nonnegative and real-valued, modulates the gray values. It is possible to verify that the direction measurements adhere to the theoretical values, even in the noisy part of the test image, where the signal-to-noise ratio, (SNR), was $2\log_2(\frac{0.3}{0.7}) \approx -2.4$ dB. The brightness of the points at the noisy part are lower in both images for two reasons. First, the $|I_{20}|$ should be weak or ideally zero because the unique direction is disturbed, and second, because the linear derivation and integration operations have effectively a bandpass character and the noise components are suppressed by the linear process. In the clean part of the signal, the brightness of the corresponding points in the left and the right images is the same. We will discuss these conclusions more quantitatively next.

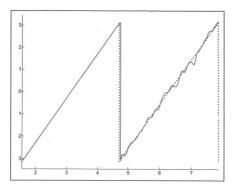 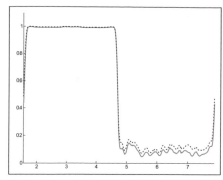

Fig. 10.17. The graphs in the *left image* represent the estimated $(\arg(I_{20}^l)$, *solid*) as well as the ideal direction angle (*dashed*) on a ring in the FM test image. The graphs on the *right* show $|I_{20}^l|$ (*solid*) and I_{11}^l (*dashed*) on the same ring

In the left graph of Fig. 10.17 we show in solid green the direction measurements, $\arg(I_{20})$, extracted along the circle passing through the points 1 to 4, see also Fig. 10.15. We show by the dotted black curve the reference direction measurements. The direction measurements agree with the reference values nearly exactly in the noise-free parts of the image, whereas they follow the reference quite well in the noisy parts of the image. The estimated significance of the direction measurements is given by the graphs of the right image. The solid curve shows $|I_{20}|$, whereas the dotted curve shows I_{11}. As predicted by the theory, we have $0 \ll |I_{20}| = I_{11}$ in the noise-free part, whereas $0 \approx |I_{20}| < I_{11}$ in the noisy part. Both $|I_{20}|$ and I_{11} are invariant to directional changes, which is manifested by the fact that they both equal a constant (one) in the noise-free part of the test image.

Likewise, in the left graph of Fig. 10.18 we display in blue the analogous measurements, but for $\arg(I_{20})$ extracted along the horizontal line joining point 6 to the center, as marked in Fig. 10.15. This curve is reasonably horizontal even in the noisy part. The constant represented by the green line shows the direction measurements extracted along the line joining point 5 to the center. Being in the noise-free part, this line adheres nearly perfectly to the reference measurements, shown as a dotted. The estimated significance of the direction measurements is given by the graphs of the right image. Here, the solid, and the dashed curves at the top represent the measurements of $|I_{20}|$, and I_{11} respectively for the clean signal, (from point 5 to the center). The corresponding measurements for the noisy part (from Point 6 to the center) are given by the solid and the dashed curves at the bottom, respectively. The direction estimation quality is assured by the condition that $|I_{20}|$ is high and is close to its upper bound, I_{11}. By contrast, when the linear symmetry in the signal is disturbed poor estimations of the direction are obtained. This is manifested by the fact that $|I_{20}|$ is close to zero while I_{11} is weak.

The structure tensor measurements are implemented by using the results of linear operators. First, the partial derivative operator $D_x + iD_y$ using a Gaussian derivative

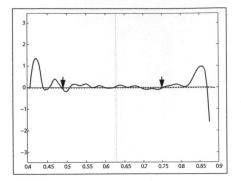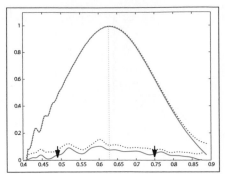

Fig. 10.18. The graphs in the *left image* represent the estimated $(\arg(I_{20}^l))$ as well as the ideal direction angle on the two radial lines shown in the FM test image. The graphs on the *right* represent $|I_{20}^l|$ and I_{11}^l on the same lines

filter with variance σ_p^2 is applied. The result is squared pointwise and smoothed by the Gaussian filter with the variance $\sigma_p^2 + \sigma_w^2$. Despite the nonlinear squaring between the two linear operators, the combined operator exhibits a bandpass character that is tuned to a particular frequency. This is seen in the right graph of Fig. 10.18, where the *tune-on frequency*, which the structure tensor is most sensitive to, has a well-distinguished peak in the solid blue curve. The tune-on frequency of this particular implementation at 0.62 radians, corresponding to a sinusoidal wave with a period of 10 pixels, has been shown in dotted black in both graphs. For convenience, even the points 1 through 4 defining the sampling ring above correspond to 0.62 radians. The tune-on frequency can be changed in a variety of ways, a straightforward one of which is by changing the derivative Gaussian variance σ_p^2. The arrows show the limits of the frequency annulus to which the filters are most sensitive, as confirmed by the graphs of $|I_{20}|$ and I_{11}.

10.12 Application Examples

We present two applications to illustrate the use of the structure tensor.

Fingerprint Recognition

In Fig. 10.19, a *minutia* detection process is visualized. A minutia in a fingerprint is a point that can be easily identified. Typically it is an end point of a ridge or a bifurcation point. The first two images visualize the original fingerprint and its enhanced version, respectively. Image III represents I_{20}/I_{11}, where the numerator and denominator at each point are obtained by direct tensor discretization as discussed above. Even here the minutiae are discernable as dark spots, indicating lack of linear symmetry. The hue represents local direction. Image IV represents the presence of linear symmetry in parabolic coordinates, a subject that we will discuss in detail in

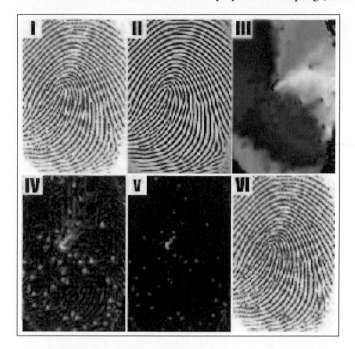

Fig. 10.19. Minutia detection process. Adapted after [76]

Chap. 11. Image V represents a fusion of the information in image III and IV by pointwise multiplication according to $(IV) \cdot (1 - |III|)$. Finally, image VI presents the automatically found minutiae. The positions of the minutiae as well as the directions between and around them are useful personal attributes in biometric person authentication. The details of this minutiae detection approach are given in [76].

Image Enhancement

Image enhancement is a valuable tool to reinforce the quality of recoverable details. In Fig. 10.20, we show the result of line reinforcement in color images [224]. The processing performs a so-called shock filltering to bias lines by means of direction estimation using the structure tensor.

10.13 Discrete Structure Tensor by Spectrum Sampling (Gabor)

We will attempt to estimate the *discrete structure tensor* by *spectrum sampling* as follows. In Sect. 9.6 we concluded that the responses of a set of Gabor filters constitute a sampled version of the local Fourier spectrum. Each filter is tuned to a particular position in the spectral domain corresponding to a particular frequency and direction with a certain support in the frequency domain. Assuming uniform sampling, the

Fig. 10.20. Image enhancement by reinforcing linear symmetry tensor directions, after [224]

number of filters determines how coarsely the Fourier spectrum is down-sampled. The more filters there are in the filter ensemble, the more the Gabor filter responses will approach an ordinary Fourier spectrum while the size of the neighborhood being analyzed will approach the size of the original image. Figure 10.21 shows a Gabor filter set which is (polar) separable in tuning directions and absolute frequencies. The tuning frequencies of the filters (marked with $\times$) along with the isoamplitude curves of the filters are given in the graph. For real images, half of the filters are not needed e.g., those drawn in cyan, because the response of a filter equals the complex conjugate of the response of the filter at the mirror site through the origin. Increasing the number of filters in the Gabor filter bank amounts to increasing the sampling resolution of the Fourier spectrum, since the filter bandwidths decrease when the frequency resolution increases.

A linearly symmetric (local) image has a power spectrum that is concentrated to a line. Depending on the direction of the image and the spectrum discretization, this line may or may not pass exactly through the tune-on frequencies, the frequency grid points, e.g., see the two drawn axes in Fig. 10.21. In the case when this axis has the same direction as the direction of one or more frequency grid points (Gabor filters), then the responses of these filters will account for a significant share of the total response power stemming from the filter ensemble. In Fig. 10.22, the Fourier transform of such an image is shown as the dashed energy concentration passing exactly through a subset of filter tuning sites. The filter tuning sites are marked with $\times$ in the figure. The magnitudes of the filter responses are modulated by the sizes of the circles placed at the corresponding frequency grid points. The responses represented by cyan circles can be deduced from their mirrored counterparts as mentioned before.

To determine the dominant direction of a local image, one could identify the Gabor filter yielding the largest magnitude and then use the tune-on direction of the filter as the direction searched for. However, the disadvantage of this approach is

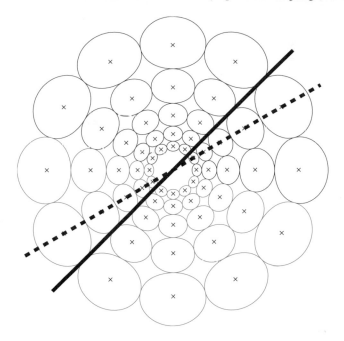

Fig. 10.21. A Gabor filter set in the Fourier domain and the Fourier transforms of two linearly symmetric images, overlayed. The *dashed* and *solid* axes represent the spectral energy concentrations of the two linearly symmetric images having gradient directions of $30°$ and $45°$ direction, respectively

an unnecessarily rough quantization of *the direction* because when the linear symmetry axis direction falls between two filter directions, two directions will signal a presence of significant response power in their respective directions (Fig. 10.23). In consequence, choosing the maximum power direction will not give better direction resolution than the direction resolution in the Gabor filter tunings. Below, by using direction tensors, we show that the limitation of the direction resolution attributable to the filter tuning resolution can be reduced significantly by utilizing the structure tensor theory.

Let the coordinates of the filter tune-on frequencies, marked by $\times$ in the example of Fig. 10.21, be given as complex numbers:

$$z_{kl} = x_{kl} + iy_{kl} \tag{10.83}$$

The indices k, l, running over all filters in the Gabor filter set, represent the direction and the absolute frequency of the filter tunings. Then the second-order complex moments of the discrete power spectrum are given by:

$$I_{20} = \sum_{k,l} z_{kl}^2 (z_{kl}^*)^2 |F^{\{k,l\}}|^2 = \sum_{k,l} z_{kl}^2 |F^{\{k,l\}}|^2 \tag{10.84}$$

and

Fig. 10.22. Gabor filter responses when an input image is linearly symmetric with a gradient direction aligned with filter tuning orientations. The sizes of the circles represent the magnitudes of the complex filters at the respective site

$$I_{11} = \sum_{k,l} z_{kl}^1 (z_{kl}^*)^1 |F^{\{k,l\}}|^2 = \sum_{k,l} |z_{kl}|^2 |F^{\{k,l\}}|^2 \qquad (10.85)$$

Using theorem 10.4, we conclude that these complex numbers are the elements of the direction tensor constructed from the discrete local spectrum. Accordingly, the direction tensor encodes the axis having the least square error when fit to the discrete power spectrum. In case the image is linearly symmetric, the error of the fit, encoded by $I_{11} - |I_{20}|$ will vanish and the $\arg(I_{20})$ will deliver the direction in double-angle representation as 2φ, with φ being the angle of the image gradient. Suggested by [90, 139], the double angle representation (to represent direction) maps an angle φ and its mirror angle $\varphi + \pi$ to the same angle. Previously, we arrived at this representation by a total least squares minimization. According to the above equations, the direction 2φ can be interpreted as an angle interpolation between fixed filter directions using the magnitude responses of a bank of filters. Because of the interpolation, the result should not suffer from the direction quantization as much when compared to using the maximum power direction. In Fig. 10.25, encoded as images, we show the direction estimations as I_{20} and I_{11}, for three tune-on directions in a *log–polar Gabor decomposition*, the details of which are discussed further below. However, it should be emphasized that the structure tensor estimation via the formulas given by Eqs. (10.84)–(10.85) are equally valid for a *Cartesian Gabor decomposition*.

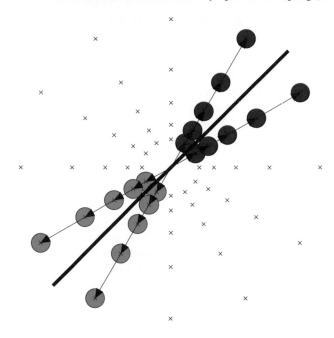

Fig. 10.23. The 2D spectrum and the Gabor filter responses when an input image is linearly symmetric with a direction between filter tune-on directions. The *crosses* represent the tune-on frequencies. The *arrows* represent the coordinate vectors, z_k, of some of these, whereas the *circles* illustrate, via their radii, the magnitudes of the respective filter responses, $|F^{\{k,l\}}|$

Structure Tensor by Log-Polar Gabor Decomposition

The sums in Eqs. (10.84)–(10.85) are taken over the entire range of Gabor filters. Assuming log–polar separable filter tune-on frequencies discussed in Sect. 9.6, one could, however, split the sum over the direction and the frequency components, as shown for I_{20}:

$$I_{20} = \sum_{l} \sum_{k} z_{kl}^2 |F^{\{k,l\}}|^2 = \sum_{l} I_{20}^l \tag{10.86}$$

with

$$I_{20}^l = \sum_{k} z_{kl}^2 |F^{\{k,l\}}|^2 \tag{10.87}$$

being the complex moment contribution from filters on a "ring" of Gabor tune-on frequency sites. Analogously, we can obtain

$$I_{11} = \sum_{l} \sum_{k} |z_{kl}|^2 |F^{\{k,l\}}|^2 = \sum_{l} I_{11}^l \tag{10.88}$$

with

$$I_{11}^l = \sum_k |z_{kl}|^2 |F^{\{k,l\}}|^2 \tag{10.89}$$

In fact, the quantities I_{20}^l and I_{11}^l are equal to I_{20} and I_{11}, respectively, if the image is bandpass filtered in such a way that the absolute frequency contents differing from $|z_{kl}|$ are suppressed. Accordingly, I_{20}^l and I_{11}^l represent the direction tensor of a specific (absolute) frequency (which is a ring with a "small" width in the spectrum) of the image. Apart from requiring less computational resources, restricting the Gabor filter–based direction tensor to a specific absolute frequency is also desirable in many situations where multi-scale analysis is needed, e.g., texture and fingerprint image analysis, since this allows one to split the analysis over an array of scales naturally. Originally suggested in [137], via a scheme that is a special case of Eq. (10.84) applied to a frequency ring, such an orientation interpolation is sufficient for the purpose of estimating the local direction for many applications. The original scheme employed quadrature mirror filters with the angular bandwidth of π i.e., the angular deviation from the tune-on direction was measured by the function $\cos^2(\theta)$ as compared to the Gaussians discussed here. Regardless of the filters shape[5] and how they tessellate the spectrum, it should be emphasized that I_{20}^l must be completed with I_{11}^l to make it a direction tensor. Otherwise I_{20} alone cannot encode both the minimum and the maximum error, apart from their difference.

Although the interpolated direction offers better direction accuracy, there is obviously a limit on the minimum number of filters one can have in the decomposition. To begin with, it is not possible to compute I_{20}^l, which consists of two real variables, and I_{11}^l, which consists of one nonnegative real variable, from just one Gabor filter, (actually two filters when mirrored). The I_{20}^l would point in the same direction, regardless of the direction of the image. Even two Gabor filter directions differing with $\frac{\pi}{2}$ (actually 4 filter sites when mirrored) are not enough because this would result in an I_{20}^l that could not possibly point in other than the two directions: 2φ and $2\varphi + \pi$ where φ is the direction of one of the two filters. Using three filters with directions separated by $\frac{\pi}{3}$ is the minimum requirement on the Gabor decomposition to yield a meaningful direction tensor.

In Fig. 10.24, this is illustrated by investigating the Gabor filter responses to an image composed of a planar sinusoid. The intersection point of three circles is marked with a small circle, and it represents the frequency coordinates of one of the two Dirac impulses composing the sinusoid image. The three circles show the frequency coordinates (the direction and absolute frequencies) at which the Dirac impulse can be placed, judging from the individual Gabor filter magnitudes. Provided that it is on such a ring, the magnitude response of the same Gabor filter is invariant to the position changes of the Dirac pulse. Only by investigating no less than three Gabor filter magnitudes, is it possible to uniquely determine the position of a Dirac impulse. This result is nonetheless not surprising, because there are three freedoms in a 2D direction tensor and these cannot be fixed by less than three independent measurements.

---

[5] Rotation invariant versions of quadrature filters have also been suggested [71].

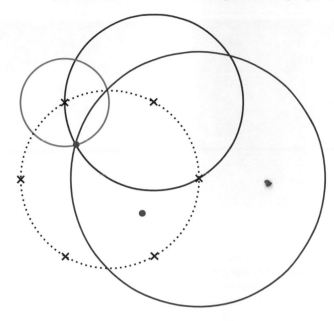

Fig. 10.24. Iso-magnitudes of three Gabor filters corresponding to an image that consists of a sinusoid (planar) wave, having the direction and the absolute frequency indicated by the pair of small circles.

In Fig. 10.25, we show on the left the quantity I_{20} computed according to Eq. (10.87) where the Gabor decomposition had three filters differing only in tune-on directions with increments of $\frac{\pi}{3}$. The hue encodes $\arg(I_{20})$, whereas the brightness encodes $|I_{20}|$. The absolute frequency represented by the radius of the ring where the tune-on frequencies of the filters are placed is 0.62π. The standard deviation of the directional Gaussians was set to $\sigma_\eta = \frac{\pi}{6}$ in the log–polar mapped frequency domain discussed in Sect. 9.6. The hue appears to change continuously in the angular direction, and the response is concentrated to a strict annulus, indicating a reasonable approximation of the direction as well as the certainty in a narrow band of frequencies. On the right we show I_{11} supporting the same visual conclusion. However, this visual evaluation is too coarse to quantify the quality of the approximation.

To this end, on the left graph of Fig. 10.26 we show a plot of $\arg(I_{20})$ along the ring passing through the points 1–4 marked in Fig. 10.15. The direction estimations follow the true direction, the piecewise linear graph, reasonably well, although the estimation accuracy is not as good as in Fig. 10.17. Likewise, the total error graphs on the right representing $|I_{20}|$ and I_{11} do not perform nearly as well as compared to Fig. 10.17. This is because on the clean part of the image $|I_{20}| \approx I_{11}$, and both should be constant and large compared to the noisy part. However, this is only approximately true, as the oscillations show. The same result is confirmed in the measurements across the lines joining the point 5 and point 6 to the center (Fig. 10.27). The image on the left shows the direction estimations in double angle representation

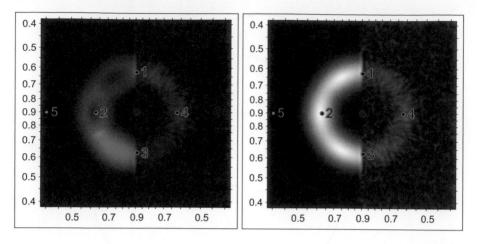

Fig. 10.25. The images illustrate the direction tensor, represented as I_{20} (*left*) and I_{11} (*right*), for Fig. 10.15 and computed by using theorem 10.4 for Eqs. (10.87)–(10.89). The hue encodes the direction, whereas the brightness represents the magnitudes of complex numbers. There were three different tune-on directions in a set of three Gabor filters

(radians). The arrows show the limits of the frequency annulus to which the Gabor filters are most sensitive, as confirmed by the graphs of $|I_{20}|$ and I_{11} in the right. The main disadvantage of this direction estimation compared to the one presented in Sect. 10.11 is due to the poor quality estimations, caused by the very coarse discretization of the spectrum in three directions. The accuracy of the direction estimation itself is, however, fully sufficient for many applications. Both direction accuracy and error estimation accuracy can be improved significantly by a nonminimalist set-up of the Gabor decomposition. This will be elaborated in the next section further. Here, we wanted to show that even with three Gabor filters it is possible to obtain reasonably accurate estimates of the direction, albeit the error estimates of the directions are not nearly as good.

Structure Tensor by Cartesian Gabor Decomposition

It is worth noting that the local spectrum must not be sampled via a separable tessellation in the orientation and the absolute frequency, although such a decomposition has advantages for pattern recognition purposes. The algorithm represented by Eqs. (10.84)–(10.85) will still hold for reasonably densely discretized power spectra, including a Gabor filter bank that covers the frequency plane in a Cartesian-separable fashion, as shown in Fig. 10.28. In this graph the tuning frequencies of the filters (marked with ×) along with the iso-amplitude curves of the filters are shown (the red and the cyan curves). For real images, half of the filters are not used, those drawn with cyan in the graph, as before because they are deducible from the other half. The dashed and solid axes show two example frequency responses at which the spectral energy will ideally be concentrated if the image is linearly symmetric having a gra-

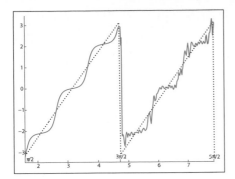
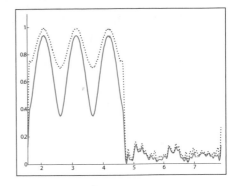

Fig. 10.26. The graphs in the *left image* represent the estimated $(\arg(I_{20}^l r), \textit{solid})$ as well as the ideal direction angle (*dashed*) on the circle passing through 1 to 4 in Fig. 10.15 when using three tune-on directions. The graphs on the *right* show $|I_{20}^l|$ (*solid*) and I_{11}^l (*dashed*) on the same ring

dient in the 45° and 30° directions respectively. We show the frequency coordinates z_{kl} (marked by $\times$), representing the tune-on frequencies of the filters. Note that the indices k and l now represent frequencies in the horizontal and vertical directions of the image instead of explicitly encoding the direction and absolute frequencies of the filters.

Marked explicitly by arrows in Fig. 10.29, we show the frequency coordinates z_{kl}. The sizes of the circles are modulated by the magnitudes of the filter responses, $|F^{k,l}|$. The magnitudes of the responses are larger when the corresponding filter tune-on frequencies are close to the solid line because we assumed that we have a linearly symmetric image, which has a spectral energy concentrated to the shown solid line.

To obtain a reasonably good estimate of the structure tensor, the filters should cover all directions, not necessarily all frequencies. This is because most images with directions consist of a wide range of frequencies. Some studies have used other tessellations which are a mixture of Cartesian and log–polar tessellation [124, 144]. Along the positive horizontal axis, they divided the frequency coordinate in exponentially increasing cell sizes. In each such cell they then placed Cartesian-separable Gaussians,

$$G_k(\omega_x, \omega_y) = \exp\left[\frac{(\omega_x - \omega_k)^2}{2\sigma_{\omega_x}^2}\right] \exp\left(\frac{\omega_y^2}{2\sigma_{\omega_y}^2}\right) \tag{10.90}$$

symmetric around the cell centers, $(\omega_k, 0)^T$ representing the horizontal tune-on directions. This set of filters is then rotated with an angular increment to obtain the Gabor filters for other directions, allowing an angular tessellation of the frequency domain. However, we emphasize that the filter functions themselves are Cartesian because they are fully symmetrical Gaussians in the frequency domain, although they are placed at frequency sites that tessellate the spectrum in a log–polar fashion. One advantage is implementation efficiency because all horizontal filters are Cartesian-separable enabling a computation by cascades of 1D filtering. Other direc-

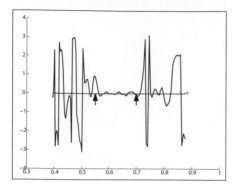 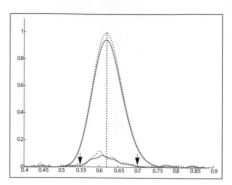

Fig. 10.27. The same as in Fig. 10.26 but the graphs on the *left* represent angle estimations on the line joining point 5 to the center (*green*) and on the line joining point 6 to the center (*blue*) marked in Fig. 10.15. In the *right* image, the two curves at the top, and the two curves at the bottom represent $|I_{20}^l|$ (*solid curves*), and I_{11}^l (*dashed curves*) for two parts of the image. The measurement pair on the top originates from the line joining point 5 to the center (clean), whereas the pair at the bottom represents the measurements for the line between point 6 and the center (noisy)

tions are evidently not Cartesian-separable, but the image can be rotated with fixed increments, and the same horizontal set of filters can be applied. The result is rotated back with the same amount. A disadvantage is that the filters, which are Gaussians, give the same weight to low and high frequencies within the frequency cell they are placed.

10.14 Relationship of the Two Discrete Structure Tensors

The structure tensor measures the second-order moment properties of the power spectrum. It estimates the most prominent direction of the image in the TLS sense and provides estimates on the quality of the fit. As we saw in Sect. 10.11, the tensor can be directly discretized or, as in Sect. 10.13, first the spectrum can be discretized via a Gabor filtering and then the structure tensor is computed for the discrete spectrum. Either case yields a discrete structure tensor that represents how well a TLS fit of a line models the image spectrum. Below we discuss the relationship of both techniques and the advantages of each.

The Gabor decomposition used in Sect. 10.13 to estimate the structure tensor sampled the spectrum at just six points (only three filters were needed though!) along a ring. According to the sampling theory discussed in Sect. 9.6, a sampled signal must be lowpass filtered before sampling to deter sampling artifacts. This is, in fact, done by the Gabor decomposition by a smoothing of the spectrum before sampling. Suppose that we have a perfectly linear symmetric image, for the sake of discussion a sinusoid in the spatial domain, with a crisp direction. This has a spectrum that has a pair of Dirac pulses. As the direction of the image changes in the spatial domain, the

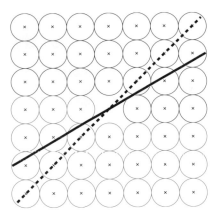

Fig. 10.28. Cartesian-separable Gabor decomposition where the × mark the filter centers and the *circles* show the iso-amplitudes of the filters midway between the grid sites. The two axes (*dashed and solid*) illustrate, overlayed, the spectra of two linearly symmetric images

pair of Dirac pulses rotates on the same ring. Accordingly, it makes sense to attempt to discretize the spectrum along a ring passing nearby the Dirac pulses. A log–polar Gabor decomposition will do the sampling, but before sampling the spectrum along the ring, the spectrum will be smeared. Thus, the Dirac pulses are smeared out in the angular direction in an amount that corresponds to the size of the window for which the Gabor filters are designed. In consequence, the more Gabor filters we use, the more the structure tensor computed by Gabor decomposition will approach the one discussed in Sect. 10.11.

In Fig. 10.30 we show on the left the quantity I_{20} which is encoded by the hue, and the brightness as obtained for the test image in Fig. 10.15. We used 18 directions, i.e., the tune-on directions differed by increments of $\frac{\pi}{18}$. The band concentration of the response is the same as before (cf. Fig. 10.25). This is to be expected because we have not changed the radial widths of the log–polar frequency Gaussians. Instead we see that the hue variation better follows the ideal hue variation shown in Fig. 10.15. On the right in Fig. 10.30 we see I_{11} modulating the brightness holding itself into an annulus, and following the brightness of the image on the left, $|I_{20}|$, closely and without oscillations. To assure ourselves, we also study the graph of these functions along the same ring as before. This is shown in Fig. 10.31, where on the left we see that the estimations follow the true directions nearly perfectly. On the right in the first half of the graph, representing the clean part of the test image, we see that the certainty, $|I_{20}|$, is nearly constant and follows I_{11} apart from a minor bias. The small bias is not surprising because even with 18 directions, the sampling density of the spectrum afforded by the Gabor decomposition is inferior to what is available in the original spectrum. The result is a smearing of the perfect energy concentration of the Dirac pulse. The dilation of the concentration is noticed by the structure tensor and is quantified as a small error in its attempt of fitting an ideal (infinitesimally narrow) line. In the noisy part of the graph, the quality of the fit is lower as expected,

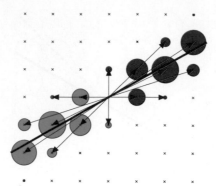

Fig. 10.29. Computing the second-order complex moments of the local power spectrum obtained via a Cartesian separable Gabor decomposition according to Eqs. (10.84)–(10.85). The sizes of the *circles*, representing the magnitudes of the filter responses become larger the closer the corresponding filter sites get to the solid axis

but the direction estimations follow the true directions quite well, nevertheless. We see a confirmation of these conclusions in the graphs of Fig. 10.32, which represent $\arg(I_{20})$ on the left, and $|I_{20}|$, I_{11} on the right, corresponding to sites along the lines joining points 5 and 6 to the center, marked in the test image of Fig. 10.15. The direction estimates are constant, as they should be and the quality of the fit as represented by the magnitudes of $|I_{20}|$ and I_{11} is high. These are close to each other in the clean signal, whereas they are much smaller and not very close to each other (relative the amplitude of $|I_{20}|$).

Which structure tensor decomposition should one then use? The direct approach is a cascade of 1D Gaussian filtering and directionally isotropic that is, free of direction-dependent artifacts. Using this technique results in far fewer operations in a sequential computer, such as a personal computer. Accordingly, if dense direction tensor maps are needed, direct sampling of the structure tensor offers computational advantages, while the accuracy of all elements of the structure tensor is virtually unaffected by the minimum number of the filters needed (three filters are needed: D_x, D_y, and a Gaussian). However, a direct steering of the absolute frequency to which the structure tensor is sensitive and the bandwidth of the frequency sensitivity range are more conveniently achievable by a log–polar mapping of the spectrum and discretization via Gabor filtering. Such a steering is possible for direct sampling of the structure tensor too, but indirectly. A Laplacian pyramidlike processing must first be applied to the image to make sure that the frequency ranges of interest are sufficiently well isolated. Also, dense structure tensors are not always needed. Some applications can successfully be developed on sparse grids, e.g., the square grid suggested in [144] or a log–polar sampling of the world [196, 216]. Tracking of humans and their identification is, for example, achievable on a doubly log–polar sampling of the image and its spectrum, i.e., on an image grid which samples the image plane in a log–polar fashion the local spectrum is sampled in a log–polar fashion too [27, 205].

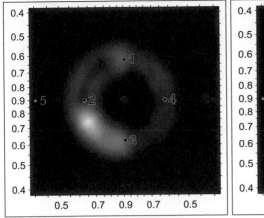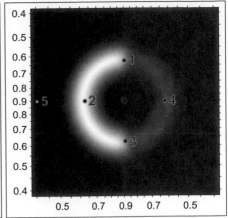

Fig. 10.30. The images illustrate the direction tensor, represented as I_{20} (*left*) and I_{11} (*right*), for Fig. 10.15 and computed by using theorem 10.4 and Eqs. (10.87)–(10.89). The hue encodes the direction, whereas the brightness represents the magnitudes of the complex numbers. There were 18 tune-on directions in the filter set

In this case, the Cartesian separability of the filters is not a serious advantage because the grid is sparse and the spectral sampling can be done by simple scalar products with Gabor filters.

10.15 Hough Transform of Lines

The Hough transform is a nonlinear filtering technique to estimate the position and direction of certain curves in a discrete image [111]. Despite its name, it is not an invertible transform in the sense of Fourier transform or the like.

The simplest Hough transform is the *Hough transform of lines*. It detects (infinitely) long lines, which we discuss in this section. To find lines, it can be imagined that a gradient filtering would be sufficient. Even in an ideal image free of noise, significantly large segments of lines may be missing for a variety of reasons. For example, lines of an object may be occluded because another object is in front of it. The Hough transform does not replace gradient filtering but starts from the result of it, to be precise the gradient image to the magnitude of which a threshold has been applied, to group together the scattered segments of the same line. A summary of the Hough transform for lines is given next.

1. The curve family of lines is modeled and parameterized. Because of its unbiased properties w.r.t. directions, one commonly used line model is:

$$\mathbf{k}^T \mathbf{r} = \cos(\phi) \cdot x + \sin(\phi) \cdot y = b \qquad (10.91)$$

Here ϕ and b are the parameters representing the direction and the closest distance of the line to the origin, marked between the points O and B in Fig. 10.33.

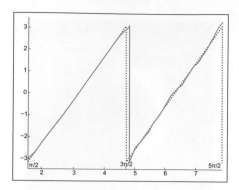 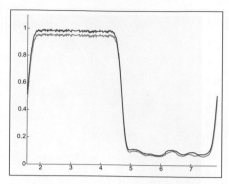

Fig. 10.31. The graphs in the *left image* represent the estimated directions ($\arg(I_{20}^l)$, *solid*) as well as the ideal direction angles (*dashed*) on the circle passing through 1 to 4 in Fig. 10.15 when using 18 tune-on directions. The graphs on the *right* show $|I_{20}^l|$ (*green*) and I_{11}^l (*magenta*) on the same ring

Every parameter pair (ϕ, b) represents a unique and infinitely long line. Likewise, every (x, y) pair defines a curve in the (ϕ, b) space.

2. The local edge strengths along with their directions are extracted by a gradient filtering,

$$|\nabla f(x, y)| = |(D_x + iD_y)f(x, y)| \qquad (10.92)$$
$$\theta(x, y) = \arg(\nabla f) = \arg[(D_x + iD_y)f] \qquad (10.93)$$

and a binary image is obtained by thresholding the edge strength, $|\nabla f|$. The result is an edge image, $\alpha(x, y)$, which is quantized to either 0 (background point), or 1 (edge point) with the local direction of the edge being $\theta(x, y)$.

3. Given an edge point (x_0, y_0) in the edge image $\alpha(x, y)$, it defines uniquely one curve in the (ϕ, b)-plane:

$$\cos(\phi) \cdot x_0 + \sin(\phi) \cdot y_0 = b \qquad (10.94)$$

because for every imaginable ϕ there is a uniquely defined b. However, only a discrete version of (ϕ, b)-plane is investigated in practice. A grid $A(\phi_k, b_k)$ is defined by having an appropriate level of quantization for ϕ and b parameters. Each point of the grid is assigned initially the value zero, $A(\phi_k, b_k)$. This artificially constructed grid is called the accumulator, because it will be used for a voting procedure storing the votes at each cell node, $A(\phi_k, b_k)$.

4. Assume that (x_0, y_0) is an edge point of the image found in $\alpha(x, y)$. Because every point in the (x, y) plane defines a curve in the (ϕ, b) plane, one votes for all points of the curve C corresponding to (x_0, y_0) in the (ϕ, b) plane. Adding the value 1 (a vote) in each of the grid cells $A(\phi_m, b_m)$ through which the curve C passes, one accomplishes the voting by repeating the procedure for all edge points of $\alpha(x, y)$. To know which cells the curve C passes in the (ϕ, b) plane, one only needs to substitute $\phi = \phi_k$ in Eq. (10.94), where ϕ_k is incremented

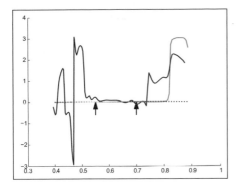 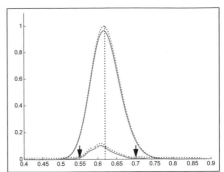

Fig. 10.32. The same as in Fig. 10.31 but the graphs on the *left* represent angle estimations on the line joining point 5 to the center (green) and on the line joining point 6 to the center (blue) marked in Fig. 10.15. The graphs on the right represent the corresponding $|I_{20}^l|$ (solid) and I_{11}^l (dashed), respectively. The measurements on the noisy line are at the bottom.

through all discrete ϕ values of the grid A established in step 3. For each such ϕ_k, one b is obtained, which is rounded off towards the closest discrete b value of the grid cells. The final $A(\phi_m, b_m)$ obtained after the voting procedure is the Hough transform for lines.

5. A long line causes a high peak in the (ϕ, b) plane, because every long line in the (x, y)-plane generates a point in the (ϕ, b) plane. Accordingly, the position of each such peak yields a specific (ϕ, b) parameter that represents an infinitely long line in the (x, y)-plane.

The Hough transform procedure above can be simplified further by voting only for one parameter cell (ϕ, b) per edge point (x, y) in step 4, the most likely direction $\phi_k \approx \theta(x, y)$ and its corresponding b value closest to a grid node. The robustness can be improved by smoothing the accumulator before finding the peaks. Accordingly, the votes cast for the line (ϕ_j, b_j) are

$$A(\phi_j, b_j) = \sum_l \delta[\theta(x_j + x_l, y_j + y_l) - \theta(x_j, y_j)] \tag{10.95}$$

where (x_l, y_l) are the edge points along the line given ϕ_j, b_j having the direction $\theta(x_j, y_j)$. The sum represents a filtering that counts the occurrence of the edges encountered along the line. The occurrence is w.r.t. edges *and* edge directions, i.e., if there is a directional conflict the value of the cast vote is zero.

Nonuniqueness of Line Directions

In step 1 we have a model of a line for every fixed pair of parameters of (φ, b). However, the model parameters and the lines do not uniquely correspond to each other because by substituting $(\phi + \pi, -b)$ in Eq. (10.91), one sees that this parameter

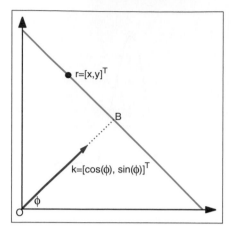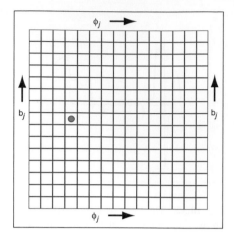

Fig. 10.33. (*left*) A common model of a line (*green*) to be used in the Hough transform of a line is given by the vector **k** and the perpendicular distance of the origin to the line, $b = |\mathbf{OB}|$. (*right*) The discretized parameter space (ϕ, b_j) is shown. Edges in the (x, y) space cast votes to cells in the (ϕ_j, b_j) space

point generates the same line as the one generated by (ϕ, b). The accumulator will then have two peaks for every line in the (x, y)-plane if no measures are taken. One technique is to add the votes of one half of the (ϕ, b) plane to the respective cells in the other half, which is equivalent to forcing $\theta(x, y)$ to the range $[0, \pi]$.

10.16 The Structure Tensor and the Hough Transform

In this section we develop the Hough voting process to see that the structure tensor averaging is a voting process too. The votes have tensor values, or equivalently complex values, jointly encoding the line strength and the line direction continuously. In case of strong coherence of local edge directions with the line model, both techniques yield identical results. However, they differ when the local edge directions are not consistent with the line model, because the tensor voting allows voting to other candidates too. In that respect, the structure tensor voting can be likened to a multiparty election that extends the single-party election, the *Hough transform* voting. The electors of the structure tensor voting are allowed to cast a vote even to the opposition party, whereas the electors of the Hough transform are only offered to be absent in case they disagree with the single party.

In Sect. 10.15, the nonuniqueness of lines w.r.t. the line direction was observed. At first, this may appear as a technical problem that can be solved by using half of the arc circle as angle parameter. However, there is a fundamental problem that is not resolved by such an approach. This is because a numerical discontinuity at an end of the interval $[0, \pi]$ must be introduced since the angles 0 and π correspond to the same line direction but differ maximally numerically. We suggest a different

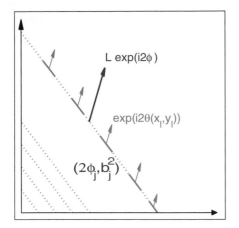
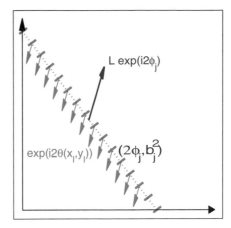

Fig. 10.34. The graphs illustrate the structure tensor voting along an infinitely long line drawn in *magenta*. (*Left*) The directions of the *green edges* are approximately consistent with the long *magenta line*, (ϕ_j, b_j). A few lines corresponding to the same ϕ_j but different b_j are shown in *magenta*. (*Right*) The coherent directions of the *green lines* are in maximal conflict with the long line. The linear symmetry tensors of the edges are shown in *green* as the complex numbers $\exp(i2\theta(x_l, y_l))$

parametrization that yields a unique representation of lines. Squaring both sides of Eq. (10.91) yields:

$$(\mathbf{k}^T \mathbf{r})^2 = \mathbf{r}^T \mathbf{k} \mathbf{k}^T \mathbf{r} = b^2 \tag{10.96}$$

where

$$\mathbf{k}\mathbf{k}^T = \begin{pmatrix} \cos^2 \phi & \sin \phi \cos \phi \\ \sin \phi \cos \phi & \cos^2 \phi \end{pmatrix} = \frac{1}{2} \left[\mathbf{I} + \begin{pmatrix} \cos 2\phi & \sin 2\phi \\ \sin 2\phi & \cos 2\phi \end{pmatrix} \right] \tag{10.97}$$

The parameters $(2\phi, b^2)$ will uniquely represent lines since $(\phi + \pi, -b)$ and (ϕ, b) map to the same parameter, the one and the same line.

The problem is thus to find how much support or votes there are along a presumed line represented by $(2\phi_j, b_j^2)$, given the observed edge points having the directions $\{\exp(i2\theta_l)\}_l$ along that line. In other words, we wish to know if a "subimage", which only consists of the long narrow line $(2\phi_j, b_j^2)$, is composed of edge elements having the same direction. The subimage is illustrated by the magenta line in Fig. 10.34. This is equivalent to investigating if the image is linearly symmetric in the subimage. A few other subimages, i.e., lines, corresponding to the same $2\phi_j$ but different b_j^2 are shown in magenta. According to the structure tensor theory, the solution to this problem is to measure I_{20} and I_{11} in the image, the long line. Accordingly, we can compute I_{20}:

$$I_{20} = \sum_l \exp(i2\theta(x_l, y_l)) \tag{10.98}$$

where $(x_l, y_l)^T$ are edge points along the presumed line. Likewise, summing the magnitudes $|\exp(i2\theta_l)| = 1$ yields:

$$I_{11} = \sum_l 1 = L \tag{10.99}$$

where L is the length of the observed line segments along the presumed line. Accordingly, the inequality

$$|I_{20}| \le I_{11} \tag{10.100}$$

holds with equality if all observed directions of the edges are are the same, or else $|I_{20}|$ will be less than I_{11}. Accordingly, the I_{20} values computed by Eq. (10.98) along a given line $(2\phi_j, b_j^2)$ are the total votes in the accumulator cell $A(2\phi_j, b_j^2)$. Alternatively, a normalized vote I_{20}/I_{11} can be cast if there are large contrast variations in the image. For a given direction $2\phi_j$, the tensor summation in Eq. (10.98) can be computed for all b_j^2s, yielding one column in the accumulator matrix A. Called orientation radiograms, such columns can encode the shape information, useful for various applications, e.g., image database retrievals [163].

In the ideal case, the inequality $|I_{20}| \le I_{11}$ will hold with equality, and the count of votes in the parameter cell $A(2\phi_j, b_j^2)$ will be identical to that obtained by use of the Hough transform. Both methods will thus deliver the same vote count in case all edge elements have the same direction as the presumed direction.

What happens if equality is not reached, i.e., some of the directions of the edges are in maximal conflict with each other? To illustrate this, we assume that half of the edge elements have directions orthogonal to the other half, i.e., we have either the direction θ_0 or $\theta_0 + \frac{\pi}{2}$, yielding the complex tensor elements:

$$I_{20} = \sum_m \exp(i2\theta_0) + \sum_n \exp(i2\theta_0 + i\pi) = \frac{L}{2} - \frac{L}{2} = 0 \tag{10.101}$$

The conflicting votes of the orthogonal directions are thus counted as negative votes, reducing the strength of the total vote for $(2\phi_j, b_j^2)$. By contrast, in the Hough transform an observed edge element with a conflicting direction is only prevented from voting. Such edge elements are hindered from reducing the accumulator votes.

In fact, the votes of the structure tensor can not only be negative, but can even be complex-valued. The argument of the complex-valued vote tells which (direction) candidate the current edge is supporting. Vote counting is a vectorial averaging, and the result, $|I_{20}|$, will be as large as I_{11} if the edge directions are collinear, or else the total vote will be reduced, possibly until $|I_{20}|$ reaches 0. In contrast to Hough voting, tensor voting allows one to fit a line even to a dashed line where the edge elements share the same direction consistently, but this common direction is different from the direction of the presumed long line, drawn in magenta in Fig. 10.34. If this is not desired, the total complex votes not agreeing with the corresponding cell labels, i.e., $2\phi_j$, can be eliminated by applying a threshold to the total vote arguments (angles).

10.17 Appendix

Proof of lemma 10.1 Noting that $\|\mathbf{k}\| = 1$, we can choose $\mathbf{u}$ so that it is orthogonal to $\mathbf{k}$. By juxtaposing these two column vectors side by side, we can construct the orthogonal matrix $\mathbf{U} = [\mathbf{k}, \mathbf{u}]$, i.e., $\mathbf{U}\mathbf{U}^T = \mathbf{1}$. Recalling that $\mathbf{r} = (x, y)^T$, and $\boldsymbol{\omega} = (\omega_x, \omega_y)^T$, the 2D Fourier transform of a linearly symmetric image then yields:

$$F(\omega_x, \omega_y) = \int \int f(x, y) \exp(-i\boldsymbol{\omega}^T \mathbf{r}) dx dy$$
$$= \int \int g(\mathbf{k}^T \mathbf{r}) \exp(-i\boldsymbol{\omega}^T \mathbf{r}) dx dy \qquad (10.102)$$

where we have substituted the 2D linearly symmetric function $f(x, y)$ with its version constructed from the 1D function via $g(\mathbf{k}^T \mathbf{r})$. We can now perform a variable substitution to the effect that $\mathbf{r}' = \mathbf{U}^T \mathbf{r}$, where $\mathbf{r}' = (x', y')^T$. The differential term $dx dy$ can be replaced by $dx' dy'$ without further consideration because $\mathbf{U}$ is orthogonal and causes the Jacobian determinant to become 1. Because of the orthogonality of $\mathbf{U}$, we also have $\mathbf{r} = \mathbf{U}\mathbf{r}'$. Consequently, the Fourier transform reduces to

$$F(\omega_x, \omega_y) = \int \int g(x') \exp(-i\boldsymbol{\omega}^T \mathbf{U}\mathbf{r}') dx' dy'$$
$$= \int \int g(x') \exp(-i\boldsymbol{\omega}^T [\mathbf{k}, \mathbf{u}]\mathbf{r}') dx' dy'$$
$$= \int \int g(x') \exp(-i[\boldsymbol{\omega}^T \mathbf{k}, \boldsymbol{\omega}^T \mathbf{u}]\mathbf{r}') dx' dy'$$
$$= \int g(x') \exp(-i\boldsymbol{\omega}^T \mathbf{k} x') dx' \int \exp(-i\boldsymbol{\omega}^T \mathbf{u} y') dy'$$
$$= G(\mathbf{k}^T \boldsymbol{\omega}) \delta(\mathbf{u}^T \boldsymbol{\omega}) \qquad (10.103)$$

∎

Proof of theorem 10.2 We will first express the error or the inertia given by Eq. (10.17) by means of complex numbers instead of vectors. To be precise, the distance defined in Eq. (10.16) can be computed by replacing the scalar product in the 2D plane with complex multiplications via

$$\mathbf{x}_1^T \mathbf{x}_2 = \Re(\widehat{\mathbf{x}}_1^* \widehat{\mathbf{x}}_2) = \frac{1}{2}[\widehat{\mathbf{x}}_1^* \widehat{\mathbf{x}}_2 + (\widehat{\mathbf{x}}_1^* \widehat{\mathbf{x}}_2)^*] \qquad (10.104)$$

where $\widehat{\cdot}$ is the operator that constructs a complex number from a real 2D vector:

$$\widehat{\mathbf{x}} = x + iy, \quad \text{when} \quad \mathbf{x} = (x, y)^T, \qquad (10.105)$$

and $\Re(\widehat{\mathbf{x}})$ equals to x, the real part of the complex number $\widehat{\mathbf{x}}$. Since $\|\mathbf{k}\| = 1$ and $\mathbf{S}$ is positive semidefinite , $\mathbf{k}$ maximizing $\mathbf{k}^T \mathbf{S} \mathbf{k}$ is the eigenvector belonging to the largest eigenvalue of $\mathbf{S}$. By using Eqs. (10.23) and Eq. (10.29) we obtain:

$$\mathbf{k}^T \mathbf{S} \mathbf{k} = \frac{1}{4\pi^2} \int \mathbf{k}^T (\nabla f_j)(\nabla^T f)\mathbf{k} d\mathbf{x} = \frac{1}{4\pi^2} \int (\mathbf{k}^T \nabla f)(\mathbf{k}^T \nabla f) d\mathbf{x}$$

$$= \frac{1}{4\pi^2} \int (\Re(\widehat{\mathbf{k}}^* \widehat{\nabla} f))^2 d\mathbf{x}$$

Now, with ordinary algebraic manipulations applied to complex numbers, we obtain

$$\mathbf{k}^T \mathbf{S} \mathbf{k} = \frac{1}{4\pi^2} \int \frac{1}{4} \left[(\widehat{\mathbf{k}}^2)^* (\widehat{\nabla} f)^2 + [(\widehat{\mathbf{k}}^2)^* (\widehat{\nabla} f)^2]^* + 2|\widehat{\nabla} f|^2 \right) d\mathbf{x}$$

$$= \frac{1}{4\pi^2} \int \frac{1}{2} |\widehat{\nabla} f|^2 d\mathbf{x} + \frac{1}{2} \Re[(\widehat{\mathbf{k}}^2)^* (\widehat{\nabla} f)^2] d\mathbf{x} \qquad (10.106)$$

where

$$\widehat{\nabla} = D_x + iD_y = \frac{\partial}{\partial x} + i\frac{\partial}{\partial y} \qquad (10.107)$$

By construction, $\widehat{\nabla} f$ and $\widehat{\mathbf{k}}$ are the complex interpretations of the real 2D vectors ∇f and $\mathbf{k}$. Since the first term of the sum in Eq. (10.106) is free of $\widehat{\mathbf{k}}$, the complex number $\widehat{\mathbf{k}}^2$ that maximizes Eq. (10.106) is the same as the one that maximizes the second term:

$$\frac{1}{4\pi^2} \int \frac{1}{2} \Re[(\widehat{\mathbf{k}}^2)^* (\widehat{\nabla} f)^2] d\mathbf{x} = \frac{1}{2} \Re \left[(\widehat{\mathbf{k}}^2)^* \frac{1}{4\pi^2} \int (\widehat{\nabla} f)^2 d\mathbf{x} \right] \qquad (10.108)$$

$$= \frac{1}{2} \Re[(\widehat{\mathbf{k}}^2)^* I_{20}] \qquad (10.109)$$

where we have used the fact that $\Re(\cdot)$ and $\int \cdot$ are linear operators that commute with each other, and I_{20} represents the complex scalar:

$$I_{20} = \frac{1}{4\pi^2} \int (\widehat{\nabla} f)^2 d\mathbf{x} = \int (\omega_x + i\omega_y)^2 |F|^2 d\boldsymbol{\omega}. \qquad (10.110)$$

The choice of the subscript in I_{20} is justified because the second integral in Eq. (10.110) is a complex moment.

Identifying the scalar product in Eq. (10.109) via Eq. (10.104) and remembering that $|\widehat{\mathbf{k}}^2| = 1$, we interpret the expression geometrically. The vector that corresponds to the complex scalar I_{20} is projected onto a line with the direction vector that corresponds to $\widehat{\mathbf{k}}^2$. The projection, a real scalar, is given by Eq. (10.109), which is to be maximized. Evidently, the direction $\widehat{\mathbf{k}}^2_{\max}$ that maximizes this projection is the same as the direction of I_{20}:

$$\widehat{\mathbf{k}}^2_{\max} = I_{20}/|I_{20}|. \qquad (10.111)$$

and the projection result is the (positive real) scalar $|I_{20}|$. Similarly, the complex number $\widehat{\mathbf{k}}^2_{\min}$ that minimizes Eq. (10.106) is given by the (negative real) scalar $-\widehat{\mathbf{k}}^2_{\max}$ yielding the negative scalar $-|I_{20}|$. Calling the (positive real) scalar expression in the first term of Eq. (10.106) as I_{11}:

$$I_{11} = \frac{1}{4\pi^2} \int |\widehat{\nabla} f|^2 dx \qquad (10.112)$$

and substituting $\widehat{\mathbf{k}}^2_{\min}$ and $\widehat{\mathbf{k}}^2_{\max}$ in Eq. (10.106) yields:

$$\lambda_{\min} = e(\mathbf{k}_{\min}) = \frac{1}{2}(I_{11} - |I_{20}|) \qquad (10.113)$$

$$\lambda_{\max} = e(\mathbf{k}_{\max}) = \frac{1}{2}(I_{11} + |I_{20}|). \qquad (10.114)$$

That is:

$$|I_{20}| = \lambda_{\max} - \lambda_{\min} \qquad (10.115)$$

$$\arg I_{20} = 2\theta_0 \qquad (10.116)$$

$$I_{11} = \lambda_{\max} + \lambda_{\min} \qquad (10.117)$$

■

11

Direction in Curvilinear Coordinates

This chapter provides a general technique for detection of patterns possessing linear symmetry, with respect to curvilinear coordinates in 2D images. Curves given by a *harmonic function pair* (HFP) will be discussed in detail. The idea is to "bend and twist" the image by means of an HFP so that the patterns can be detected by the same formalism that we developed for the structure tensor. This will lead us to the concepts of *coordinate transformations* (CT) and *generalized structure tensor* (GST), which will be discussed further from the viewpoint of pattern recognition. Since very intricate patterns can be described by such CTs, the technique is a general toolbox for pattern detection. We will also develop a unifying concept for geometric shape quantitation and detection, to the effect that the generalized structure tensor becomes an extension of the generalized Hough transform, with the additional property that it is also capable of handling negative votes as well as complex-valued votes. In the generalized structure tensor theory, the detection of intricate target objects is equivalent to a problem of symmetry detection in the HFP coordinate system. The generalized structure tensor does not necessitate explicit coordinate transformations. Instead, via Lie operators, the "bending and twisting" occurs in the complex-valued filters implicitly once for all, rather than being performed explicitly on every image to be recognized.

11.1 Curvilinear Coordinates by Harmonic Functions

Let $\xi(x, y)$ be a *harmonic function,* that is, its partial derivatives of the first two orders are continuous and it satisfies the *Laplace equation:*

$$\Delta \xi = \frac{\partial^2 \xi}{\partial x^2} + \frac{\partial^2 \xi}{\partial y^2} = 0. \tag{11.1}$$

Due to the linearity of Laplace's equation, linear combinations of harmonic functions are also harmonic. If two harmonic functions ξ and η satisfy the *Cauchy–Riemann equations:*

$$\frac{\partial \xi}{\partial x} = \frac{\partial \eta}{\partial y}, \quad \frac{\partial \xi}{\partial y} = -\frac{\partial \eta}{\partial x} \tag{11.2}$$

then η is said to be the *conjugate harmonic function* of ξ. Equivalently, the pair (ξ, η) is said to be a *harmonic function pair* (HFP). Conversely, if two functions with continuous second-order partial derivatives satisfy Eq. (11.2), then both are harmonic, i.e., fulfill the Laplace equation. This is seen by applying $\frac{\partial}{\partial x}$ and $\frac{\partial}{\partial y}$, respectively, to the Cauchy–Riemann equations, which yield:

$$\frac{\partial^2 \xi}{\partial x^2} = \frac{\partial^2 \eta}{\partial x \partial y}, \quad \frac{\partial^2 \xi}{\partial y^2} = -\frac{\partial^2 \eta}{\partial y \partial x} \tag{11.3}$$

where we have $\frac{\partial^2 \eta}{\partial x \partial y} = \frac{\partial^2 \eta}{\partial y \partial x}$ because of the continuity assumption on the second-order derivatives. Accordingly, the conjugate harmonic function of a known harmonic function, ξ, is found by solving the Laplace equation using the Cauchy–Riemann equations as boundary conditions. These equations stipulate that the gradients of a harmonic function pair are orthogonal to each other at every point, i.e., they are *locally orthogonal*.

An analytic function is generally a complex function and is characterized by the fact that it has complex derivatives of all orders. It can be shown that the imaginary part of any analytic function is the conjugate harmonic function of the real part. Without loss of generality, we can assume both ξ and η to be single-valued by imposing proper restrictions. Then by definition Eq. (11.2), an HFP curve pair,

$$\xi_0 = \xi(x, y) \tag{11.4}$$
$$\eta_0 = \eta(x, y) \tag{11.5}$$

has orthogonal gradients at the same point. For nontrivial $\xi(x, y)$ and $\eta(x, y)$, Eqs. (11.4) and (11.5) define a *coordinate transformation* (CT) which is invertible almost everywhere.

Let an image be represented by the real function $f_1(x, y)$. Another representation of the image f_2 can be obtained by means of a CT using the HFP,

$$f_1(x(\xi, \eta), y(\xi, \eta)) = f_2(\xi, \eta) \tag{11.6}$$

As before, the term image will refer to a subimage.

Definition 11.1. *The image $f(\xi, \eta)$ is said to be linearly symmetric in the coordinates (ξ, η) if there exists a one-dimensional function g such that*

$$f(\xi, \eta) = g(a\xi + b\eta) \tag{11.7}$$

for some real constants a and b. Here $\xi(x, y), \eta(x, y)$ is a HFP, and the symmetry direction vector, (a, b), has its length normalized to unity, i.e., $\sqrt{a^2 + b^2} = 1$.

The notion "linearly symmetric in (ξ, η)" is, in analogy with Chap. 10, motivated by the fact that the isocurves of such images are *parallel lines* in the $\xi\eta$ coordinates. Likewise, such images have a high concentration of their spectral power along a line through the origin.

Starting with the trivial unity transformation we give examples of CTs for pattern families that can be modeled and detected by the toolbox to be presented.

Fig. 11.1. The HFPs used in Examples 11.1 (*top*) and 11.2 (*bottom*), respectively

Example 11.1. *Let $w(z)$ be defined as the identity coordinate transformation:*

$$w(z) = z = \xi(x,y) + i\eta(x,y) = x + iy$$

Since w is an analytic function in z, (x,y) is a HFP. For illustration, we let the one-dimensional function g be

$$g(\tau) = (1 + \cos \tau)/2 \qquad (11.8)$$

while bearing in mind that g can be any 1D function. The argument τ is replaced by $a\xi + b\eta$ to generate the family of isocurves defined by this transformation. Figure 11.1 (top) illustrates the two basis patterns, $g(\xi) = $ constant, and $g(\eta) = $ constant respectively. The linear combinations of these two patterns generate new patterns, all of which belong to the same family of curves, those that are linearly symmetric w.r.t. the Cartesian x and y, which we studied in Chap. 10.

Example 11.2. *By using the same g as in Example 11.1, we can illustrate the transformation defined by*

$$w(z) = \log z = \xi(x,y) + i\eta(x,y) = \log \sqrt{x^2 + y^2} + i\tan^{-1}(x,y)$$

which is analytic everywhere except at the origin. We assume the principal branch as the value set of w to avoid multiple-valued functions. Figure 11.1 (bottom) shows

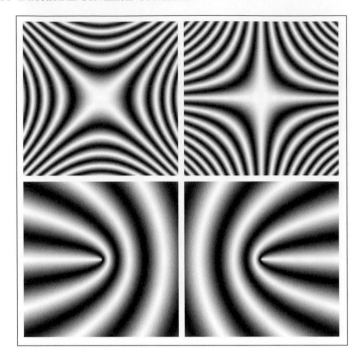

Fig. 11.2. The HFPs used in Examples 11.3 (*top*) and 11.4 (*bottom*), respectively

the basis pair of this transformation. The local orthogonality can be seen by super-imposing the two figures so that the origins coincide. The linear combinations of the basis pair, $a\xi + b\eta$, generate the family of logarithmic spirals. Some members of this family are displayed in Fig. 11.3. We note that the sign of $a \cdot b$ determines the chirality of the spirals, i.e., whether they are twisted to the left or to the right. By measuring the direction angle of the vector (a, b), it will be possible to tell apart a left-handed pattern from a right-handed pattern, as well as whether a pattern is circular or star shaped, without actually knowing the gray levels of the pattern (g) in advance.

Example 11.3. *We use the analytic function z^2 to obtain the HFP ξ, η*

$$w(z) = z^2 = \xi(x, y) + i\eta(x, y) = x^2 - y^2 + i2xy \qquad (11.9)$$

which is illustrated in Fig. 11.2 (top). The generated pattern family, $a\xi + b\eta$ corresponds to rotated versions of a basis pattern. The asymptotes of the generated hyperbolic patterns are orthogonal and the direction of the cross is given by the direction of (a, b), which, as will be discussed further below, can be used to detect "crosslike" junctions.

Example 11.4. *The analytic function $\sqrt{z}$:*

$$w(z) = \sqrt{z} = \xi(x,y) + i\eta(x,y) = \sqrt{r}\exp\left(i\frac{\varphi}{2}\right) = \sqrt{r}\cos\left(\frac{\varphi}{2}\right) + i\sqrt{r}\sin\left(\frac{\varphi}{2}\right)$$

generates the basis patterns that are illustrated in Fig. 11.2 (bottom). The pattern family generated by this pair consists of rotated versions of one of the basis patterns; that is, the rotation angle corresponds to the direction of the vector (a,b).

11.2 Lie Operators and Coordinate Transformations

Here we summarize the essential parts of the coordinate transformation theory using differential operators to detect symmetric pattern families. We will briefly present the Lie operators that perform small (infinitesimal) CTs, which are all equivalent to translations in HFP coordinates.

We start by performing translations of ξ and η, which are coordinates that an arbitrary image f will later be represented by, beginning with ξ. The translated[1] coordinates are marked with $'$ and yield:

$$T_1(\xi, \eta, \epsilon_1) = \begin{cases} \xi' = \xi + \epsilon_1, \\ \eta' = \eta. \end{cases} \tag{11.10}$$

Two successive translations are equivalent to a single translation which can be obtained by using the parameter combination rule:

$$\phi(\epsilon_1, \delta) = \epsilon_1 + \delta \tag{11.11}$$

where ϕ is analytic with respect to both of its arguments and fulfills the group axioms with $\epsilon = 0$ being the identity element of the group. These properties make T_1 a *one-parameter Lie group of transformations* [31]. With each Lie group of transformation an *infinitesimal generator* is associated. In this case, this is[2] $D_\xi = \frac{\partial}{\partial \xi}$. When applied to f the CT results in a translation of the isocurves of f along the basis vector, $\hat{\xi}$, which is the *curvilinear basis* related to ξ. D_ξ applied to any function whose isocurves consist of $\xi(x,y) = \lambda$ delivers the tangent fields of ξ. An arbitrary amount of translations in the $\hat{\xi}$ direction can be obtained by applying the exponential form of D_ξ:

$$f(\xi', \eta') = \left(1 + \epsilon_1 D_\xi + \frac{\epsilon_1^2}{2!} D_\xi^2 + \cdots\right) f(\xi, \eta) = \exp(\epsilon_1 D_\xi) f(\xi, \eta) \tag{11.12}$$

which is a Taylor expansion of $f(\xi + \epsilon_1, \eta)$ around (ξ, η). The CT (ξ', η') is completely determined by D_ξ's actions on the ξ, η coordinates. As a special case, if the isocurves of f are given by $\xi = $ constant, i.e., $f(\xi, \eta) = g(\xi)$ for some g, then we

---

[1] The symbol $'$ represents new coordinates after a CT is applied to original coordinates in this chapter.

[2] In studies of CTs, it is common to use the notation $\mathcal{L}_1$ for D_ξ, and $\mathcal{L}_2$ for D_η, where $\mathcal{L}$ is a mnemonic for the "Lie operator".

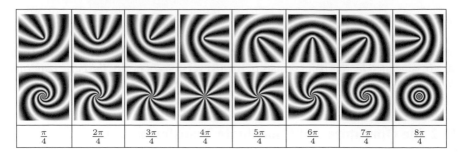

Fig. 11.3. The *top row* shows the rotated patterns generated by $g(z) = \sqrt{z}$, i.e., when $n = -1$ in Eq. (11.87), for various angles between the parameters. The *second row* displays the corresponding curves for $g(z) = \log(z)$, i.e., when $n = -2$. The *third row* is φ, which represents the parameter ratios used to generate the patterns. The change of φ represents a geometric rotation on the *top row*, whereas it represents a change of bending in the second row

have $D_\xi f = g'(\xi)$, leaving the *isocurve families* of f invariant. The *isocurve* families obtained for various fixed λs by the equation $\xi(x, y) = \lambda$ are *integral curves* of D_ξ, that is, an isocurve maps to another isocurve, $\xi(x, y) + \epsilon_1 = \lambda$, which is within the same family. A more powerful invariance is obtained if $f(\xi, \eta) = g(\eta)$ for some g. This yields $D_\xi f = 0$, that is, the isocurves $\eta(x, y) = \lambda$, are invariant. In this case one isocurve maps onto itself.

Similarly, $T_2(\xi, \eta, \epsilon_2)$, which translates the η coordinate, is a one-parameter Lie group of transformation:

$$T_2(\xi, \eta, \epsilon_2) = \begin{cases} \xi' = \xi, \\ \eta' = \eta + \epsilon_2, \end{cases} \tag{11.13}$$

with the corresponding infinitesimal generator

$$D_\eta = \frac{\partial}{\partial \eta} \tag{11.14}$$

As before, one can reconstruct T_2 by means of the operator $\exp(\epsilon_2 D_\eta)$.

We can formally define a new infinitesimal operator:

$$D_\zeta = \cos\theta D_\xi + \sin\theta D_\eta \tag{11.15}$$

which is linear in D_ξ, D_η, and expect to have a one parameter group of transformations $T(\xi, \eta, \epsilon)$ corresponding to it. This expectation can be realized precisely because D_ξ and D_η are tangent vector fields of a coordinate basis by construction.[3] A finite motion along the integral curves of D_ζ can be obtained by successive exponentiations (Taylor series):

---

[3] For general vector fields represented by $\frac{\partial}{\partial \lambda}$ and $\frac{\partial}{\partial \mu}$, in order for a coordinate basis generating them via tangents to exist, the vector fields must commute and be independent. Luckily, D_ξ and D_η] always commute if ξ, η are HFPs.

$$\exp(\epsilon D_\zeta) = \exp(\epsilon \cos\theta D_\xi)\exp(\epsilon \sin\theta D_\eta)$$
$$= \exp(\epsilon \sin\theta D_\eta)\exp(\epsilon \cos\theta D_\xi) \qquad (11.16)$$

Paraphrasing, D_ξ and D_η together act as an operator basis pair for *any* translation along a "line" (which is a linear combination of ξ, η) [31], which in turn makes D_ζ a classical *directional gradient*, but in the curvilinear coordinates (ξ, η). The corresponding one-parameter Lie group of transformation is given by

$$T(\xi, \eta, \epsilon_1) = \begin{cases} \xi' = \xi + \epsilon \cos\theta, \\ \eta' = \eta + c \sin\theta, \end{cases} \qquad (11.17)$$

with θ being a "directional" constant, characterizing a unique family of curves, which are also the invariants of the operator D_ζ, Eq. (11.15).

$$-\xi\sin\theta + \eta\cos\theta = \text{constant} \qquad (11.18)$$

The converse is also true, i.e., Eq. (11.18) uniquely represents D_ζ. In Fig. 11.3 we show two curve families generated by using Eq. (11.18) for various θs. The bottom row shows $\log(x+iy) = \xi(x,y) + i\eta(x,y)$, while the top row shows $\sqrt{x+iy} = \xi(x,y) + i\eta(x,y)$. Notice that when changing θ we generate the members of the same pattern family, whereas when changing the ξ, η pair we generate completely new families of patterns.

11.3 The Generalized Structure Tensor (GST)

We wish to know how well an arbitrary image can be described by means of analytic curves. In our approach ξ and η are known a priori and are conjugate harmonic pairs, whereas neither θ nor the gray values are known. We will attempt to fit a family of curves as defined by Eq. (11.18) to an arbitrary image by finding its "closest member" to the image. We do this by minimizing the following error or energy:

$$e(\theta) = \int |D_\zeta f(\xi,\eta)|^2 d\xi d\eta = \int |(\cos\theta D_\xi + \sin\theta D_\eta)f(\xi,\eta)|^2 d\xi d\eta \quad (11.19)$$

with respect to θ. Since we do not know if the image is really a member of the space linearly spanned by the ξ, η pair, the next best thing is to find a member of this family closest to our image and see if the error is small enough. Notice that $e(\theta)$ is a norm (in the sense of $\mathcal{L}^2$), and from this it follows that when $e(\theta)$ is zero for some θ, the image f fulfills $D_\zeta f(x,y) = 0$, almost for all x, y. By using Eq. (11.12), Eq. (11.19) can be seen as the total error in a small translation. Furthermore, the error is the total least square error as in the ordinary linear symmetry error function discussed in Sect. 10.10 except that the coordinates are now the curve pair ξ, η.

Example 11.5. *To fix the ideas, we return to Example 11.2 and note that the operators $\frac{\partial}{\partial\xi}$ and $\frac{\partial}{\partial\eta}$ have the curves*

$$\eta = \tan^{-1}(x, y) = constant \qquad and \qquad \xi = \log(\sqrt{x^2 + y^2}) = constant \quad (11.20)$$

as invariants. Because $D_\xi = \frac{\partial}{\partial \xi}$ acts as a zoomer and $D_\eta = \frac{\partial}{\partial \eta}$ as a rotator, the direction given by $\theta_{\min}$ represents the amount of scaling versus rotation leaving f unchanged (in practice, least changed). The $e(\theta_{\min})$ contains information as to whether $\theta_{\min}$, which represents the symmetry "direction", is or is not significant.

The basis of the recognition is that $e(\theta)$ is small for some θ. We define

$$k(\theta) \stackrel{\triangle}{=} (\cos \theta, \sin \theta)^T \tag{11.21}$$

and we rewrite the error function in quadratic form:

$$e(\theta) = \int k(\theta)^T \begin{pmatrix} \frac{\partial f}{\partial \xi} \\ \frac{\partial f}{\partial \eta} \end{pmatrix} \left(\frac{\partial f}{\partial \xi}, \frac{\partial f}{\partial \eta} \right) k(\theta) d\xi d\eta = k(\theta)^T \mathbf{S} k(\theta) \tag{11.22}$$

where $\mathbf{S}$ is a matrix defined as follows:

Definition 11.2. *The matrix*

$$\mathbf{S} = \begin{pmatrix} \int \frac{\partial f}{\partial \xi} \frac{\partial f}{\partial \xi} d\xi d\eta, & \int \frac{\partial f}{\partial \xi} \frac{\partial f}{\partial \eta} d\xi d\eta \\ \int \frac{\partial f}{\partial \xi} \frac{\partial f}{\partial \eta} d\xi d\eta, & \int \frac{\partial f}{\partial \eta} \frac{\partial f}{\partial \eta} d\xi d\eta \end{pmatrix} \tag{11.23}$$

is called the generalized structure tensor (GST).

Applying Parseval–Plancherel identity to the elements of matrix $\mathbf{S}$ gives

$$\mathbf{S} = \begin{pmatrix} \mu_{20}(|F(\omega_\xi, \omega_\eta)|^2), & \mu_{11}(|F(\omega_\xi, \omega_\eta)|^2) \\ \mu_{11}(|F(\omega_\xi, \omega_\eta)|^2), & \mu_{02}(|F(\omega_\xi, \omega_\eta)|^2) \end{pmatrix} \tag{11.24}$$

Hence, elements of $\mathbf{S}$ correspond to the second-order moments of the power spectrum $(|F(\omega_\xi, \omega_\eta)|^2)$, which is the Fourier transform of image $f(\xi, \eta)$ taken in the $\xi\eta$ coordinates. It should be emphasized that the Fourier transform of f w.r.t. the (ξ, η) coordinates is not the same as when taken w.r.t. the (xy) coordinates.

We know from Chap. 10 that the second-order moments can also be represented by the second-order complex moments, I_{pq} with $p + q = 2$, see Eq. 10.36, so that we can conveniently use another matrix as defined next.

Definition 11.3. *The complex representation of the generalized structure tensor is given as :*

$$\mathbf{Z} = \frac{1}{2} \begin{pmatrix} I_{11}(|F(\xi, \eta)|^2) & -iI_{20}(|F(\xi, \eta)|^2) \\ iI_{20}(|F(\xi, \eta)|^2)^* & I_{11}(|F(\xi, \eta)|^2) \end{pmatrix} \tag{11.25}$$

where

$$\begin{aligned} I_{20} &= \mathbf{S}(1, 1) - \mathbf{S}(2, 2) + i2\mathbf{S}(1, 2) \\ I_{11} &= \mathbf{S}(1, 1) + \mathbf{S}(2, 2) \end{aligned} \tag{11.26}$$

with $\mathbf{S}(k, l)$ being the elements of $\mathbf{S}$ according to Eq. (11.23).

It follows from the above that the building blocks of $\mathbf{Z}$ are obtained as follows:

$$I_{20}(|F(\omega_\xi, \omega_\eta)|^2) = \int (D_\xi f(\xi, \eta) + i D_\eta f(\xi, \eta))^2 d\xi d\eta \qquad (11.27)$$

and

$$I_{11}(|F(\omega_\xi, \omega_\eta)|^2) = \int |D_\xi f(\xi, \eta) + i \mathcal{L}_\eta f(\xi, \eta)|^2 d\xi d\eta \qquad (11.28)$$

Again, the real moments μ_{20}, μ_{02}, and μ_{11}, as well as the complex moments I_{20} and I_{11} are all taken w.r.t *harmonic coordinates* as in Eqs. (11.23), Eq. (11.27), and Eq. (11.28). We must still find a way to obtain them directly by using Cartesian coordinates. In different coordinate systems the representation of the Lie operators and the linear symmetry operator may look quite different despite the identical physical effect when applied to a function. The representation of the Lie operators in the Cartesian coordinates can be obtained by using the chain rule:

$$D_\xi = x_\xi \frac{\partial}{\partial x} + y_\xi \frac{\partial}{\partial y} = \frac{\xi_x}{\xi_x^2 + \xi_y^2} \frac{\partial}{\partial x} + \frac{\xi_y}{\xi_x^2 + \xi_y^2} \frac{\partial}{\partial y} = \frac{\nabla^T \xi}{\|\nabla \xi\|^2} \nabla \qquad (11.29)$$

$$D_\eta = x_\eta \frac{\partial}{\partial x} + y_\eta \frac{\partial}{\partial y} = -\frac{\xi_y}{\xi_x^2 + \xi_y^2} \frac{\partial}{\partial x} + \frac{\xi_x}{\xi_x^2 + \xi_y^2} \frac{\partial}{\partial y} = \frac{\nabla_\perp^T \xi}{\|\nabla \xi\|^2} \nabla \qquad (11.30)$$

where the definitions $\xi_x = \frac{\partial \xi(x,y)}{\partial x}$ and $\nabla_\perp^T \xi = (-\xi_y, \xi_x)$ are used for simplicity. Moreover, the partial derivatives of ξ and η are obtained by inverting the Jacobian[4] of T, and using the Cauchy–Riemann equations, Eq. (11.2):

$$\frac{\partial^2 (x, y)}{\partial \xi \partial \eta} = \begin{pmatrix} x_\xi & x_\eta \\ y_\xi & y_\eta \end{pmatrix} = \left(\frac{\partial^2 (\xi, \eta)}{\partial x \partial y} \right)^{-1} = \begin{pmatrix} \xi_x & \xi_y \\ \eta_x & \eta_y \end{pmatrix}^{-1} = \frac{1}{\xi_x^2 + \xi_y^2} \begin{pmatrix} \xi_x & -\xi_y \\ \xi_y & \xi_x \end{pmatrix}$$

We note that D_η does not explicitly depend on partial derivatives of η with respect to x or y, which of course is the consequence of the HFP assumption that binds the gradients of ξ and η together.

Example 11.6. *For illustration we go back to Example 11.2, and see that the corresponding infinitesimal operators are found by using Eqs. (11.29) and (11.30):*

$$D_\xi = x \frac{\partial}{\partial x} + y \frac{\partial}{\partial y}$$

$$D_\eta = -y \frac{\partial}{\partial x} + x \frac{\partial}{\partial y}$$

These are well-known scaling and rotation operators from differential geometry. By applying D_ξ to the coordinate pair x, y we obtain an infinitesimal increment which

---

[4] Represented by the symbol $\frac{\partial^2 (\xi, \eta)}{\partial \partial y}$, a Jacobian of a 2D CT $(x, y)^T \rightarrow (\xi, \eta)^T$ is defined as the matrix $\begin{pmatrix} \frac{\partial \xi}{\partial x} & \frac{\partial \xi}{\partial y} \\ \frac{\partial \eta}{\partial x} & \frac{\partial \eta}{\partial y} \end{pmatrix}$.

we can add to x, y to obtain the new coordinates $(x', y')^T$. This will yield an infinitesimal scaling:

$$\begin{cases} D_\xi x = x \\ D_\xi y = y \end{cases} \quad \Rightarrow \quad \begin{cases} x' = x + x dx \\ y' = y + y dy \end{cases} \tag{11.31}$$

Similarly, a small rotation is obtained as

$$\begin{cases} D_\eta x = -y \\ D_\eta y = x \end{cases} \quad \Rightarrow \quad \begin{cases} x' = x - y dx \\ y' = y + x dy \end{cases} \tag{11.32}$$

$$D_\eta x = -y \Rightarrow x' = x - y dx, \quad \text{and} \quad D_\eta y = x \Rightarrow y' = y + x dy. \tag{11.33}$$

We extend the definition of the ILST introduced in Chap. 10 to harmonic coordinates

Definition 11.4. *The infinitesimal linear symmetry tensor w.r.t. the harmonic coordinates ξ, η is defined as*

$$\daleth_{\xi,\eta}(f)(\xi,\eta) = [D_\xi f(\xi,\eta) + i D_\eta f(\xi,\eta)]^2 \tag{11.34}$$

The subscripts in $\daleth_{\xi,\eta}$ represent the derivation variables, i.e., they remind that the two derivations of the operator are taken with respect to ξ and η variables, respectively.

Lie operators can be translated to Cartesian coordinates even if they were initially formulated in canonical coordinates. Accordingly, we can also translate $\daleth_{\xi,\eta}$, which consists of the square of a Lie operator applied to an image, as follows:

$$\daleth_{\xi,\eta}(f)(\xi,\eta) = (D_\xi f + i D_\eta f)^2 = [(\nabla\xi + i\nabla_\perp\xi)^T \nabla f]^2 / \|\nabla\xi\|^4. \tag{11.35}$$

However, since

$$\nabla\xi + i\nabla_\perp\xi = \begin{pmatrix} \xi_x - i\xi_y \\ i(\xi_x - i\xi_y) \end{pmatrix} = (\xi_x - i\xi_y)\begin{pmatrix} 1 \\ i \end{pmatrix} \tag{11.36}$$

where $\xi_x - i\xi_y$ is complex-valued (scalar function), we have the result

$$\daleth_{\xi,\eta}(f)(\xi,\eta) = \frac{(\xi_x - i\xi_y)^2}{|\xi_x - i\xi_y|^4}\left[(1,i)\begin{pmatrix} f_x \\ f_y \end{pmatrix}\right]^2 = \frac{(\xi_x - i\xi_y)^2}{|\xi_x - i\xi_y|^4}(f_x + i f_y)^2$$

$$= \frac{\daleth_{\xi,y}^*(\xi)(x,y)}{|\daleth_{x,y}^*(\xi)(x,y)|^2}\daleth_{x,y}(f)(x,y). \tag{11.37}$$

which we restate in the following lemma.

Lemma 11.1. *Under a harmonic conjugate basis change given by ξ, η, the ILST changes basis according to:*

$$\daleth_{\xi,\eta}(f)(\xi,\eta) = \frac{\daleth_{x,y}^*(\xi)(x,y)}{|\daleth_{x,y}^*(\xi)(x,y)|^2} \cdot \daleth_{x,y}(f)(x,y) \tag{11.38}$$

♦

Notice that the basis change resulted in a decoupling of the terms according to their dependency on coordinates. The first term depends only on ξ, computable without a knowledge of f, whereas the second term is dependent only on the image f, computable without a knowledge of ξ. For convenience, in what follows we drop the mnemonic subscripts of $⅂_{x,y}$ when it is clear from the context that the derivations are w.r.t. x, y. Consequently, I_{20} is given by

$$I_{20} = \int ⅂_{\xi,\eta}(f)(\xi,\eta)d\xi d\eta \tag{11.39}$$

$$= \int \frac{⅂^*(\xi)(x,y)}{|⅂^*(\xi)(x,y)|^2} ⅂(f)(x,y) \det\left(\frac{\partial\xi(x,y)}{\partial x \partial y}\right) dx dy \tag{11.40}$$

$$= \int \frac{⅂^*(\xi)(x,y)}{|⅂^*(\xi)(x,y)|} ⅂(f)(x,y) dx dy \tag{11.41}$$

$$= \int w^{20*}(x,y) ⅂(f)(x,y) dx dy \tag{11.42}$$

with

$$w^{20}(x,y) = \frac{⅂(\xi)(x,y)}{|⅂(\xi)(x,y)|} \tag{11.43}$$

Here the Cauchy–Riemann equations applied to the functional determinant:

$$\det\left(\frac{\partial\xi(x,y)}{\partial x \partial y}\right) = \begin{vmatrix} \xi_x & \xi_y \\ -\xi_y & \xi_x \end{vmatrix} = \xi_x^2 + \xi_y^2 = |⅂^*(\xi)(x,y)| \tag{11.44}$$

have been used. Accordingly, we have the following result:

Lemma 11.2. *The complex moment I_{20} of the power spectrum in harmonic coordinates can be estimated in Cartesian coordinates as follows*

$$I_{20} = \int w^{20*}(x,y) ⅂(f)(x,y) dx dy \tag{11.45}$$

where

$$w^{20}(x,y) = \frac{⅂(\xi)(x,y)}{|⅂(\xi)(x,y)|} \tag{11.46}$$

◆

In Eq. (11.45), we note that I_{20} is obtained by projecting the ILST of the image in Cartesian coordinates on the kernel w^{20}, which consists of the ILST of the target curve pattern, ξ, except for a magnitude normalization. This is a scalar product between a function, which only depends on the image and is independent of the target pattern family ξ, and a kernel that encodes the directional information of the pattern family. Consequently, once the ILST of the image is available it can be tested for matching a multitude of pattern families by changing the corresponding kernel w^{20} without recalculating the ILST of the image. This is not surprising as the ILST of the image represents "universal" information for pattern matching, namely the local edges and the double of their directions [90, 137].

Similarly, I_{11} which is the average ILST magnitude, can be obtained. We state this result as a lemma.

Lemma 11.3 (Energy conservation). *The sum of the maximum and the minimum error is independent of the coordinate system chosen for symmetry investigation of the image* f:

$$I_{11} = e(\theta_{\max}) + e(\theta_{\min}) = \int |\daleth_{\xi,\eta}(f)(\xi,\eta)| d\xi d\eta$$

$$= \int |\daleth(f)(x,y)| dx dy = \int \left(\frac{\partial f}{\partial x}\right)^2 + \left(\frac{\partial f}{\partial y}\right)^2 dx dy. \qquad (11.47)$$

◆

This suggests the interesting property[5] that I_{11} can be computed even without knowledge of ξ. This is because I_{11}, which is the upper bound of $|I_{20}|$ such that $|I_{20}| \leq I_{11}$ attains the upper bound with equality if and only if the image has linear symmetry w.r.t. ξ, η curve family, is independent of the ξ, η. Accordingly, it follows that the upper bound of $|I_{20}|$ is the same for all harmonic CTs, I_{11} computed for f w.r.t. the (ordinary Cartesian) linear symmetry. An image cannot be both linear symmetric w.r.t. (ordinary Cartesian) lines obtained as the linear combinations of $\xi = x$ and $\eta = y$ and w.r.t. another harmonic curve family at the same time. This means that it is possible to construct independent shape properties of f by measuring I_{20}s w.r.t. different CTs, ξ, η.

In summary, the elements of the GST can be found by filtering $\daleth(f)$ as stated in the following theorem [18]:

Theorem 11.1 (Generalized structure tensor). *The structure tensor theorem holds in harmonic coordinates. In particular, the second-order complex moments determining the minimum inertia axis of the power spectrum,* $|F(\omega_\xi, \omega_\eta)|^2$, *can be obtained in the (Cartesian) spatial domain as:*

$$I_{20} = (\lambda_{\max} - \lambda_{\min})e^{i2\varphi_{\min}} = \iint (\omega_\xi + i\omega_\eta)^2 |F|^2 d\omega_\xi d\omega_\eta \qquad (11.48)$$

$$= \iint \daleth_{\xi,\eta}(\xi,\eta) d\xi d\eta = \iint \frac{\daleth_{x,y}^*(\xi)}{|\daleth_{x,y}^*(\xi)|} \daleth_{x,y}(f) dx dy \qquad (11.49)$$

$$= \iint \exp(i \arg \daleth_{x,y}^*(\xi)) \daleth_{x,y}(f) dx dy \qquad (11.50)$$

$$I_{11} = \lambda_{\max} + \lambda_{\min} = \iint (\omega_\xi + i\omega_\eta)(\omega_\xi - i\omega_\eta)|F|^2 d\omega_\xi d\omega_\eta \qquad (11.51)$$

$$= \iint |\daleth_{\xi,\eta}(f)| d\xi d\eta = \iint |\daleth_{x,y}(f)| dx dy \qquad (11.52)$$

The quantities $\lambda_{\min}$, $\varphi_{\min}$, *and* $\lambda_{\max}$ *are, respectively, the minimum inertia, the direction of the minimum inertia axis, and the maximum inertia of the power spectrum of the harmonic coordinates,* $|F(\omega_\xi, \omega_\eta)|^2$.

◆

---

[5] Note that this conclusion is valid for the continuous representation. In the discrete case the integral still needs a ξ-dependent kernel as will be discussed further below.

We emphasize that it is the coordinate transformation that determines what I_{20} and I_{11} represent and detect. Central to the generalized structure tensor is the harmonic function pair $\xi(x, y)$ and $\eta(x, y)$, which creates new coordinate curves to represent the points of the 2D plane. An image $f(x, y)$ can always be expressed by such a coordinate pair $\xi(x, y)$ and $\eta(x, y)$ as long as the transformation from (x, y) to (ξ, η) is one-to-one and onto. The deformation by itself does not create new gray tones, i.e., no new function values of f are created, but rather it is the isogray curves of f that are deformed. The harmonic coordinate transformations deform the appearance of the target patterns to make the detection process mathematically more tractable. In the principle suggested by theorem 11.1, these transformations are not applied to an image because they are implicitly encoded in the utilized complex filters. The deformations occur only in the idea, when designing the detection scheme and computing the filters.

11.4 Discrete Approximation of GST

For computation of the generalized structure tensor parameters, the ILST of the image f,

$$\daleth(f)(x, y) = (D_x f + i D_y f)^2 \tag{11.53}$$

is needed. This is, however, the same complex image that is used to compute the ordinary structure tensor, discussed in Eq. (10.79) and in theorem 10.4. Accordingly, provided that the image is densely sampled to a sufficient degree, the ILST image can be obtained through:

$$\daleth(f)(x, y) = (D_x f(x, y) + i D_y f(x, y))^2 \big|_{(x,y)=(x_j, y_j)} = [(f_{x_j} + i f_{y_j})]^2 \tag{11.54}$$

where (x_j, y_j) is a point on the grid on which the original discrete image f is defined. Below we discuss how to estimate the GST elements on the discrete Cartesian grid, (x_j, y_j).

Case 1: Estimation of I_{20} with known analytic expression of $\xi(x, y)$

Assuming that an analytic expression of $\xi(x, y)$ is known explicitly, it follows from this that $\daleth(\xi)(x, y)$ is known. An approximation of I_{20} can be obtained by substituting the reconstructed (from its samples) $\daleth(f)$,

$$\daleth(f)(x, y) = \sum_j \psi(x - x_j, y - y_j) \daleth(f)(x_j, y_j) \tag{11.55}$$

into Eq. (11.45)

$$
\begin{aligned}
I_{20} &= \sum_j \daleth(f)(x_j, y_j) \int \psi(x - x_j, y - y_j) [\frac{\daleth(\xi)(x, y)}{|\daleth(\xi)(x, y)|}]^* dx dy \\
&= \sum_j \daleth(f)(x_j, y_j)(w_j^{20})^*
\end{aligned}
\tag{11.56}
$$

where w_j^{20} is given by:

$$w_j^{20} = \int \psi(x - x_j, y - y_j) \exp(i2 \tan^{-1}(\xi_x, \xi_y)) dx dy \qquad (11.57)$$

Here, we observe that the discrete kernel of I_{20}, which is w_j^{20}, is obtained by projecting the continuous kernel, Eq. (11.45),

$$w^{20}(x, y) = \frac{\daleth(\xi)(x, y)}{|\daleth(\xi)(x, y)|} = \exp(i2 \tan^{-1}(\xi_x, \xi_y)) \qquad (11.58)$$

onto the space of band-limited signals. We note that the continuous kernel has modulus 1 except at $\nabla \xi(x, y) = 0$, where $\daleth(\xi)$ is undefined. At these points, w^{20} can safely be assumed to be 0, as the values of the integrand on a set of points with zero measure do not affect I_{20}. Technically, Eq. (11.57) is a lowpass filtering followed by discretization, which is also known as perfect sampling. Thus, Eq. (11.56) is essentially a matching of the direction of the basis tangent vector field with the tangent vector field of the image. This observation will prove to be useful in Sect. (11.4). Naturally, the closed form of the integral in Eq. (11.57) is not possible to obtain for most ξs. However, w_j^{20} can be computed numerically and off line, e.g. [19], for pattern recognition purposes. In one important case, when $\xi = x$, though, Eq. (11.57) can be derived analytically and reduces to $w_j^{20} = 1$ which gives the ordinary structure tensor kernel for the image. Assuming that we will need I_{20} for a local image, w_j^{20} will however, be a window function, e.g., a Gaussian. There are other nontrivial $\xi(x, y)$s yielding analytically tractable kernels, those with polynomial derivatives. We will discuss this class further in Sect. 11.7.

Is it really worth computing w_j^{20} exactly through Eq. (11.57) to obtain a useful approximation of I_{20}? The answer to this question depends on the application at hand. The computation of w_j^{20} through Eq. (11.57) and then substituting it in Eq. (11.56) yields robust approximations of I_{20} since the weight zero is automatically given to the appropriate points of the kernel at the same time as all kernel coefficients vary smoothly. If the digitized image f_j, represents a small neighborhood, it might be worth computing w_j^{20} in the aforementioned "orthodox" fashion (i.e., by projection on the band-limited functions), as this yields less biased estimates. However, this may not be worth doing if the number of singularity points is negligible compared to the total number of the image points, since the bias of the singularity points will be negligible. In this case one can use the approximation:

$$w_j = \begin{cases} \exp[i2 \tan^{-1}(\xi_x, \xi_y)]|_{x=x_j, y=y_j}, & \text{if } \nabla \xi(x_j, y_j) \neq 0 \, ; \\ 0, & \text{if } \nabla \xi(x_j, y_j) = 0. \end{cases} \qquad (11.59)$$

Case 2: Estimation of I_{20} with unknown analytic expression of ξ

Often a digital image of a prototype pattern $\tilde{\xi}_j$ is all that is known, and one would like to know whether $\tilde{\xi}_j$ occurs in a discrete image f_j or not. One cannot assume that isocurves of the prototype are sampled isocurves of harmonic functions, i.e., the isocurves fulfill the Laplacian equation, as this is difficult to verify for an arbitrary digital prototype, and it is not very helpful when $\tilde{\xi}$ is not strictly harmonic. The tilde in $\tilde{\xi}_j$ is used in order to emphasize that we do not know whether $\tilde{\xi}$ is harmonic or not. We can exploit the fact that the computation of I_{20} is essentially a matching between the ILST image and the sampled version of the normalized ILST of the prototype. As an algorithm implementing the discrete ILST operator according to Eq. (11.54) is assumed to exist, we can apply such an algorithm to $\tilde{\xi}_j$ to obtain:

$$\daleth(\tilde{\xi})(x_j, y_j) = [(\tilde{\xi}_{x_j} + i\tilde{\xi}_{y_j})]^2, \tag{11.60}$$

We can then proceed as if $\tilde{\xi}_j$ is a sampled harmonic function and find the kernel w_j^{20} as

$$w_j^{20} = \begin{cases} \exp(i2\tan^{-1}(\tilde{\xi}_{x_j}, \tilde{\xi}_{y_j})) & \text{if } \nabla\xi(x_j, y_j) \neq 0; \\ 0, & \text{if } \nabla\xi(x_j, y_j) = 0. \end{cases} \tag{11.61}$$

Accordingly, in analogy with Eq. (11.56), I_{20} can be computed as

$$I_{20} = \sum_j \daleth(f)(x_j, y_j)(w_j^{20})^* \tag{11.62}$$

The classical alternative to the case in this subsection is to directly match (correlate) the two digital images, f_j and $\tilde{\xi}_j$, without filtering them through the ILST operator. However, matching the ILST image with an appropriate kernel has certain advantages. In the ILST approach it is the edges of f_j and not the gray values which are aligned in case of match. As a consequence we can expect a high localization. However, this is only a byproduct; the main advantage is the complex voting process and its rich interpretability, as will be discussed in Sect. 11.5.

Case 3: Estimation of I_{11}

According to Eq. (11.47), I_{11} is obtained as

$$I_{11} = \int |\daleth(f(x, y))| \cdot 1 dxdy = \int |\daleth(f)(x, y)||w^{20}(x, y)| dxdy \tag{11.63}$$

Consequently, the continuous kernel of I_{11} is

$$w^{11} = |w^{20}| \tag{11.64}$$

By using the reconstructed $|\daleth(f)|$, and in analogy with Eq. (11.56),

$$\bar{I}_{11} = \sum_j |\daleth(f)(x_j, y_j)|\bar{w}_j^{11} \tag{11.65}$$

where the discrete kernel,

$$\bar{w}_j^{11} = \int \psi(x - x_j, y - y_j)|w^{20}(x, y)| dxdy \tag{11.66}$$

is the projection of the continuous kernel $w^{11}(x, y)$ onto the space of band-limited signals. Using the kernel $\bar{w}_j^{11}$ in Eq. (11.65) yields a biased estimate of I_{11} as shown below. The discrete kernel $\bar{w}_j^{11}$ is the perfect sampling of $w^{11}(x, y)$. The magnitudes of the two kernels fulfill the inequality $|w_j^{20}| \leq \bar{w}_j^{11}$ which is a weaker relationship than the equality relationship in the continuous case, Eq. (11.64). Consequently, when $|w_j^{20}| < \bar{w}_j^{11}$ even for one kernel coefficient j, we get

$$|I_{20}| = |\sum_j \daleth(f)(x_j, y_j) w_j^{20*}| \leq \sum_j |\daleth(f)(x_j, y_j)||w_j^{20}|$$
$$< \sum_j |\daleth(f)(x_j, y_j)|\bar{w}_j^{11} = \bar{I}_{11} \tag{11.67}$$

This implies that $|I_{20}|$ will not attain the value $\bar{I}_{11}$ even if the image is linearly symmetric in harmonic coordinates. By using the triangle inequality, it can be shown that $|I_{20}|$ will attain the upper bound only when the discrete ILST of the image and the kernel coefficients are collinear. This behavior is similar to the continuous case, as can be seen by applying the triangle inequality to Eq. (11.45) and comparing the result to (11.47). In order to avoid the bias introduced by the discretization process, we will use the discrete kernel,

$$w_j^{11} = |w_j^{20}| \tag{11.68}$$

instead of Eq. (11.66) to compute

$$I_{11} = \sum_j |\daleth(f)(x_j, y_j)||w_j^{20}| \tag{11.69}$$

where w^{20} is assumed to be available through either of the processes described in Eqs. (11.57), (11.59), and (11.61).

11.5 The Generalized Hough Transform (GHT)

When an analytic expression for a target curve or a collection of curves is not available, provided that there is a discrete version of the target, $\tilde{\xi}$, it is still possible to detect it in discrete images. However, in this section we consider the alternative approach, the *generalized Hough transform* (GHT). In the subsequent section, we will establish that GST is a GHT, with the additional capability to recognize antitargets *during* the target recognition. We will dwell only on the case when an analytic expression for the target curve is not available. The discussion when the target curve is analytically available is analogous and is omitted.

The chief tool to achieve machine recognition of general curves in images is the GHT [12, 111] which has been extensively studied [54, 118, 126, 177]. GHT is popular for its robustness, because even when the occurrence of a target curve is only partial in an image, for some reason, including occlusion, the method can find the

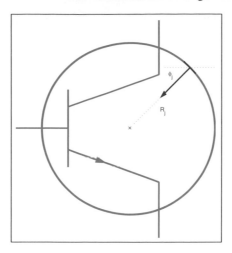

Fig. 11.4. A collection of edges describing an object to be identified is shown in *green*. An edge with the direction ϕ_j and the distance R_j to the reference point, marked with a ×, are shown in *magenta*. The latter information is used by GHT to encode the position of the prototype. All edge matches occuring in the target and the image result in a vote cast to the corresponding reference point in the image

presence of target curves. It consists of a table look-up indexed by the direction of the target curve and a voting procedure, [12, 111]. The table entries are discrete edge directions of the edges of the prototype. To each tabulated direction corresponds a list of edge positions, expressed relative to a unique reference point (origin) of the prototype. There is a direction associated with each contour edge, both in the image and in the prototype. Typically the contour points and their directions are obtained after applying a threshold to the gradient-filtered image and the quantized gradient directions. This is because edges participating in the GHT poll are binary, i.e., they either exist with a unique direction, or they do not. Given the table, one can easily construct the *contours of the prototype,* or inversely construct the table, given the contours of the prototype. The contours of a transistor, an example prototype, its reference point, an edge, and its (ϕ_j, R_j) parameters used in a GHT table are shown in Fig. 11.4. Accordingly, the GHT direction table is equivalent to a 2D template containing the contour points and their directions. The GHT procedure consists of a 2D nonlinear filtering identical to sliding a template over the contour-direction image and counting the direction matches per template position. Thus, the vote accumulator, A, is two-dimensional and has usually the same resolution as the image itself in this process. However, it may be necessary to estimate more than two parameters, e.g., translation, rotation, and scale. This can be handled by the 2D accumulators too, provided that one can fix the other parameters by use of a numerical optimization procedure, such as gradient descent. The peaks in the accumulator A will indicate the positions where the template presence is most likely. The accumulator votes are given by

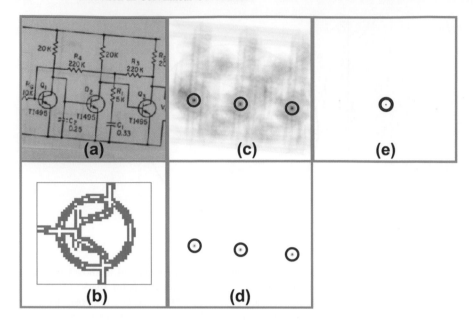

Fig. 11.5. Electronic circuit and transistor detection. *a,b.* the original and the prototype. *c,d.* $|I_{20}|$ and I_{11}. *e.* GHT accumulator with $2\pi/256$ angle resolution (from [22]). The *circle* is inserted to show the location of the peak

$$A(x_j, y_j) = \sum_l \delta(\theta(x_j + x_l, y_j + y_l) - \phi(x_l, y_l)) \qquad (11.70)$$

where l runs over the edge elements of the prototype, and $\phi(x_l, y_l)$ is the direction of the prototype edge. The $\theta(x_j + x_l, y_j + y_l)$ corresponds to the direction of the edge at the position $(x_j + x_l, y_j + y_l)$ in the contour image, whenever there is an edge pixel, or $i\infty$. That θ is defined to be $i\infty$ is purely symbolic and serves to make θ different from ϕ, because expected contours modeled by the prototype may be missing in the contour image. In such instances this will generate a zero contribution (vote) from the δ-function. The array $A(x_j, y_j)$ represents the vote accumulator and can be constructed on the same grid as the image itself. Figure 11.5 shows the result of detection of a transistor in an electronic circuit picture.

11.6 Voting in GST and GHT

The parameter estimation of the generalized structure tensor is more than a correlation of edge magnitudes. It carefully takes into account the directions of the edges, too. Here we show that the GST detection extends GHT, to a complex GHT in which the votes are allowed to assume complex values.

The GHT accumulator, $A(x_j, y_j)$, can be compared to an estimation of I_{20} at local images around all image points,

$$I_{20}(x_j, y_j) = \sum_l \daleth(f)(x_j + x_l, y_j + y_l)w^{20*}(x_l, y_l)$$

$$= \sum_l \exp[i2\theta(x_j + x_l, y_j + y_l) - i2\phi(x_l, y_l)] \qquad (11.71)$$

The latter assumes that the same unitary gradient of the image and the unitary gradient of the prototype as those used in the computation of A are employed to obtain $\daleth(f)(x_j, y_j)$ and $w^{20}(x_j, y_j)$, via Eq. (11.61), and Eq. (11.62). Consistent with Eq. (11.70), $\theta = i\infty$ whenever it is not defined, i.e., when a pixel position does not represent an edge pixel position, the vote contribution of that point to I_{20} is reduced to 0. While $A(x_j, y_j)$ is the count of the positive matches only, $|I_{20}|$ is the positive matching score *adjusted downwards* with the amount of negative matches. At the reference point, when the image is the same as the prototype itself, we obtain the maximum match with both of the techniques:

$$A = L, \qquad I_{20} = L, \qquad (11.72)$$

where L represents the number of edge pixels in the prototype. When 100% of the edge directions mismatch maximally i.e., when all prototype directions are orthogonal to the image directions, we obtain:

$$A = 0 \qquad I_{20} = -L \qquad (11.73)$$

while if 50% of the edge directions match perfectly and 50% mismatch maximally we obtain,

$$A = \frac{L}{2}, \qquad I_{20} = 0. \qquad (11.74)$$

In computing $I_{20}(x_j, y_j)$ the scores of the positions with contradictory gradient directions will be reduced compared to the GHT scores, $A(x_j, y_j)$. The GHT is without score reduction since a score of $A(x_j, y_j)$ is only allowed to increase (in case of gradient direction match) or it is unchanged (in case of mismatch). For GHT this is a necessity, because negative scores would not be meaningful in case they were given to empty accumulators, while in the case of I_{20} this is allowed as this simply corresponds to a vote for a pattern which is locally orthogonal to the prototype, *antiprototype*. Clearly, the computation of I_{20} is a voting process in which not only negative but also complex votes are allowed.

Formally, nonprototype ILST images can be generated by a phase shift of the prototype ILST image:

$$\exp[i\varphi_0]\exp(i2\tan^{-1}(\tilde{\xi}_{x_j}, \tilde{\xi}_{y_j})) \qquad (11.75)$$

But to which real patterns $\varphi_0 \neq 0$ corresponds, as this new ILST image is a purely synthetic construct, is not obvious. This is because a phase shift of the prototype ILST may result in tangent fields that are not always imaginable or intelligible by visual inspection of Eq. (11.75). To give every φ_0 an exact meaning, i.e., to find a ζ_j that approximates $\tilde{\xi}_j$, an estimation should be made by numerical methods. For

z^{-1}	$z^{-0.5}$	$\log(z)$	$z^{0.5}$	z^1	$z^{1.5}$	z^2	$z^{2.5}$
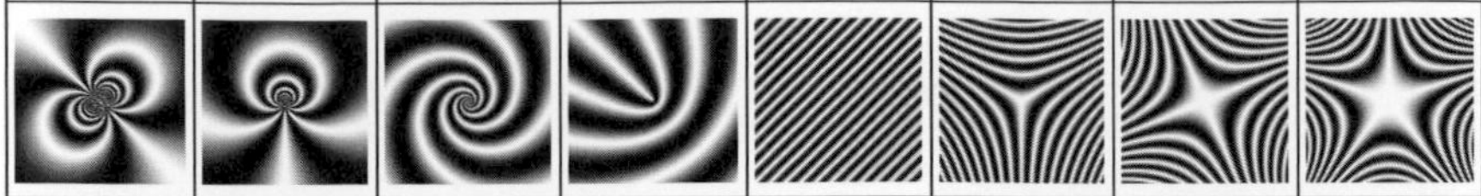							
$\Gamma\{-4,\sigma^2\}$	$\Gamma\{-3,\sigma^2\}$	$\Gamma\{-2,\sigma^2\}$	$\Gamma\{-1,\sigma^2\}$	$\Gamma\{0,\sigma^2\}$	$\Gamma\{1,\sigma^2\}$	$\Gamma\{2,\sigma^2\}$	$\Gamma\{3,\sigma^2\}$
$n=-4$	$n=-3$	$n=-2$	$n=-1$	$n=0$	$n=1$	$n=2$	$n=3$

Fig. 11.6. The *top row* shows the harmonic functions, Eqs. (11.87), that generate the patterns in the *second row*. The isocurves of the images are given by a linear combination of the real and the imaginary parts of the harmonic functions on the *top* according to Eq. (11.88) with a constant parameter ratio, i.e., $\varphi = \tan^{-1}(a,b) = \frac{\pi}{4}$. The *third row* shows the filters that are tuned to detect these curves for any φ, while the *last row* shows the symmetry order of the filters

pattern recognition purposes, however, this will not be necessary given the control possibility the I_{11} estimation offers. If, for an image for which $0 \ll |I_{20}| \approx I_{11}$, $\varphi_0 = \arg I_{20} \neq 0$ is obtained, the nonprototype is known in reality too (as the gradients come from a real image). Thus, in practice only when $|I_{20}| \ll I_{11}$ may pose interpretation difficulties of $\arg I_{20}$, in which case no member of this class is a good fit to the data anyway.

Both the Hough accumulator value A and the value of I_{20} will be maximal and identical to each other when there is maximal match between the prototype and the image edges (i.e in GST terms when $\arg(I_{20}) = 0$ and $|I_{20}| = I_{11}$). However, because of the complex votes, the two GST measurements, I_{20} and I_{11}, additionally offer detection of other prototypes not detected by the Hough transform, e.g., the antiprototype when $\arg(I_{20}) = \pi$ and $|I_{20}| = I_{11}$. We summarize our findings in the following lemma.

Lemma 11.4. *The GST is GHT with complex votes. Except for a possible vote reduction due to edge direction mismatch, the GHT accumulator value A is equivalent to I_{20} in the 0 radian direction, i.e., GST will only sharpen the accumulator peaks of the GHT. The GST parameter I_{20} in other directions than 0 radian, along with I_{11}, can detect and identify other prototypes not detected by GHT.*

♦

11.7 Harmonic Monomials

Here we will discuss a specific harmonic function class with member functions having direction fields that are monomials of z. As will be shown, this class of harmonic function families is easily found analytically while they constitute computationally powerful models to process symmetric patterns in images.

Assuming $z = x + iy$, we will study those

$$g(z) = \xi(x,y) + i\eta(x,y) \tag{11.76}$$

such that

$$\daleth^*(\xi)(x,y) = (D_x\xi - iD_y\xi)^2 = z^n, \qquad \text{with} \qquad n = 0, \pm1, \pm2, \cdots . \quad (11.77)$$

This study is motivated because, to estimate I_{20} via theorem 11.1, $\daleth^*(\xi)$ is needed[6] and it makes sense to know what kind of curve families the ILST fields are projected on,

$$e^{i\,\arg([(D_x - iD_y)\xi]^2)} = e^{i\,\arg([\frac{dg}{dz}]^2)} = e^{i\,\arg(z^n)} = \frac{z^n}{|z^n|} = e^{in\,\arg(x+iy)}(11.78)$$

First, we establish a relationship between the operator[7] $D_x - iD_y$ and complex derivatives as follows:

$$(D_x - iD_y)\Re[g(z)] = D_x[\Re g(z)] - iD_y[\Re g(z)] \qquad (11.79)$$

$$= \Re[D_x g(z)] - i\Re[D_y g(z)] \qquad (11.80)$$

$$= \Re\left(\frac{dg}{dz}\frac{dz}{dx}\right) - i\Re\left(\frac{dg}{dz}\frac{dz}{dy}\right) \qquad (11.81)$$

$$= \Re\left(\frac{dg}{dz}\right) - i\Re\left(\frac{dg}{dz}i\right) \qquad (11.82)$$

$$= \Re\left(\frac{dg}{dz}\right) - i\Re\left[i\Re\left(\frac{dg}{dz}\right) - \Im\left(\frac{dg}{dz}\right)\right] \qquad (11.83)$$

$$= \Re\left(\frac{dg}{dz}\right) + i\Im\left(\frac{dg}{dz}\right) = \frac{dg}{dz} \qquad (11.84)$$

Thus, we obtain:

$$\daleth^*(\xi(x,y)) = (\frac{dg(z)}{dz})^2 = z^n, \qquad \text{with} \qquad n = 0, \pm1, \pm2, \cdots . \quad (11.85)$$

and establish

$$g(z) = z^{\frac{b}{2}}, \qquad \text{with} \qquad n = 0, \pm1, \pm2 \cdots \qquad (11.86)$$

as a solution. We integrate $\frac{dg}{dz} = z^{\frac{n}{2}}$ to obtain the real and imaginary parts of g,

$$g(z) = \begin{cases} \frac{1}{\frac{n}{2}+1}z^{\frac{n}{2}+1}, & \text{if } n \neq -2; \\ \log(z), & \text{if } n = -2. \end{cases} \qquad (11.87)$$

The scheme discussed in Sects. 11.3 and 11.4 detects the patterns that are generated by real and imaginary parts of $g(z)$. Such patterns are shown in Fig. 11.6 by gray modulation:

$$s(a\xi + b\eta) = \cos(a\Re[g(z)] + b\Im[g(z)]) \qquad (11.88)$$

The 1D function $s(t) = \cos(t)$ is chosen for illustration purposes. The filters that are tuned to detect the isocurves $a\xi + b\eta$ are not sensitive to s, but to the angle

$$\varphi = \tan^{-1}(a, b). \qquad (11.89)$$

---

[6] Since ξ is harmonic, to study η does not represent freedom of choice but is determined as soon as ξ is given.

[7] The properties of such linear operators will be discussed in further detail in Sect. 11.9

11.8 "Steerability" of Harmonic Monomials

In Fig. 11.6, the angle is fixed to $\varphi = \frac{\pi}{4}$, and n is varied between -4 and 3. Each n represents a separate isocurve family. By changing φ and keeping n fixed, the parameter pair (a, b) is rotated to (a', b'). Except for the patterns with $n = -2$, which we will come back to next, this results in rotating the isocurves, since for $n \neq -2$ and $g(z) = z^{\frac{n}{2}+1}$ we have:

$$a'\xi + b'\eta = \Re[(a' - ib')(\xi + i\eta)] = \Re[(a' - ib')g(z)] \tag{11.90}$$

$$= \Re[(a - ib)e^{i\varphi}z^{\frac{n}{2}+1}] = \Re[(a - ib)g(ze^{i\frac{1}{\frac{n}{2}+1}\varphi})] \tag{11.91}$$

$$= a\xi' + b\eta' \tag{11.92}$$

Here ξ' and η' are rotated versions of the harmonic pair ξ and η, so that $g(z \exp^{i\varphi_0}) = \xi' + i\eta'$ for some φ_0. The top row of Fig. 11.3 displays the curves generated by Eq. (11.88), for increasing values of φ in $(a, b) = (\cos(\varphi), \sin(\varphi))$, to illustrate that a coefficient rotation results in a pattern rotation.

When $n = -2$, we obtain the isocurves via the function

$$g(x + iy) = \log(|x + iy|) + i \arg(x + iy)$$

which is special in that it represents the only case when a change of the ratio between a and b does not result in a rotation of the image pattern. Instead, changing the angle φ bends the isocurves,

$$\cos(\varphi) \log(|x + iy|) + \sin(\varphi) \arg(x + iy) = \text{constant} \tag{11.93}$$

That is, the spirals become "tighter" or "looser" until the limit patterns, circles, and radial patterns, corresponding to infinitely tight and infinitely loose spirals, are reached.

The relationships (11.90)-(11.92) show that the isocurves of the patterns that are modeled by harmonic monomials are obtained as a linear combination of the non-rotated isocurves, except $g(z) = \log(z)$. Yet half of these patterns fail to fulfill the steerability condition [75, 180]. The steerability condition foresees that the angular Fourier series expansion function, f, must have a limited number of elements to allow steering by linear weighting of basis elements. In our case, $g(z) = z^{\frac{n}{2}+1}$ does not satisfy the steerability condition on angular band-limitedness when n is odd since it is not possible to expand odd powers of a square root with a limited number of (integer) angular frequencies. The same holds evidently for the isocurve family represented by the CT $\log(z)$, the members of which cannot be rotated by changing the linear coefficients. Consequently, the patterns with odd n or with $n = -2$ cannot be generated by weighted sums of a low number of *steerable functions*. In turn, this makes it impossible to detect the mentioned patterns by correlating the *original gray images* with steerable filters. Yet, these patterns can be accurately generated by *analytic functions*.[8] They can even be detected by steerable filters, as will be shown

---

[8] Analytic functions are harmonic, but they do not necessarily meet the steerablity condition of [75, 180].

in Sect. 11.10, though not by gray image correlations as these filters were originally intended for, but via GST field correlations. We can conclude that the angular band-limitedness condition is a sufficient but not a necessary condition for steering the rotations of 2D patterns or for their detection.

11.9 Symmetry Derivatives and Gaussians

In this section we describe a set of tools that will be useful when recognizing intricate patterns that certain harmonic curve families and the Lie operators are capable representing.

Definition 11.5 (Symmetry derivative). *We define the first symmetry derivative as the complex partial derivative operator:*

$$D_x + iD_y = \frac{\partial}{\partial x} + i\frac{\partial}{\partial y} \tag{11.94}$$

The symmetry derivative resembles the ordinary gradient in 2D. When it is applied to a scalar function $f(x, y)$, the result is a complex field instead of a vector field. Consequently, the first important difference is that it is possible to take the (positive integer or zero) powers of the symmetry derivative, e.g.,,

$$(D_x + iD_y)^2 = (D_x^2 - D_y^2) + i(2D_xD_y) \tag{11.95}$$
$$(D_x + iD_y)^3 = (D_x^3 - 3D_xD_y^2) + i(3D_x^2D_y - D_y^3) \tag{11.96}$$
$$\cdots$$

Second, being a complex scalar, it is even possible to exponentiate the result of the symmetry derivative, i.e., $(D_x + iD_y)^n f$, to yield nonlinear functionals:

$$[(D_x + iD_y)^n f]^m$$

Had it not been for the two mentioned exponentiation properties, there would not be any real reason to introduce the symmetry derivatives concept, because the ordinary gradient operator, ∇, would be sufficient in practice. As will be seen, however, the symmetry derivatives posses properties that make them elegant tools when modeling and detecting intricate patterns.

The operator $(D_x + iD_y)^n$ will be defined as the *nth order symmetry derivative* since its invariant patterns (those that vanish under the linear operator) are highly symmetric. In an analogous manner, we define, for completeness, *the first conjugate symmetry derivative* as $D_x - iD_y = \frac{\partial}{\partial x} - i\frac{\partial}{\partial y}$ and the *nth conjugate symmetry derivative* as $(D_x - iD_y)^n$. We will, however, only dwell on the properties of the symmetry derivatives. The extension of the results to conjugate symmetry derivatives are straightforward.

Definition 11.6. *We apply the pth symmetry derivative to the Gaussian and define the function* $\Gamma^{\{p,\sigma^2\}}$ *as*

$$\Gamma^{\{p,\sigma^2\}}(x,y) = (D_x + iD_y)^p \frac{1}{2\pi\sigma^2} e^{-\frac{x^2+y^2}{2\sigma^2}} \qquad (11.97)$$

with $\Gamma^{\{0,\sigma^2\}}$ *being the ordinary Gaussian.*

Theorem 11.2. *The differential operator* $D_x + iD_y$ *and the scalar* $\frac{-1}{\sigma^2}(x+iy)$ *operate on a Gaussian in an identical manner:*

$$(D_x + iD_y)^p \Gamma^{\{0,\sigma^2\}} = (\frac{-1}{\sigma^2})^p (x+iy)^p \Gamma^{\{0,\sigma^2\}} \qquad (11.98)$$

$\blacklozenge$

The theorem, proved in the Appendix (Sect. 11.13), reveals an invariance property of the Gaussians w.r.t. symmetry derivatives. We compare the second-order symmetry derivative with the classical Laplacian, also a second-order derivative operator, to illustrate the analytical consequences of the theorem. The Laplacian of a Gaussian,

$$(D_x^2 + D_y^2)\Gamma^{\{0,\sigma^2\}} = (-\frac{2}{\sigma^2} + \frac{x^2+y^2}{\sigma^4})\Gamma^{\{0,\sigma^2\}} \qquad (11.99)$$

can obviously not be obtained by a mnemonic replacement of the derivative symbols D_x with x and D_y with y in the Laplacian operator. As the Laplacian already hints, with an increased order of derivatives, the resulting polynomial factor, e.g.,, the one on the right-hand side of Eq. (11.99), will resemble less and less the polynomial form of the derivation operator. Yet, it is such a form invariance that the theorem predicts when symmetry derivatives are utilized. By using the linearity of the derivation operator, the theorem can be generalized, see Sect. 11.13, to any polynomial as follows:

Lemma 11.5. *Let the polynomial* Q *be defined as* $Q(q) = \sum_{n=0}^{N-1} a_n q^n$. *Then*

$$Q(D_x + iD_y)\Gamma^{\{0,\sigma^2\}}(x,y) = Q(\frac{-1}{\sigma^2}(x+iy))\Gamma^{\{0,\sigma^2\}}(x,y) \qquad (11.100)$$

$\blacklozenge$

That the Fourier transformation of a Gaussian is also a Gaussian has been known and exploited in information sciences. It turns out that a similar invariance is valid for symmetry derivatives of Gaussians, too.

A proof of the following theorem is omitted because it follows by observing that derivation w.r.t. x corresponds to multiplication with $i\omega_x$ in the Fourier domain and applying Eq. (11.98). Alternatively, theorem 3.4 of [208] can be used to establish it.

Theorem 11.3. *The symmetry derivatives of Gaussians are Fourier transformed on themselves, i.e.,*

$$\mathcal{F}[\Gamma^{\{p,\sigma^2\}}] = \iint \Gamma^{\{p,\sigma^2\}}(x,y)e^{-i\omega_x x - i\omega_y y}dxdy$$

$$= 2\pi\sigma^2 (\frac{-i}{\sigma^2})^p \Gamma^{\{p,\frac{1}{\sigma^2}\}}(\omega_x,\omega_y) \qquad (11.101)$$

◆

We note that, in the context of prolate spheroidal functions [204], and when constructing rotation-invariant 2D filters [52], it has been observed that the (integer) symmetry order n of the function $h(\rho)\exp(in\theta)$, where h is a one-dimensional function, ρ and θ are polar coordinates, is preserved under the Fourier transform. Accordingly, the Fourier transforms of such functions are: $H_0[h(\rho)](\omega_\rho)$, where H_0 is the Hankel transform (of order 0) of h. However, a further precision is needed as to the choice of the function family h, to render it invariant to Fourier transform.

Another analytic property that can be used to construct efficient filters by cascading smaller filters or simply to gain further insight into steerable filters or rotation-invariant filters is the addition rule under the convolution. This is stated in the following theorem and is proved in the Appendix.

Theorem 11.4. *The symmetry derivatives of Gaussians are closed under the convolution operator so that the order and the variance parameters add under convolution:*

$$\Gamma^{\{p_1,\sigma_1^2\}} * \Gamma^{\{p_2,\sigma_2^2\}} = \Gamma^{\{p_1+p_2,\sigma_1^2+\sigma_2^2\}} \qquad (11.102)$$

◆

11.10 Discrete GST for Harmonic Monomials

We discussed above how analytic functions $g(z)$ generate curve families via a linear combination of their real, $\xi = \Re[g(z)]$, and imaginary, $\eta = \Im[g(z)]$ parts. The curve families generated by the real part are locally orthogonal to those of the imaginary part. Here we discuss a specific family. The next lemma, a proof of which is given in the Appendix, makes use of the symmetry derivatives of Gaussians to represent and to sample the generalized structure tensor. Sampled functions are denoted as f_k, i.e., $f_k = f(x_k,y_k)$, as before.

Lemma 11.6. *Consider the analytic function $g(z)$ with $\frac{dg}{dz} = z^{\frac{n}{2}}$, and let n be an integer, $0, \pm 1, \pm 2 \cdots$. Then the discretized filter $\Gamma_k^{\{n,\sigma_2^2\}}$ is a detector for patterns generated by the curves $a\Re[g(z)] + b\Im[g(z)] = $ constant, provided that a shifted Gaussian is assumed as an interpolator and the magnitude of a symmetry derivative of a Gaussian acts as a window function. The discrete scheme*

$$I_{20}(|F(\omega_\xi,\omega_\eta)|^2) = C_n\Gamma_k^{\{n,\sigma_2^2\}} * (\Gamma_k^{\{1,\sigma_1^2\}} * f_k)^2 \qquad (11.103)$$

$$I_{11}(|F(\omega_\xi,\omega_\eta)|^2) = C_n|\Gamma_k^{\{n,\sigma_2^2\}}| * |\Gamma_k^{\{1,\sigma_1^2\}} * f_k|^2 \qquad (11.104)$$

where $0 \leq n$ and C_n is a real constant, estimates the direction parameter $\tan^{-1}(a, b)$ *as well as the error via* I_{20} *and* I_{11} *according to theorem 11.1. For* $n < 0$ *the following scheme yields the analogous estimates:*

$$I_{20}(|F(\omega_\xi, \omega_\eta)|^2) = C_n \Gamma_k^{*\{n, \sigma_2^2\}} * (\Gamma_k^{\{1, \sigma_1^2\}} * f_k)^2 \qquad (11.105)$$

$$I_{11}(|F(\omega_\xi, \omega_\eta)|^2) = C_n |\Gamma_k^{*\{n, \sigma_2^2\}}| * |\Gamma_k^{\{1, \sigma_1^2\}} * f_k|^2 \qquad (11.106)$$

where $\Gamma^{*\{n, \sigma_2^2\}} = (\Gamma^{\{n, \sigma_2^2\}})^*.$

◆

We note that the parameter C_n is constant w.r.t. to (x_l, y_l) and has no implications to applications because it can be assumed to have been incorporated to the image the filter is applied to. In turn, this amounts to a uniform scaling of the gray-value gamut of the original image. There are two parameters employed by the suggested scheme that control filter sizes: σ_1, which is the same as in the ordinary structure tensor, determining how much of the high frequencies are assumed to be noise; and σ_2, representing the size of the neighborhood.

Equations (11.103) and (11.105) can be implemented via separable convolutions since the filters $\Gamma_k^{\{n, \sigma_2^2\}}$ are separable for all n. The same goes for Eqs. (11.104) and (11.106), provided that n is even. Consequently, for even n, both I_{20} and I_{11} can be computed with 1D filters. For odd n, only $|\Gamma_k^{\{n, \sigma_2^2\}}|$ is not separable. For such patterns, while I_{20} can be computed by the use of 1D filters, the computation of I_{11} will need one true 2D convolution or an inexact approximation of it obtainable, e.g., by the singular value decomposition of the 2D filter (Sect. 15.3). Alternatively, the computational costs can still be kept small by working with small σ_2 and Gaussian pyramids.

Lemma 11.6 assumes that there is a window function whose purpose is to limit the estimation of I_{20} and I_{11} to a neighborhood around the current image point. Apart from $n = 0$, straight line patterns, the local gradient direction $\arg\left(\frac{dg}{dz}\right) = \arg(z^{\frac{n}{2}})$ is not well-defined in the origin for patterns generated by Eq. (11.87). This is visible in Fig. 11.6. The factor $|x + iy|^n$ in the window function is consequently justified, since it suppresses the origin as information provider for $n \neq 0$. Figure 11.6 shows that the filters that are suggested by the lemma for various patterns vanish at the origin except for the (Gaussian) one used for straight line extraction.

As mentioned, for $n \neq 0$ the origins of the target patterns are singular. Because of this, with increased $|n|$, the continuous image of such a pattern becomes increasingly difficult to discretize in the vicinity of the origin, to the effect that their discrete images will have an appearance less faithful to the underlying continuous image near the origin. As a consequence, approximating such patterns with band-limited or other regular functions, a necessity for accurate approximation of the integrals representing I_{20} and I_{11}, will be problematic because the singularity at the origin is barely or not at all accounted for already at the original discrete image. This can be achieved by signal theoretically correct sampling [63], e.g., when the square of an image on a discrete grid is needed, then the discrete image must be assured to

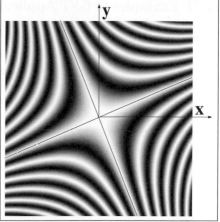

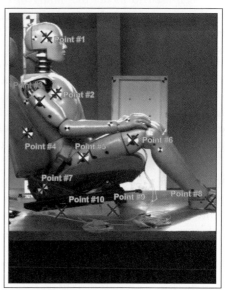

Fig. 11.7. Detecting cross markers and their direction in crash tests (from [23]). (*Top, left*) The ideal model of a cross with $\varphi = \frac{\pi}{8}$. (*Top, right*) The *gray tones* change linearly with the function $\sin(2\varphi)(x^2 - y^2) - \cos(2\varphi)2xy$, where $\varphi = \frac{\pi}{8}$. (*Bottom, left*) The first frame of an image sequence with the tracked forward (o, *thick curve*) and backward (·, *thin curve*) trajectory of the cross attached to the head. (*Bottom, right*) The automatically identified crosses

have been oversampled, before pixelwise squaring is applied. Oversampling can be effectively implemented by separable filters and pyramids [43].

11.11 Examples of GST Applications

Example 11.7. *In vehicle crash tests, the test event is filmed with a high-speed camera to quantify the impact of various parameters on human safety by tracking markers. A common marker is the "cross", which allows, to quantitate the planar position of an object as well as its planar rotation (Fig. 11.7). Markers have to be tracked across numerous frames (in the order of hundreds to thousands). The tracking has to be fast and robust in that the markers should not be lost from frame to frame. The rotations and translations of the objects are not constant due to the large accelerations/decelerations, while severe light conditions are common between two frames (e.g., imperfect flash synchronization). A symmetry tracker using hyperbolas, i.e.,*

$$n = 2 \quad \Rightarrow \quad g(z) = z^2 = x^2 - y^2 + i2xy \qquad (11.107)$$

has been used to model the family of cross markers. Because of this, the cross-markers were detected and simultaneously their rotation angles were quantitated automatically. The image at the bottom, left shows the original superimposed trajectory of the head, whereas the image on the right shows the identified crosses [23]. Alternatively, the candidate cross markers could have been detected by lack of linear symmetry in Cartesian coordinates, but then the rotation angles of the crosses would have to be identified separately. This is because a lack of a symmetry in any coordinate system does not provide the angle informationa since only the presence of an explicitly modeled curve family has a well-defined direction whereas the lack of a modeled curve family cannot provide the direction information, e.g. [97].

Example 11.8. *In biometric authentication, alignment of two fingerprints without extraction of minutiae[9] has gained increased interest, e.g., [108], since this improves the subsequent person authentication (minutiae-based or not) performance substantially. Besides improved accuracy, this helps to reduce the costly combinatorial match of fingerprint minutiae. Recently, silicon-based imaging sensors have become cheaply available. However, because sensor surfaces are decreasing, in order to accommodate them to portable devices, e.g., mobile phones, the delivered images of the fingerprints are also small. In turn this results in fewer minutiae points that are available to consumer applications, which is an additional reason for why nonminutiae-based alignment techniques in biometric authentication have gained interest. A high automation level of accurate fingerprint alignment is desirable independent of which matching technique is utilized. For robustness and precision, an automatical identification of two standard landmark types: core and delta (Fig. 11.8) has been proposed [169]. These can be modeled and detected by symmetry derivative filters in a scheme based on lemma 11.6. Naturally, the used coordinate transformations are different than the one modeling the cross markers. Furthermore, the detection is performed within a Gaussian pyramid scheme to improve the SNR. The real and imaginary parts of the analytic functions, i.e.,*

---

[9] Typically a minutia point is the end of a line or the bifurcation point of two lines.

Fig. 11.8. The "+" and the "□" represent the delta and core (parabolalike) points that have been automatically extracted in several fingerprint images. *Top, left* and *bottom, left* are two fingerprints of the same finger that differ significantly in quality. The certainties are 0.84 (+) and 0.64 (□) in *top, left*. The certainties are 0.73 (□), 0.22 ("+") and 0.10 (O) in *bottom, left*

$$n = -1 \quad \Rightarrow \quad g(z) = z^{\frac{1}{2}} \tag{11.108}$$

were used to model core, whereas

$$n = 1 \quad \Rightarrow \quad g(z) = z^{\frac{3}{2}} \tag{11.109}$$

were used for delta, see also Figs. 11.6 and 11.8.

Example 11.9. *Figure 11.9 shows the result of GST to locate letters for an optical character recognition system [186], in Sinhala script. Letters of a script encoding a natural language can be more effectively modeled by GST than direct gray values because occlusion and noise affects the directional information less than gray values. In this use of GST the analytic expressions of the pattern isocurves are not known, but they are deducted from the structure tensor variations of the reference characters,*

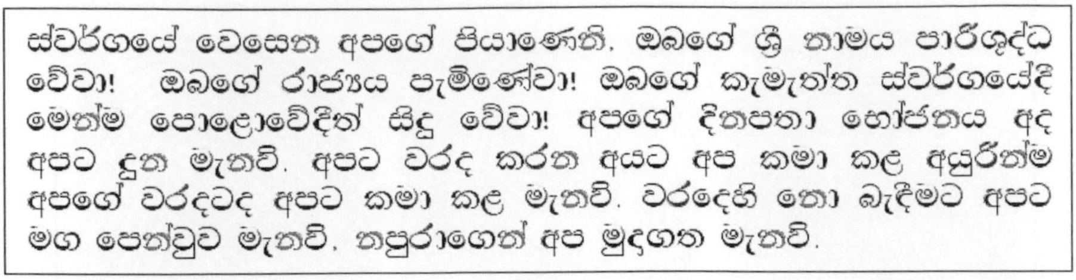

ස්වර්ගයේ වෙසෙන අපගේ පියාණනි. ඔබගේ ශ්‍රී නාමය පාරිශුද්ධ වේවා! ඔබගේ රාජ්‍යය පැමිණේවා! ඔබගේ කැමැත්ත ස්වර්ගයේද මෙන්ම පොළොවෙහිත් සිදු වේවා! අපගේ දිනපතා හෝජනය අද අපට දුන මැනවි. අපට වරද කරන අයට අප කමා කළ අයුරින්ම අපගේ වරදටද අපට කමා කළ මැනවි. වරදෙහි තො බැඳීමට අපට මග පෙන්වුව මැනවි. තපුරාගෙන් අප මුදගත මැනවි.

Fig. 11.9. The illustration of a segmentation-free OCR for Sinhala text (Sri Lanka), implemented by use GST, after [186]. The *points* are painted *colored* at the places where a support for a GST character template has been found

in analogy with Figs. 11.4 and 11.5. Similarly, Fig. 11.10 illustrates the recognition of Ethiopian script.

Example 11.10. *Two-dimensional gel* electrophoresis *is an important tool of proteomics, e.g., in search of new drugs. The result is 2D data that can be viewed as images (Fig. 11.11). These contain spots that are maps of a large number of proteins, requiring automatic image processing for efficient analysis. Quantification of individual proteins and tracing changes in expression between gels require accurate protein detection. One can approach the problem by modeling circles via the monomial family corresponding to $n = -2$. A set of computationally cheap and robust symmetry derivative features can implement a detector that is particularly tuned to circles. In the same figure, the color image represents I_{20} using the CT implied by $\log z$. The red hue represents the presence of circles. Brightness represents the certainty. Using this model as one of the features, an efficient classification and segmentation can be achieved [182].*

11.12 Further Reading

In computer vision, normally one does not locate an edge or compute the direction of a line by correlation with multiple templates consisting of rotated edges, incremented with a small angle. However, multiple correlations with the target pattern rotated in increments are commonly used to detect other shapes. This approach is also used to estimate the direction of such shapes. The number of rotations of the template can be fixed a priori or, as in [30], dynamically. The precision of the estimated direction is determined by the number of rotated templates used or by the amount of computations allowed. Although such techniques yield a generally good precision when estimating the affine parameters, which include target translation and rotation, in certain applications a beforehand undetermined number of iterations may not be possible or be desirable due to imposed restrictions that include hardware and software resources. Furthermore, the precision and/or convergence properties remain satisfactory as long as the reference pattern and the test pattern do not violate the

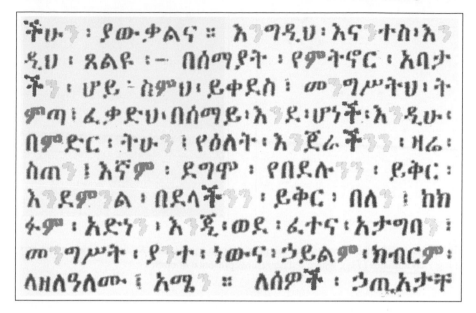

Fig. 11.10. Recognition of characters in Ethiopian text. The *colors* encode the labels of some identified target characters

image constancy hypothesis severely. In other words, if the image gray tones representing the same pattern differ nonlinearly and significantly between the reference and the test image, then a good precision or a convergence may not be achieved.

An early exception to the "rotate and correlate" approach is the pattern recognition school initiated by Hu [112], who suggested the moment-invariant signatures to be computed on the original image, which was assumed to be real-valued. Later, Reddi [188] suggested the magnitudes of complex moments to efficiently implement the moment invariants of the spatial image, mechanizing the derivation of them. The complex moments contain the rotation angle information directly encoded in their arguments as was shown in [25]. An advantage they offer is a simple separation of the direction parameter from the model evidence, i.e., by taking the magnitudes of the complex moments one obtains the moment invariants, which represent the evidence. The linear rotation-invariant filters suggested by [51, 228] or the steerable filters [75, 180] are similar to filters implementing complex moments. With appropriate radial weighting functions, the rotation-invariant filters can be viewed as equivalent to Reddi's complex moment filters, which in turn are equivalent to filters implementing Hu's geometric invariants. From this viewpoint, the suggestions of [2, 199] are also related to the computation of complex moments of a gray image, and hence deliver correlates of Hu's geometric invariants. Despite their demonstrated advantage in the context of real images, it is, however, not a trivial matter to model *isocurve families* embedded in gray images by correlating complex moment filters with gray

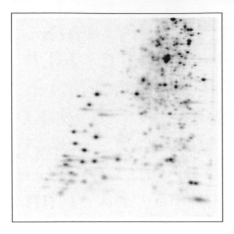 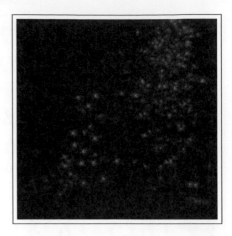

Fig. 11.11. (*Left*) An electrophoresis image of protein expression. (*Right*) the linear symmetry in $\log z$ coordinates (I_{20}), where red represents presence of circles

images. By contrast, this is possible when (GST) tensor-valued images are correlated with (GST) tensor valued filters.

Sharing similar integral curves with the generalized structure tensor, the Lie operators of [72, 107] should be mentioned. However, these studies have not provided tools on how to estimate the parameters of the integral curves, e.g., the direction and the estimation error.

Granlund and Knutsson first showed that convolving complex images by complex filters can result in detection of intricate patterns [140]. In the follow-up studies [17, 22, 23, 129] the GST theory, which achieved isocurve detection and identification by use of analytic models and tensor fields, was introduced. Hansen [94] studied the structure tensor when it is extended to curves and surfaces in three and higher dimensional spaces.

We have used a biometric person authentication application to illustrate the theory. A review of biometric recognition with brief summaries of the main schools are given in [175].

11.13 Appendix

Proof of theorem 11.2 and lemma 11.5

We prove the theorem by induction.

1. Eq. (11.98) holds for $p = 1$.
2. Assume (induction) that Eq. (11.98) is true for $p = p_0$, for some $p_0 \geq 1$.
3. Then

$$\Gamma^{\{p_0+1,\sigma^2\}}(x,y) =$$

$$= \frac{1}{2\pi\sigma^2}(D_x + iD_y)(D_x + iD_y)^{p_0}\exp(-\frac{x^2+y^2}{2\sigma^2})$$

$$= \frac{1}{2\pi\sigma^2}(D_x + iD_y)\left[\left(\frac{-x}{\sigma^2} + i\frac{-y}{\sigma^2}\right)^{p_0}\exp\left(-\frac{x^2+y^2}{2\sigma^2}\right)\right] \quad (11.110)$$

where we indicated by brackets [ ] the terms on which the differential operators act, is obtained by applying the induction assumption (point 2). Consequently, and by using the chain rule as well as the linearity of the partial differential operators, we obtain:

$$\Gamma^{\{p_0+1,\sigma^2\}}(x,y)$$

$$= \frac{1}{2\pi\sigma^2}\left[(D_x + iD_y)\left(\frac{-x}{\sigma^2} + i\frac{-y}{\sigma^2}\right)^{p_0}\right]\exp\left(-\frac{x^2+y^2}{2\sigma^2}\right)\cdots$$

$$+ \frac{1}{2\pi\sigma^2}\left(\frac{-x}{\sigma^2} + i\frac{-y}{\sigma^2}\right)^{p_0}\left[(D_x + iD_y)\exp\left(-\frac{x^2+y^2}{2\sigma^2}\right)\right] \quad (11.111)$$

By repeated applications of the algebraic rules that govern the differential operators, we obtain:

$$\Gamma^{\{p_0+1,\sigma^2\}}(x,y)$$

$$= \frac{1}{2\pi\sigma^2}\left[D_x\left(\frac{-x}{\sigma^2} + i\frac{-y}{\sigma^2}\right)^{p_0} + iD_y\left(\frac{-x}{\sigma^2} + i\frac{-y}{\sigma^2}\right)^{p_0}\right]\exp\left(-\frac{x^2+y^2}{2\sigma^2}\right)\cdots$$

$$+ \sigma^2\left(\frac{-x}{\sigma^2} + i\frac{-y}{\sigma^2}\right)^{p_0}\left(\frac{-x}{\sigma^2} + i\frac{-y}{\sigma^2}\right)\exp\left(-\frac{x^2+y^2}{2\sigma^2}\right)$$

$$= \frac{1}{2\pi\sigma^2}\left(p_0\left(\frac{-1}{\sigma^2}\right)\left(\frac{-x}{\sigma^2} + i\frac{-y}{\sigma^2}\right)^{p_0-1} + i^2 p_0\left(\frac{-1}{\sigma^2}\right)\left(\frac{-x}{\sigma^2} + i\frac{-y}{\sigma^2}\right)^{p_0-1}\right)\cdots$$

$$\times \exp\left(-\frac{x^2+y^2}{2\sigma^2}\right) + \frac{1}{2\pi\sigma^2}\left(\frac{-x}{\sigma^2} + i\frac{-y}{\sigma^2}\right)^{p_0+1}\exp\left(-\frac{x^2+y^2}{2\sigma^2}\right)$$

$$= \frac{1}{2\pi\sigma^2}\left(\frac{-x}{\sigma^2} + i\frac{-y}{\sigma^2}\right)^{p_0+1}\exp\left(-\frac{x^2+y^2}{2\sigma^2}\right) \quad (11.112)$$

Consequently, when it holds for $p = p_0$, Eq. (11.98) will also hold for $p = p_0+1$.

Now we turn to the general case, lemma 11.5.

$$Q(D_x + iD_y)\frac{1}{2\pi\sigma^2}\exp(-\frac{x^2+y^2}{2\sigma^2}) =$$

$$= \frac{1}{2\pi\sigma^2}[\sum_{n=0}^{N-1} a_n(D_x + iD_y)^n]\exp(-\frac{x^2+y^2}{2\sigma^2})$$

and obtain

$$Q(D_x + iD_y)\frac{1}{2\pi\sigma^2}\exp(-\frac{x^2 + y^2}{2\sigma^2}) = \tag{11.113}$$

$$= \frac{1}{2\pi\sigma^2}\sum_{n=0}^{N-1}a_n(\frac{-x}{\sigma^2} + i\frac{-y}{\sigma^2})^n\exp(-\frac{x^2 + y^2}{2\sigma^2})$$

$$= Q(\frac{-x}{\sigma^2} + i\frac{-y}{\sigma^2})\frac{1}{2\pi\sigma^2}\exp(-\frac{x^2 + y^2}{2\sigma^2}) \tag{11.114}$$

in which the linearity of derivative operators and Eq. (11.98) have been used.   ∎

Proof of theorem 11.4

We can ignore nearly all steps below to conclude that the theorem holds, provided that the following conditions (to be proven) are granted: (i) convolution and symmetry derivatives are distributive, i.e., $(D_x + iD_y)[f * g] = [(D_x + iD_y)f] * g = f * [(D_x + iD_y)g]$, and (ii) the Gaussian convolutions are variance-additive, i.e., $\Gamma^{\{0,\sigma_1^2\}} * \Gamma^{\{0,\sigma_2^2\}} = \Gamma^{\{0,\sigma_1^2+\sigma_2^2\}}$.

Convolution in the Fourier transform domain reduces to an ordinary multiplication:

$$\Gamma^{\{p_1,\sigma_1^2\}} * \Gamma^{\{p_2,\sigma_2^2\}} \leftrightarrow \mathcal{F}[\Gamma^{\{p_1,\sigma_1^2\}}]\mathcal{F}[\Gamma^{\{p_2,\sigma_2^2\}}] \tag{11.115}$$

By using Eq. (11.101) we note that

$$\mathcal{F}[\Gamma^{\{p_1,\sigma_1^2\}}](\omega_x, \omega_y) = 2\pi\left(\frac{-i}{\sigma_1^2}\right)^{p_1}\sigma_1^2\Gamma^{\{p_1,\frac{1}{\sigma_1^2}\}}(\omega_x, \omega_y) \tag{11.116}$$

Due to Eq. (11.98), we can write:

$$\mathcal{F}[\Gamma^{\{p_1,\sigma_1^2\}}] \cdot \mathcal{F}[\Gamma^{\{p_2,\sigma_2^2\}}] \tag{11.117}$$

$$= 2\pi\left(\frac{-i}{\sigma_1^2}\right)^{p_1}\sigma_1^2\Gamma^{\{p_1,\frac{1}{\sigma_1^2}\}}(\omega_x, \omega_y) \cdot 2\pi\left(\frac{-i}{\sigma_2^2}\right)^{p_2}\sigma_2^2\Gamma^{\{p_2,\frac{1}{\sigma_2^2}\}}(\omega_x, \omega_y)$$

$$= 2\pi\sigma_1^2\left(\frac{-i}{\sigma_1^2}\right)^{p_1}(-\sigma_1^2)^{p_1}(\omega_x + i\omega_y)^{p_1}\frac{1}{2\pi\sigma_1^2}\exp\left(-\sigma_1^2\frac{\omega_x^2 + \omega_y^2}{2}\right)\cdots$$

$$\times 2\pi\sigma_2^2\left(\frac{-i}{\sigma_2^2}\right)^{p_2}(-\sigma_2^2)^{p_2}(\omega_x + i\omega_y)^{p_2}\frac{1}{2\pi\sigma_2^2}\exp\left(-\sigma_2^2\frac{\omega_x^2 + \omega_y^2}{2}\right)$$

$$= \left(\frac{-i}{\sigma_1^2}\right)^{p_1}\left(\frac{-i}{\sigma_2^2}\right)^{p_2}(-\sigma_1^2)^{p_1}(-\sigma_2^2)^{p_2}(\omega_x + i\omega_y)^{p_1+p_2}\exp\left(-\frac{\omega_x^2 + \omega_y^2}{2\frac{1}{\sigma_1^2+\sigma_2^2}}\right)$$

Here, we have used the definition of $\Gamma^{\{p,\frac{1}{\sigma}\}}$ to obtain the Gaussian terms. Cancelling the redundant terms allows us to write the product as:

$$\mathcal{F}[\Gamma^{\{p_1,\sigma_1^2\}}] \cdot \mathcal{F}[\Gamma^{\{p_2,\sigma_2^2\}}] = (i)^{p_1+p_2}(\omega_x + i\omega_y)^{p_1+p_2} \exp\left(-\frac{\omega_x^2 + \omega_y^2}{2\frac{1}{\sigma_1^2+\sigma_2^2}}\right)$$

$$= \frac{2\pi(\sigma_1^2 + \sigma_2^2)}{2\pi(\sigma_1^2 + \sigma_2^2)}(i)^{p_1+p_2}(\omega_x + i\omega_y)^{p_1+p_2} \exp\left(-\frac{\omega_x^2 + \omega_y^2}{2\frac{1}{\sigma_1^2+\sigma_2^2}}\right) \frac{(-\sigma_1^2 - \sigma_2^2)^{p_1+p_2}}{(-\sigma_1^2 - \sigma_2^2)^{p_1+p_2}}$$

$$= 2\pi(\sigma_1^2 + \sigma_2^2)\left(\frac{i}{-\sigma_1^2 - \sigma_2^2}\right)^{p_1+p_2} \Gamma^{\{p_1+p_2, \frac{1}{\sigma_1^2+\sigma_2^2}\}}(\omega_x, \omega_y) \qquad (11.118)$$

Remembering, Eq. (11.115) we now inverse Fourier transform Eq. (11.118) by using Eq. (11.101) and obtain:

$$\Gamma^{\{p_1,\sigma_1^2\}} * \Gamma^{\{p_2,\sigma_2^2\}} = \Gamma^{\{p_1+p_2,\sigma_1^2+\sigma_2^2\}} \qquad (11.119)$$

∎

Proof of lemma 11.6

We use theorem 11.1 to estimate I_{20}, for which $(D_x - iD_y)\xi$ is needed.[10] We write the coordinates as a complex variable $z = x + iy$ and remember their relationship between symmetry derivatives and complex derivatives that was derived in Eq. (11.84).

$$(D_x - iD_y)\Re[g(z)] = \frac{dg}{dz} \qquad (11.120)$$

But $\frac{dg}{dz} = z^{\frac{n}{2}}$, so that we can obtain the complex exponential as:

$$e^{i \arg\{[(D_x - iD_y)\xi]^2\}} = e^{i \arg([\frac{dg}{dz}]^2)} = e^{i \arg(z^n)} = \frac{z^n}{|z^n|} = e^{in \arg(x+iy)}$$

Consequently, the expression I_{20} in equation (11.50) of theorem 11.1 reduces to

$$I_{20} = \iint ((D_\xi + iD_\eta)f)^2 d\xi d\eta = \iint e^{in \arg(x+iy)}[(D_x + iD_y)f]^2 dx dy$$

We assume that $[(D_x + iD_y)f]^2$ is discretized on a Cartesian grid and use a Gaussian as interpolator[11] to reconstruct it from its samples:

$$h(x, y) = 2\pi\sigma_1^2 \sum_k h_k \Gamma^{\{0,\sigma_1^2\}}(x - x_k, y - y_k) \qquad (11.121)$$

Here h_k represents the samples of $h(x, y)$, and the constant $2\pi\sigma_1^2$ normalizes the maximum of $\Gamma^{\{0,\sigma_1^2\}}$ to 1. We include the window function $K_n|x+iy|^n \Gamma^{\{0,\sigma_2^2\}}(x, y)$, where K_n is the constant[12] that normalizes the maximum of the window function to

---

[10] Because ξ is harmonic, η does not represent freedom of choice, but is determined as soon as ξ is given.

[11] This is possible by using a variety of interpolation functions, not only Gaussians. For example, the theory of band-limited functions allows such a reconstruction via the Sinc functions but also other functions have been discussed along with Gaussians, see interpolation and scale space reports of [141, 152, 221].

[12] At $|x + iy| = (n\sigma_2^2)^{1/2}$, the window $K_n|x + iy|^n \Gamma^{\{0,\sigma_2^2\}}(x, y)$ attains the value 1 when $K_n = 2\pi\sigma_2^2(\frac{e}{n\sigma_2^2})^{n/2}$.

1, into $[(D_x + iD_y)f]^2$, to estimate Eq. (11.121) in a neighborhood:

$$I_{20} = \iint K_n |x + iy|^n e^{in \arg(x+iy)} \Gamma^{\{0,\sigma_2^2\}}(x,y)[(D_x + iD_y)f]^2 dxdy \quad (11.122)$$

By assuming $0 \leq n$ and substituting Eq. (11.121) in Eq. (11.122), we obtain

$$\frac{I_{20}(x',y')}{(2\pi\sigma_1^2)K_n} = \quad (11.123)$$

$$\iint |x + iy|^n e^{in \arg(x+iy)} \Gamma^{\{0,\sigma_2^2\}}(x,y) \left[\sum_k h_k \Gamma^{\{0,\sigma_1^2\}}(x - x_k, y - y_k) \right] dxdy$$

Noting that

$$|x + iy|^n e^{in \arg(x+iy)} \Gamma^{\{0,\sigma_2^2\}}(x,y)$$
$$= (x + iy)^n \Gamma^{\{0,\sigma_2^2\}}(x,y) = (-\sigma_2^2)^n \Gamma^{\{n,\sigma_2^2\}}(x,y) \quad (11.124)$$

where we used the definition of $\Gamma^{\{n,\sigma_2^2\}}$ in Eq. (11.97) and applied theorem 11.2, we can estimate I_{20} on a Cartesian grid:

$$\frac{I_{20}(x',y')}{(2\pi\sigma_1^2)K_n(-\sigma_2^2)^n}$$

$$= \sum_k h_k \iint \Gamma^{\{n,\sigma_2^2\}}(x,y)\Gamma^{\{0,\sigma_1^2\}}(x' - x - x_k, y' - y - y_k)dxdy$$

$$= \sum_k h(x_k, y_k)(\Gamma^{\{n,\sigma_2^2\}} * \Gamma^{\{0,\sigma_1^2\}})(x' - x_k, y' - y_k)$$

$$= \sum_k h(x_k, y_k)\Gamma^{\{n,\sigma_1^2+\sigma_2^2\}}(x' - x_k, y' - y_k) \quad (11.125)$$

Here Eq. (11.125) is obtained[13] by utilizing theorem 11.4. Equation (11.125) can be computed on a Cartesian discrete grid by the substitution $(x',y') = (x_l, y_l)$, yielding an ordinary discrete convolution:

$$I_{20}(x_l, y_l) = C_n \left(h * \Gamma^{\{n,\sigma_1^2+\sigma_2^2\}} \right)(x_l, y_l) \quad (11.126)$$

with $C_n = (2\pi\sigma_1^2)K_n(-\sigma_2^2)^n$.

The result for $n < 0$ is straightforward to deduce by following the steps after Eq. (11.125) in an analogous manner. Likewise, the scheme of I_{11} is obtained by following the same idea as for I_{20}. ∎

---

[13] We note that in the proof of theorem 11.4, theorem 11.3 is needed, so that all of the theorems of this paper are actually utilized in the proof of this lemma.

12

Direction in ND, Motion as Direction

In this chapter we extend our discussion of the direction estimation problem from a 2D setting in Cartesion and curvilinear coordinates to an estimation problem in N dimensions (ND). We start the discussion in 3D. To the extent that we can recognize the solutions from treatment of 2D, the solutions of 3D problems related to direction estimation will be inherited from their 2D analogues. This will provide an opportunity to focus on specific issues of direction estimation not present in 2D. In turn, the experience from 3D will help us draw conclusions on the problem of direction estimation in ND.

12.1 The Direction of Hyperplanes and the Inertia Tensor

Whether or not a 3D image has a direction can be studied in the Fourier transform domain[1] in analogy with Chap. 10, although the computations will be carried out in the spatial domain. Let f be a positive real function defined on E_3 with F being its 3D Fourier transform. As before, we will sometimes call f an image and its values as gray values, although the values of f can represent a variety of physical properties in applications. They can, for example, represent the light intensities as observed on a video camera sensor to form image sequences, $f(x, y, t)$, or the absorption of X-ray in an organic tissue to form X-ray tomography images, $f(x, y, z)$. The function f will still be called an image, even if it represents a local image.

Intuitively, if f has a "direction" then this has to be related to the locus of f's isogray values, points which have the same function values. In 3D the loci of isogray values are more complex to describe than in 2D because f's being constant can be achieved along a 1D curve, the simplest of which is a line, or along a 2D surface, the simplest of which is a plane. First, we will discuss the *simple manifolds*, namely the

---

[1] An equivalent formulation in the spatial domain utilizing the Lie operators [31] as is done in Sect. 11.2, is possible in 3D Cartesian coordinates. We prefer the frequency-domain derivation for its straightforward geometric interpretation in terms of the habitual line and plane fitting.

lines and planes embedded in 3D such that f does not change value. Subsequently, we will show that the problems of describing the direction when isovalues are planes or lines are mathematically the same. We will first be concerned with the case of planar loci since these have some notational advantages. We define them directly for the general ND case, from which the 3D case follows directly.

Definition 12.1. *The image $f(\mathbf{r})$, where $\mathbf{r} \in E_N$, is called* linearly symmetric *if the loci of its isogray points constitute hyperplanes that have a common direction, i.e., there exists a scalar function of one variable g such that*

$$f(\mathbf{r}) = g(\mathbf{k}^T \mathbf{r}) \tag{12.1}$$

for some vector $\mathbf{k} \in E_N$. The direction of the linear symmetry is $\pm \mathbf{k}$.

It should be noted that although $g(x)$ is a one-dimensional function, $g(\mathbf{k}^T \mathbf{r})$ is a function defined on ND. The way they are defined assures that the linearly symmetric images have the same gray values at all points $\mathbf{r}$ satisfying

$$\mathbf{k}^T \mathbf{r} = \text{constant} \tag{12.2}$$

which is the equation of a hyperplane embedded in ND having the normal $\mathbf{k}$. The hyperplane itself is an $(N-1)$-dimensional surface (which is embedded in ND). This corresponds to an ordinary plane (surface) in E_3 when $N = 3$.

Example 12.1. *A step edge, defined in 3D as:*

$$f(\mathbf{r}) = \begin{cases} 1, & \text{if } \mathbf{k}_0^T \mathbf{r} \geq 0; \\ 0, & \text{otherwise;} \end{cases} \tag{12.3}$$

is equal to $\chi(\mathbf{k}_0^T \mathbf{r})$, where χ is the usual one-dimensional step function:

$$\chi(x) = \begin{cases} 1, & \text{if } x \geq 0; \\ 0, & \text{otherwise.} \end{cases}$$

Thus f defined in Eq. (12.3) is linearly symmetric.

We can Fourier transform a linearly symmetric image defined in 3D, by following steps analogous to Eqs. (10.102)–(10.103):

$$F(\boldsymbol{\omega}) = \mathcal{F}(g(\mathbf{k}^T \mathbf{r}))(\boldsymbol{\omega}) = G(\mathbf{k}^T \boldsymbol{\omega})\delta(\mathbf{u}_1^T \boldsymbol{\omega})\delta(\mathbf{u}_2^T \boldsymbol{\omega}) \tag{12.4}$$

In Fig. 12.1, the dashed triangular plane shows a linearly symmetric image in the 3D space having the coordinate axes represented by three black basis vectors. The red basis vectors show the corresponding axes in the 3D Fourier transform domain, whereas the vector in black shows $\mathbf{k}$, the axis on which all the energy of the Fourier transform of the image is concentrated. Note that $\mathbf{k}$ is orthogonal to the plane. Naturally, this can be extended to ND, a result which we state as a lemma.

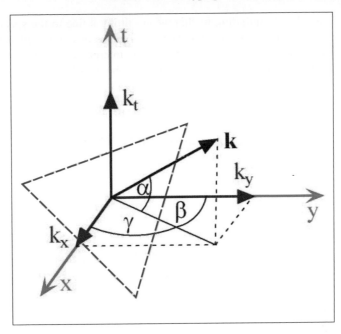

Fig. 12.1. The *green axes* show the 3D Euclidean space in which a linearly symmetric image is defined, the *dashed triangular plane*. The *red axes* show the corresponding 3D Fourier transform domain axes in which a concentration of the energy to the axis represented by the *vector* **k** occurs

Lemma 12.1. *A linearly symmetric image,* $f(\mathbf{r}) = g(\mathbf{k}_0^T \mathbf{r})$, *has a Fourier transform concentrated to a line through the origin:*

$$F(\boldsymbol{\omega}) = G(\boldsymbol{\omega}^T \mathbf{k}_0) \delta(\boldsymbol{\omega}^T \mathbf{u}_1) \delta(\boldsymbol{\omega}^T \mathbf{u}_2) \cdots \delta(\boldsymbol{\omega}^T \mathbf{u}_{N-1})$$

where $\mathbf{k}_0, \mathbf{u}_1 \cdots \mathbf{u}_{N-1}$ *are orthonormal vectors in* E_N, *and* δ *is the Dirac distribution. G is the one-dimensional Fourier transform of g.*

$\blacklozenge$

The lemma states that the function $g(\mathbf{k}_0^T \mathbf{r})$, which is in general a "spread" function, is compressed to a line, even for functions defined on spaces with higher dimension than two. To detect linearly symmetric functions is consequently the same as to check whether or not the energy is concentrated to a line in the Fourier domain. To simplify the discussion, we assume that f is defined on E_3 and we attempt to fit an axis to F, through the origin of the corresponding 3D Fourier transform domain.

$$\min_{\|\mathbf{k}\|=1} e^{\mathbf{L}}(\mathbf{k}) = \int_{E_3} (d^{\mathbf{L}}(\boldsymbol{\omega}, \mathbf{k}))^2 |F(\boldsymbol{\omega})|^2 dE_3 \qquad (12.5)$$

where $\boldsymbol{\omega} = (\omega_1, \omega_2, \omega_3)^T$ represents the frequency coordinates corresponding to the spatial coordinates $\mathbf{r} = (x_1, x_2, x_3)^T$. The dE_3 is equal to $d\omega_1 d\omega_2 d\omega_3$, and the superscript **L** is a label reminding us that we are fitting a line to the spectrum. The line-fitting process is the same as identifying iso-values of parallel planes in the **r** domain. If the energy of the Fourier transform is interpreted as the mass density, then $e^{\mathbf{L}}(\mathbf{k})$ is the inertia of a mass with respect to the axis **k**. Because the integral Eq. (12.5) is a norm, i.e.,

$$e^{\mathbf{L}}(\mathbf{k}) = \langle d^{\mathbf{L}}F, d^{\mathbf{L}}F \rangle = \|d^{\mathbf{L}}F\|^2 \qquad (12.6)$$

$e^{\mathbf{L}}(\mathbf{k}_{\min})$ vanishes if and only if F is concentrated to a line.

The distance function is given by:

$$
\begin{aligned}
(d^{\mathbf{L}}(\boldsymbol{\omega}, \mathbf{k}))^2 &= \|\boldsymbol{\omega} - (\boldsymbol{\omega}^T\mathbf{k})\mathbf{k}\|^2 \\
&= \left(\boldsymbol{\omega} - (\boldsymbol{\omega}^T\mathbf{k})\mathbf{k}\right)^T \left(\boldsymbol{\omega} - (\boldsymbol{\omega}^T\mathbf{k})\mathbf{k}\right)
\end{aligned}
$$

Using matrix multiplication rules and remembering that $\boldsymbol{\omega}^T\mathbf{k}$ is a scalar and identical to $\mathbf{k}^T\boldsymbol{\omega}$ and $\|\mathbf{k}\|^2 = \mathbf{k}^T\mathbf{k} = 1$, the *quadratic form*:

$$(d^{\mathbf{L}}(\boldsymbol{\omega}, \mathbf{k}))^2 = \mathbf{k}^T\left(\mathbf{I}\boldsymbol{\omega}^T\boldsymbol{\omega} - \boldsymbol{\omega}\boldsymbol{\omega}^T\right)\mathbf{k}$$

is obtained. Thus Eq. (12.5) is expressed as

$$e^{\mathbf{L}}(\mathbf{k}) = \mathbf{k}^T\mathbf{J}\mathbf{k} \qquad (12.7)$$

with

$$\mathbf{J} = \begin{pmatrix} \mathbf{J}(1,1) & \mathbf{J}(1,2) & \mathbf{J}(1,3) \\ \mathbf{J}(2,1) & \mathbf{J}(2,2) & \mathbf{J}(2,3) \\ \mathbf{J}(3,1) & \mathbf{J}(3,2) & \mathbf{J}(3,3) \end{pmatrix}$$

where $\mathbf{J}(i,j)$'s are given by

$$\mathbf{J}(i,i) = \int_{E_3} \sum_{j \neq i} \omega_j^2 |F(\mathbf{r})|^2 dE_3 \qquad (12.8)$$

and

$$\mathbf{J}(i,j) = -\int_{E_3} \omega_i\omega_j |F(\boldsymbol{\omega})|^2 dE_3 \quad \text{when} \quad i \neq j. \qquad (12.9)$$

Notice that the matrix **J** is symmetric per construction. The minimization problem formulated in Eq. (12.5) is solved by **k** corresponding to the least eigenvalue of the inertia matrix, **J**, of the Fourier domain [231]. All eigenvalues are real and non-negative and the smallest eigenvalue is the minimum of $e^{\mathbf{L}}$. The matrix **J** contains sufficient information to allow computation of the optimal **k** in the TLS error sense Eq. (12.5). As is its 2D counterpart, even this matrix is a tensor because it represents a physical property, *inertia*.

The obtained direction will be unique if the least eigenvalue has the multipicity[2] 1. When the multiplicity of the least eigenvalue is 2, there is no unique axis **k**, but

---

[2] An $N \times N$ matrix has N eigenvalues. If two eigenvalues are equal then that eigenvalue has the multiplicity 2. If three eigenvalues are equal then the multiplicity is 3, and so on.

plenty of them by which the image can be described as $g(\mathbf{k}^T\mathbf{r})$. Representing an important "degenerate" case, the energy in the Fourier domain will then be distributed in such a way that there is a 2D plane (instead of an axis) containing an infinite number of axes that give the (same) least square error. This plane is defined by the subspace generated by linear combinations of the eigenvectors belonging to the least eigenvalue, which has the multiplicity 2. The dimension of eigenvector subspaces is always equal to the multiplicity of the eigenvalue to which the eigenvectors belong, whereas the eigenvectors belonging to different eigenvalues are always orthogonal. This is due to the fact that $\mathbf{J}$ is positive semidefinite and symmetric by definition, Eqs. (12.5) and (12.7). Accordingly, such a 3×3 matrix always has three orthogonal eigenvectors, where those possibly belonging to the same eigenvalue will not be unique. This conclusion can be naturally extended to N dimensions.

Lemma 12.2. *An* inertia tensor $\mathbf{J}$ *in* N-*D has* N *orthogonal eigenvectors with the reservation that those belonging to the same eigenvalue, with a multiplicity 2 or larger, are not unique.*

$\blacklozenge$

12.2 The Direction of Lines and the Structure Tensor

Here, we will discuss what the degenerate solutions (the nonunique eigenvectors) of the spectral line-fitting problem corresponds to in the spatial domain. We investigate the issue first for 3D spectra. Assuming that the multiplicity of the least eigenvalue is 2, the energy will be concentrated to a 2D plane, spanned by the eigenvectors of the least eigenvalue. What sense will this plane make when the intention was to fit a line through the origin and what we found is a plane (instead of a line)?

Suppose that we were intending to fit the (complex) spectral function F a plane under the condition that it had to pass through the origin. Representing the normal of the plane by $\mathbf{k}$, we would then minimize

$$\min_{\|\mathbf{k}\|=1} e^{\mathbf{P}}(\mathbf{k}) = \int_{E_3} (d^{\mathbf{P}}(\boldsymbol{\omega}, \mathbf{k}))^2 |F(\boldsymbol{\omega})|^2 dE_3 \qquad (12.10)$$

where the $\mathbf{P}$ reminds us that we are attempting to fit a plane and $d^{\mathbf{P}}$ is the Euclidean distance between the point $\boldsymbol{\omega}$, and the plane with the normal $\mathbf{k}$ in the spectrum,

$$(d^{\mathbf{P}}(\boldsymbol{\omega}, \mathbf{k}))^2 = (\boldsymbol{\omega}^T \mathbf{k})^2 = \mathbf{k}^T \boldsymbol{\omega}\boldsymbol{\omega}^T \mathbf{k} \qquad (12.11)$$

Compared to line-fitting, the main difference is in the distance function $d^{\mathbf{P}}(\boldsymbol{\omega}, \mathbf{k})$, which is now a projection of the point $\boldsymbol{\omega}$ on the normal vector $\mathbf{k}$, the shortest distance to the plane. Thus Eq. (12.10) reduces to minimizing

$$e^{\mathbf{P}}(\mathbf{k}) = \mathbf{k}^T \mathbf{S}\mathbf{k} \qquad (12.12)$$

where

$$\mathbf{S}(i, j) = \int_{E_3} \omega_i \omega_j |F(\boldsymbol{\omega})|^2 dE_3 \qquad (12.13)$$

Comparing Eq. (12.13) with Eqs. (12.8) and (12.9) yields an invertible algebraic relationship between the matrices $\mathbf{S}$ and $\mathbf{J}$, arising from line-fitting and plane-fitting problems, respectively.

$$\mathbf{J} = \text{Trace}(\mathbf{S})\mathbf{I} - \mathbf{S} \qquad (12.14)$$

Here "Trace($\mathbf{S}$)" is the *trace of a matrix* that is the sum of all eigenvalues of $\mathbf{S}$. It can be computed conveniently by summing up $\mathbf{S}$'s diagonal elements. Like $\mathbf{J}$, the matrix $\mathbf{S}$ is also positive semidefinite, and the solution to the plane-fitting problem is given by the least eigenvalue of $\mathbf{S}$ and its corresponding eigenvector(s). The matrix $\mathbf{S}$ defined by Eq. (12.13) is the *structure tensor* in n dimensions. Both of the matrices $\mathbf{S}$ and $\mathbf{J}$ are tensors because they encode physical qualities, i.e., the scatter and the inertia along with the extremal axes of a mass distribution. The scatter and the inertia remain the same, regardless the coordinate frames they are measured in. The axes do not change relative the mass distribution. Relative to coordinate frames not attached to the mass distribution, their representation varies up to a rotation, which is a viewpoint transformation.

Lemma 12.3. *The tensors $\mathbf{J}$ and $\mathbf{S}$ have common eigenvectors, that is,*

$$\mathbf{J}\mathbf{u} = \lambda'\mathbf{u} \quad \Leftrightarrow \quad \mathbf{S}\mathbf{u} = \lambda\mathbf{u} \qquad (12.15)$$

with

$$\lambda' = \text{Trace}(\mathbf{S}) - \lambda \qquad (12.16)$$

♦

The lemma is a consequence of the fact that $\mathbf{J}$ and $\mathbf{S}$ commute but can also be proven immediately by utilizing the relationship Eq. (12.14) and operating with $\mathbf{J}$ on $\mathbf{u}$, which is assumed to be an eigenvector of $\mathbf{S}$.

According to the lemma, fitting a line to F or fitting a plane to F can be achieved by a quadratic form using the same tensor, i.e., either of $\mathbf{S}$ or $\mathbf{J}$. Then, the following lemma holds, too.

Lemma 12.4. *Let the spectrum be 3D and that the eigenvalues of the structure tensor $\mathbf{S}$ are enumerated in ascending order, $0 \le \lambda_3 \le \lambda_2 \le \lambda_1$. If and only if*

$$0 = \lambda_3 < \lambda_2 \qquad (12.17)$$

the FT values are zero outside of a plane *through the origin. The normal of this plane is given by $\mathbf{u}_3$, the least significant eigenvector of $\mathbf{S}$. On the plane, the FT values equal the 2D FT of the values of f collected from any plane with normal $\mathbf{u}_3$. Similarly, if and only if*

$$0 = \lambda_3 = \lambda_2 < \lambda_1 \qquad (12.18)$$

the FT is concentrated to a line *through origin. The direction of this line is given by $\mathbf{u}_1$, the most significant eigenvector of $\mathbf{S}$. The FT values on the plane are given by the 1D FT of f lying on any of the lines with the direction $\mathbf{u}_1$.*

♦

The first part of the lemma predicts that in case a 2D image $g(x, y)$ is translated as we move along the the third dimension t, the FT values are concentrated to a plane. This will be discussed in detail in Sect. 12.6 as this lemma will be useful to quantitate motion in image sequences, such as TV images. In such images the direction of the plane will be interpreted as velocity. Mathematically, the reach of the lemma is, however, not restricted only to the (x, y, t)-type images, where t is time, but any type of 3D images such as *magneto resonance images*, or *X-ray tomography*, which are images defined on an xyz coordinate frame, there all coordinates have the same physical interpretation, the length. The direction in such images will mean the surface direction of anatomic structures, such as the surface of the gray matter in the brain, or the direction of a blood vessel.

A generalization of lemma 12.4 to higher dimensions, where we have the function $F(\boldsymbol{\omega})$ with $\boldsymbol{\omega} \in E_N$ and $N > 3$, can be done by fitting a hyperplane to F. The (weighted) average distance to the hyperplane having the normal $\mathbf{k}$ is then

$$\|\mathbf{k}^T \boldsymbol{\omega} F\|^2 = \mathbf{k}^T (\int \boldsymbol{\omega} \boldsymbol{\omega}^T |F|^2 d\boldsymbol{\omega}) \mathbf{k} = \mathbf{k}^T \mathbf{S} \mathbf{k} \qquad (12.19)$$

where the weights are given by the spectral energy $|F|^2$. This is minimized by the least significant eigenvector of $\mathbf{S}$, being the scatter tensor of F (or the structure tensor of f). In the ideal situation of a perfect hyperplane fit, the least eigenvalue must be zero, whereas the remaining $(N - 1)$ eigenvalues must be nonzero. However, more than 1 eigenvalue could be zero too, i.e., using the sorting convention

$$0 \le \lambda_N \le \lambda_{N-1} \cdots \le \lambda_1 \qquad (12.20)$$

for labelling the eigenvalues, and calling the multiplicity of the null eigenvalue n, we will have

$$0 = \lambda_N = \lambda_{N-1} \cdots = \lambda_{N-(n-1)} < \lambda_{N-n} \qquad (12.21)$$

Then, the (range) value space of $\mathbf{S}$ will be a hyperplane with the dimension $N - n$. We summarize this in the following theorem:

Theorem 12.1. *Let*

$$\mathbf{S} = \int \boldsymbol{\omega} \boldsymbol{\omega}^T |F|^2 d\boldsymbol{\omega} \qquad (12.22)$$

be the scatter tensor of the function $F(\boldsymbol{\omega})$, where $\boldsymbol{\omega} \in E_N$, and $2 \le N$. Then, if and only if the null space[3] of $\mathbf{S}$ has the multiplicity n, with n being an integer, and $0 < n < N$, F is concentrated to a hyperplane with dimension $N - n$, through the origin. The hyperplane is given by the points satisfying $\mathbf{S}\boldsymbol{\omega} = 0$, and the values of F on this plane are given by the $(N - n)$-dimensional FT of f lying on any hyperplane spanned by the null space of $\mathbf{S}$.

♦

---

[3] The space represented by the eigenvectors belonging to the eigenvalue zero is the null space: $\mathbf{S}\boldsymbol{\omega}=0$. Regardless of its dimension, the null space always includes the vector $\boldsymbol{\omega} = 0$.

12.3 The Decomposition of the Structure Tensor

To study the decomposition in terms of its subspaces, we discuss the structure tensor in 3D, i.e., $\mathbf{S}$ is 3×3. According to the *spectral theorem* of positive semidefinite operators [136], the structure tensor in ND can be written as a weighted sum of its eigenvector subspaces. In 3D this yields,

$$\mathbf{S} = \lambda_1 \mathbf{u}_1 \mathbf{u}_1^T + \lambda_2 \mathbf{u}_2 \mathbf{u}_2^T + \lambda_3 \mathbf{u}_3 \mathbf{u}_3^T \tag{12.23}$$

which is also called the *spectral decomposition*. As theorem 12.1 suggests, the dimension of the null space of $\mathbf{S}$ determines if the tensor has "succeeded" to fit a plane, a line, or none of these to the image. Accordingly, the following three cases can be distinguished. The analogy to 2D, see Section 10.7, is discernable, but there are now three basic cases in 3D, instead of 2.

The spectral line: This is the case when λ_2 equals the least significant eigenvalue:

$$0 = \lambda_3 = \lambda_2 < \lambda_1 \tag{12.24}$$

Accordingly, the null space and the line will be given in parametric form by

$$\mathbf{S}\omega = \mathbf{0}, \quad \text{when} \quad \omega = \alpha \mathbf{u}_3 + \beta \mathbf{u}_2, \tag{12.25}$$
$$\mathbf{S}\omega \neq \mathbf{0}, \quad \text{when} \quad \omega = \gamma \mathbf{u}_1, \tag{12.26}$$

where α, β and γ are arbitrary real scalars. Alternatively, the line is given by solving for ω in the underdetermined system of equations:

$$\mathbf{u}_3^T \omega = 0 \tag{12.27}$$
$$\mathbf{u}_2^T \omega = 0 \tag{12.28}$$

which is the same as searching for the space perpendicular to the null space.

The spectral plane: This is the case when

$$0 = \lambda_3 < \lambda_2 \tag{12.29}$$

The null space and the plane will be given in parametric form by

$$\mathbf{S}\omega = \mathbf{0}, \quad \text{when} \quad \omega = \alpha \mathbf{u}_3, \tag{12.30}$$
$$\mathbf{S}\omega \neq \mathbf{0}, \quad \text{when} \quad \omega = \beta \mathbf{u}_2 + \gamma, \mathbf{u}_1 \tag{12.31}$$

where α, β, and γ are arbitrary real scalars. The plane can also be obtained by writing down the condition for ω to be perpendicular to the null space:

$$\mathbf{u}_3^T \omega = 0 \tag{12.32}$$

The balanced directions: This is the case when all three eigenvalues are equal:

$$0 < \lambda_3 = \lambda_2 = \lambda_1 \tag{12.33}$$

Accordingly this case has a nullspace with dimension 0. There is a point symmetry in the distribution of the energy to the effect that no direction is different than other directions.

We define the tensors

$$\begin{aligned} \mathbf{U}_1 &= \mathbf{u}_1 \mathbf{u}_1^T, \\ \mathbf{U}_2 &= \mathbf{u}_2 \mathbf{u}_2^T, \\ \mathbf{U}_3 &= \mathbf{u}_3 \mathbf{u}_3^T, \end{aligned} \tag{12.34}$$

and note that these are orthogonal in the sense of Eq. (3.45) and therefore constitute a basis. Accordingly, Eq. (12.23) yields:

$$\mathbf{S} = \lambda_1 \mathbf{U}_1 + \lambda_2 \mathbf{U}_2 + \lambda_3 \mathbf{U}_3 \tag{12.35}$$

which can be interpreted as an orthogonal expansion of the structure tensor, and the eigenvalues are the coordinates encoding the tensor $\mathbf{S}$ in terms of the basis. Inversely, we could try to synthesize an $\mathbf{S}$ by varying the coordinates, but we would then need to obey

$$0 \le \lambda_3 \le \lambda_2 \le \lambda_1 \tag{12.36}$$

This means that the coordinates cannot be chosen independent of each other in a synthesis process. To simplify the synthesis, we could define a new set of coordinates that are independent of each other as follows:

$$\begin{aligned} 0 &\le \lambda_1' = \lambda_1 - \lambda_2 \\ 0 &\le \lambda_2' = \lambda_2 - \lambda_3 \\ 0 &\le \lambda_3' = \lambda_3 \end{aligned} \qquad \Leftrightarrow \qquad 0 \le \boldsymbol{\lambda}' = \mathbf{C}\boldsymbol{\lambda} \tag{12.37}$$

with

$$\boldsymbol{\lambda}' = \begin{pmatrix} \lambda_1' \\ \lambda_2' \\ \lambda_3' \end{pmatrix}, \qquad \mathbf{C} = \begin{pmatrix} 1 & -1 & 0 \\ 0 & 1 & -1 \\ 0 & 0 & 1 \end{pmatrix}, \qquad \boldsymbol{\lambda} = \begin{pmatrix} \lambda_1 \\ \lambda_2 \\ \lambda_3 \end{pmatrix} \tag{12.38}$$

The new coordinates $\boldsymbol{\lambda}'$ simply encode the increments between the subsequent eigenvalues and can therefore be chosen freely, as long as they are positive or zero. The structure tensors must thus be chosen in a cone bounded by a certain basis tensor corresponding to the new coordinates. To find the new basis we only need to follow the standard rule of linear algebra. Given the coordinate transformation matrix $\mathbf{C}$, the basis tranformation matrix equals $\mathbf{C}^{-1}$:

$$[\mathbf{U}_1' \mathbf{U}_2' \mathbf{U}_3'] = [\mathbf{U}_1 \mathbf{U}_2 \mathbf{U}_3] \mathbf{C}^{-1} \tag{12.39}$$

so that we have

$$\mathbf{C}^{-1} = \begin{pmatrix} 1 & 1 & 1 \\ 0 & 1 & 1 \\ 0 & 0 & 1 \end{pmatrix} \qquad \Rightarrow \qquad \begin{aligned} \mathbf{U}_1' &= \mathbf{U}_1 \\ \mathbf{U}_2' &= \mathbf{U}_1 + \mathbf{U}_2 \\ \mathbf{U}_3' &= \mathbf{U}_1 + \mathbf{U}_2 + \mathbf{U}_3 \end{aligned} \tag{12.40}$$

Consequently, the structure tensor decomposition, Eq. (12.23), can always be rearranged in terms of independent coordinates as

$$\mathbf{S} = (\lambda_1 - \lambda_2)\mathbf{u}_1\mathbf{u}_1^T + (\lambda_2 - \lambda_3)(\mathbf{u}_1\mathbf{u}_1^T + \mathbf{u}_2\mathbf{u}_2^T) +$$
$$+ \lambda_3(\mathbf{u}_1\mathbf{u}_1^T + \mathbf{u}_2\mathbf{u}_2^T + \mathbf{u}_3\mathbf{u}_3^T) \qquad (12.41)$$

$$(12.42)$$

This rearrangement allows a direct interpretation of the eigenvalues in terms of our three fundamental cases. The line case is the dominating structure when $0 \approx \lambda_2 \ll \lambda_1$, i.e., the first term is largest. The plane case is the dominating structure when $0 \approx \lambda_3 \ll \lambda_2$, i.e., the second term is largest. The balanced direction case is the dominating structure when $0 \ll \lambda_3 \approx \lambda_2 \approx \lambda_1$, i.e., the last term is largest. Accordingly,

$$\begin{aligned} \lambda_1' &= C^{\mathbf{L}} = \lambda_1 - \lambda_2 \\ \lambda_2' &= C^{\mathbf{P}} = \lambda_2 - \lambda_3 \qquad \text{with} \\ \lambda_3' &= C^{\mathbf{B}} = \lambda_3 \end{aligned} \qquad \begin{aligned} \mathbf{U}_1' &= \mathbf{u}_1\mathbf{u}_1^T \\ \mathbf{U}_2' &= \mathbf{u}_1\mathbf{u}_1^T + \mathbf{u}_2\mathbf{u}_2^T \\ \mathbf{U}_3' &= \mathbf{u}_1\mathbf{u}_1^T + \mathbf{u}_2\mathbf{u}_2^T + \mathbf{u}_3\mathbf{u}_3^T = \mathbf{I} \end{aligned} \qquad (12.43)$$

where the three coordinates, $C^{\mathbf{L}}$, $C^{\mathbf{P}}$, $C^{\mathbf{B}}$, can be used as a certainty or saliency for $\mathbf{S}$ representing a line, a plane, and a balanced directions structure. That $\mathbf{u}_1\mathbf{u}_1^T + \mathbf{u}_2\mathbf{u}_2^T + \mathbf{u}_3\mathbf{u}_3^T$ equals the identity matrix $\mathbf{I}$ in the last row follows from the fact that $\mathbf{I}$ can be expanded in the orthonormal basis $\mathbf{U}_j = \mathbf{u}_j\mathbf{u}_j^T$ by using the scalar product, Eq. (3.45):

$$\langle \mathbf{U}_j, \mathbf{I} \rangle = \sum_{k,l} \mathbf{u}_j(k)\mathbf{u}_j(l)\delta(k - l) = \sum_k |\mathbf{u}_j(k)|^2 = 1 \qquad (12.44)$$

Following a similar procedure, these results can be generalized to the spectral decomposition of the an $N \times N$ structure tensor as follows:

Lemma 12.5 (Structure tensor decomposition). *Assuming that $\lambda_j, \mathbf{u}_j$ constitute the eigenvalue and eigenvector pairs of the symmetric positive definite matrix $\mathbf{S}$, the spectral decomposition of $\mathbf{S}$ yields*

$$\mathbf{S} = \sum_j \lambda_j \mathbf{u}_j\mathbf{u}_j^T = \sum_j \lambda_j'\mathbf{U}_j' \qquad (12.45)$$

where

$$\begin{aligned} \lambda_1' &= \lambda_1 - \lambda_2, \\ \lambda_2' &= \lambda_2 - \lambda_3, \\ &\cdots \qquad\qquad\qquad \text{with} \\ \lambda_N' &= \lambda_{N-1} - \lambda_N, \\ \lambda_N' &= \lambda_N, \end{aligned} \qquad \begin{aligned} \mathbf{U}_1' &= \mathbf{u}_1\mathbf{u}_1^T, \\ \mathbf{U}_2' &= \mathbf{u}_1\mathbf{u}_1^T + \mathbf{u}_2\mathbf{u}_2^T, \\ &\cdots \\ \mathbf{U}_{N-1}' &= \mathbf{u}_1\mathbf{u}_1^T + \mathbf{u}_2\mathbf{u}_2^T + \cdots \mathbf{u}_{N-1}\mathbf{u}_{N-1}^T, \\ \mathbf{U}_N' &= \mathbf{u}_1\mathbf{u}_1^T + \mathbf{u}_2\mathbf{u}_2^T + \cdots \mathbf{u}_N\mathbf{u}_N^T = \mathbf{I}. \end{aligned}$$
$$(12.46)$$

For $j < N$, λ_j', represents the certainty for $\mathbf{S}$ to represent a j-dimensional hyperplane, whereas λ_N' represents the certainty for $\mathbf{S}$ to represent a perfectly balanced structure.

◆

It is worth noting that in image analysis applicatons, all directions may not have the same physical meaning. This implies that distances in the N-dimensional space on which f, and thereby $\mathbf{S}$ is defined may lack physical relevance, although directions may still be meaningful. For example, in a TV sequence, the first two dimensions are length, whereas the third dimension is time, e.g., see Figs. 12.2 and 12.1. Depending on what unit has been chosen for the time (seconds or hours) and the length (meter or km) and whether or not the points are in the same image frame, the interpretation of the Euclidean length between two points in the 3D spatio–temporal sequence will be different. However, the direction of lines and planes will represent the velocity, a physically meaningful quantity. We will discuss the spatio–temporal direction below.

12.4 Basic Concepts of Image Motion

Visual motion analysis is a vital processing element of the mammalian visual systems, which include human vision. Even still image processing, which would normally be handled by still image analysis tools in computer vision, are handled via motion analysis pathways of the brain. For example, even when analyzing a painting on the wall, the human eyes perform saccades, whereby the motif is brought to an artificial motion on the retina to the effect that the relevant visual field is analyzed by the cells of the brain that are motion-direction, and/or spatial-direction sensitive. While the current understanding of mammalian vision still leaves room for discussion on the reasons why biological vision has gone to 3D signal analysis to solve 2D problems, it is beyond doubt that motion is a very valuable and powerful feature. By contrast, it is fair to say that much of the development efforts in computer vision have been steered towards still image analysis tools essentially because of the limitations in computational resources, even if the power of motion in image analysis is not a disputed issue. However, because of the rapid developments in computer and communication technologies, this argument is swiftly being suppressed as a raison d'etre.

The motion of a body can be observed by an eye or by artificial imaging equipment. In this section we will discuss the translational motion of a local patch, as this is a good starting point to approximate even more complex types of motions, e.g., those that can bring an image into rotation. We discuss the motion of a surface patch of a 3D object moving relative to a camera in a small time interval. The velocity of the motion is a 3D vector. When the moving patch is observed by a camera (on a 2D plane), the light reflections from the surface are projected to the image plane. Also, the 3D velocity vector is projected to the image plane, now becoming a 2D vector field. The vector field representing the translations of the moving object surface patches is known as the *motion field*. Whether or not the motion field is actually identifiable by a vision system from the observed projections of the object surface depends on many factors, such as the surface color, its texture, the optical properties of the material when interacting with light, and the illumination. The observable version of the motion field is known as *optical flow*. In many circumstances the optical

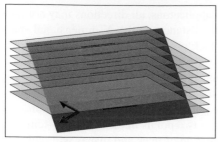 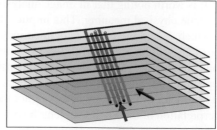

Fig. 12.2. (*Left*) The uniform motion of a line in the image plane generates the *magenta plane* in the 3D spatio–temporal signal. (*Right*) The uniform motion of a set of points, shown in *black*, generates a set of parallel lines, drawn in *magenta*

flow field will be a good approximation of the motion field, although there are numerous circumstances in which the approximation will be poor. In terms of a rotating sphere, we will discuss the distinction between the two fields and the identification of the motion from images of the motion further in Sect. 12.12. Below we will summarize the essentials of the translational motion assuming that the motion field equals to the optical flow and vice versa.

We will study the motion by modeling it as a 2D image patch that has been brought into motion according to a model, e.g., a translation or an affine motion. When doing this we also assume that the following constraint is satisfied by the spatio–temporal image observations:

Definition 12.2 (BCC). *The* brightness constancy constraint *(BCC) is satisfied if the colors of the spatio–temporal image points representing the same 3D points, remain unchanged throughout the spatio–temporal image.*

The simplest motion model is the translational model. In the following we discuss how the 2D content influences the computation of the translational motion. The observability of translation depends on the directional content of the 2D pattern brought into such a motion. Leaving out the nonobservable translation of a constant (gray value or color), there are two fundamental classes that can be discerned:

- The translation of linearly symmetric 2D images, i.e., the linear symmetry component of the 2D structure tensor is nonzero whereas the balanced direction component is zero.
- The translation of 2D images that have more than one directions, i.e., the balanced direction component of the 2D structure tensor is nonzero.

Assuming a continuum of image frames, i.e., a continuous 3D volume, the motion of a line generates *isosurfaces*[4] that are equal to a tilted plane. If there are more parallel lines in motion there will be more layers of planes that will be parallel to each other, like a cheeseburger. The more the motion plane tilts, the faster the motion

---

[4] An isosurface is the set of points for which $f(x, y, t) = $ constant.

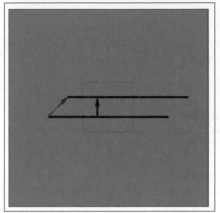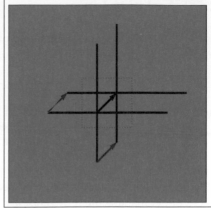

Fig. 12.3. (*Left*) The translational motion of a linearly symmetric patch. (*Right*) The translational motion of a patch containing two distinct directions. The actual and the observable translations are drawn as *magenta* and *black* vectors, respectively

is perceived. This situation is shown in Fig. 12.2, left, where the normal vector of the motion plane (magenta) is drawn in blue and the (normal) motion vector of the line is drawn in black.

By contrast, the motion of a point generates a line in the 3D continuous image. This situation is shown in Fig. 12.2, right, where dots translate upwards in the image plane, as represented by the black arrow. The result is many parallel lines, reminiscent of a bundle of spaghetti.

The two fundamental translation types are not new types of phenomena. They are not even specific to the physics of motion. Indeed, these occur as a consequence of physical quantities that happen to be equivalent to or deduced from the structure tensor. In particular, the motion vectors in the image plane are uniquely determined by the normal of the plane, $\mathbf{k}$, if the 2D image patch is linearly symmetric (a 2D property). Likewise, the 2D optical flow is uniquely determined by the direction of the generated 3D line, $\mathbf{k}$, if the 2D structure tensor of the image patch has a balanced direction tensor component that is nonzero.[5] In both cases the vector $\mathbf{k}$ is a 3D vector and the optical flow is insensitive to its direction, i.e., $-\mathbf{k}$ is as good as $\mathbf{k}$ when it comes to represent, the velocity. Accordingly, the same tensor,

$$\mathbf{k}\mathbf{k}^T \tag{12.47}$$

can represent both types of translational motion conveniently.

---

[5] This is equivalent to saying that the image patch is nontrivial (nonconstant) and it lacks linear symmetry.

12.5 Translating Lines

Assuming that the normal of the tilting plane is $\mathbf{k}$, how is the 2D normal flow obtained from this 3D vector? In this case we have a linearly symmetric image in 2D,

$$g(\cos(\theta)x_0 + \sin(\theta)y_0) \tag{12.48}$$

where $g(\tau)$ is a one-dimensional function, the vectors

$$\mathbf{s}_0 = (x_0, y_0)^T, \quad \mathbf{a} = (\cos(\theta), \sin(\theta))^T \tag{12.49}$$

represent the coordinates of an arbitrary point in the image plane, and the normal of the line(s) defining the linearly symmetric image, respectively. The 2D image in Eq. (12.48) is manufactured by replacing the argument of the 1D function with the "equation" of a line:

$$\cos(\theta)x_0 + \sin(\theta)y_0 = \tau \tag{12.50}$$

Clearly, the gray value does not change as long as we are on a certain line, i.e., the x_0, y_0 pair that satisfies Eq. (12.50), and therefore the notion of line is justified when speaking about linearly symmetric images. We take this CT reasoning one step further and translate one of the lines in the pattern g.

We can assume that the position vector $\mathbf{s}_0 = (x_0, y_0)^T$ represents a (spatial) point in the image plane at the time instant $t = 0$ and we wish to move it with the velocity $v\mathbf{a}$, where $\mathbf{a}$ is the direction of the velocity ($\|\mathbf{a}\| = 1$) and v is the absolute speed. A velocity in the direction orthogonal to $\mathbf{a}$, i.e., the line moves "along itself", will not be observable. This problem is also known as the *aperture problem*. That is why only in the direction $\mathbf{a}$ can a motion be observed in a linearly symmetric image, Eq. (12.48). It may not be the true motion, but it is the only motion that we can observe. Accordingly, after time t, a point on the line can be assumed to have moved to the position

$$\mathbf{s}(t) = (x(t), y(t))^T = \mathbf{s}_0 + vt \cdot \mathbf{a} \tag{12.51}$$

so that $\mathbf{s}_0 = \mathbf{s} - vt \cdot \mathbf{a}$, although we should be aware that the point may have actually moved to any place along the line. The vector $v\mathbf{a}$ is called the *normal image velocity* or *normal optical flow*. Substituting, the gray-values expression in Eq. (12.48) yields a spatio–temporal image (sequence) in which the lines of g move with the same velocity:

$$g(\mathbf{a}^T \mathbf{s}_0) = g(\mathbf{a}^T \mathbf{s}(t) - vt\mathbf{a}^T\mathbf{a}) = g(\mathbf{a}^T \mathbf{s} - vt) = g(\tilde{\mathbf{k}}^T \mathbf{r}) \tag{12.52}$$

Here we have defined the new variables $\tilde{\mathbf{k}}$ and $\mathbf{r}$ as the spatial variables augmented with the temporal variables $-v$ and t, respectively.

$$\tilde{\mathbf{k}} = [\mathbf{a}^T | - v]^T \in E_3, \quad \mathbf{r} = [\mathbf{s}^T | t]^T \in E_3 \tag{12.53}$$

However, Eq. (12.52) represents a linearly symmetric 3D image having parallel planes as isosurfaces. The vector $\tilde{\mathbf{k}}$ is thus equal to the normal of the plane

$\mathbf{k}^T \mathbf{r} = $ constant that will be fit by the most significant eigenvector of the 3×3 structure tensor of $f(x, y, t)$. Notice that the first two elements of $\tilde{k}$ are normalized to have length 1. Accordingly, given that one of its first two elements is nonnil, the normal vector $\mathbf{k}$ is related to $\tilde{k}$ as

$$\tilde{\mathbf{k}} = \frac{\mathbf{k}}{\sqrt{k_x^2 + k_y^2}} \tag{12.54}$$

Obtained via such a normalization from the most significant eigenvector of the structure tensor, the first two elements of $\tilde{k}$ will then equal to $\mathbf{a}$:

$$\mathbf{a} = (\frac{k_x}{\sqrt{k_x^2 + k_y^2}}, \frac{k_y}{\sqrt{k_x^2 + k_y^2}})^T \tag{12.55}$$

and the third element will be equal to the speed : v

$$v = -\frac{k_t}{\sqrt{k_x^2 + k_y^2}} \tag{12.56}$$

to the effect that the velocity or the *normal optical flow* will be given by $v\mathbf{a}$

$$\mathbf{v} = -v\mathbf{a} = -\frac{k_t}{k_x^2 + k_y^2}(k_x, k_y)^T \tag{12.57}$$

12.6 Translating Points

It is possible to estimate an unambigious optical flow for certain 2D image patterns in motion. We discuss the necessary and sufficient conditions further below, whereas for now we will be served well enough if we assume the existence of isocurves in the 3D spatio–temporal image that is parallel lines (not planes!). This should enable us to track where a point moves to with higher precision than the translating lines. Such images are obtained, for example, when we have images like a sand pattern that translates. Because of lemma 12.4, we know that the 3D Fourier transform of such translating 2D patterns is concentrated to a plane. How can the optical flow be obtained from the common direction of the 3D lines, $\mathbf{k}$, that the 2D pattern undergoing a group translation generates, see Fig. 12.2, right? We discuss an answer below.

We assume that $g(x, y)$ is a 2D pattern that contains trackable points at the time instant $t = 0$. Translating a point $(x, y)^T$ in the image plane x, y with the velocity $\mathbf{v} = (v_x, v_y)^T$ to another point $(x', y')^T$ is achieved by

$$\begin{aligned} x' &= x + tv_x \\ y' &= y + tv_y \\ t' &= t \end{aligned} \tag{12.58}$$

which is a linear CT, $\mathbf{r} \rightarrow \mathbf{r}'$. Accordingly, a 2D image $g(x, y)$ is translated by substituting the above CT in the gray image,

$$g(x,y) = g(x' - t'v_x, y' - t'v_y) = f(x', y', t') \tag{12.59}$$

This yields a spatio–temporal image function f defined on E_3 that contains a volume of image frames originating from the continuously varying time, t'. Accordingly, the CT is written in matrix form as:

$$\mathbf{r}' = \mathbf{A}\mathbf{r}, \quad \text{with} \quad \mathbf{r} = (x, y, t), \quad \mathbf{A} = \begin{pmatrix} 1 & 0 & v_x \\ 0 & 1 & v_y \\ 0 & 0 & 1 \end{pmatrix}. \tag{12.60}$$

Using $\mathbf{s}' = (x', y')^T$, and $\mathbf{s} = (x, y)^T$ for convenience, the inverse of this CT is easily found by observing that $\mathbf{s}' = \mathbf{s} + t\mathbf{v}$, Eq. (12.58),

$$\mathbf{s} = \mathbf{s}' - t\mathbf{v} = \mathbf{s}' - t'\mathbf{v} = \begin{cases} x = x' - t'v_x \\ y = y' - t'v_y \end{cases} \Rightarrow \mathbf{A}^{-1} = \begin{pmatrix} 1 & 0 & -v_x \\ 0 & 1 & -v_y \\ 0 & 0 & 1 \end{pmatrix} \tag{12.61}$$

Now, the FT of $f(x', y', t')$ is

$$\begin{aligned} F(k_x, k_y, k_t) &= \int f(\mathbf{r}') \exp(-i\mathbf{k}^T\mathbf{r}')d\mathbf{r}' \\ &= \int g(x' - t'v_x, y' - t'v_y)\exp(-i\mathbf{k}^T\mathbf{r}')d\mathbf{r}' \\ &= \int g(x, y)\exp(-i\mathbf{k}^T\mathbf{A}\mathbf{r})|J(\mathbf{r}', \mathbf{r})|d\mathbf{r} \\ &= \int g(x, y)\exp(-i\mathbf{k}'^T\mathbf{r})|J(\mathbf{r}', \mathbf{r})|d\mathbf{r} \\ &= \int g(x, y)\exp(-i(k_x'x + k_y'y))\exp(-ik_t't)d\mathbf{s}dt \\ &= G(k_x', k_y')\delta(k_t') = G(k_x, k_y)\delta(k_xv_x + k_yv_y + k_t) \tag{12.62} \end{aligned}$$

where $\mathbf{k}' = (k_x', k_y', k_t')^T$ and $|J(\mathbf{r}', \mathbf{r})|$, the determinant of the Jacobian, are

$$\mathbf{k}'^T = \mathbf{k}^T\mathbf{A}, \quad \text{and} \quad |J(\mathbf{r}, \mathbf{r}')| = \det\left[\begin{pmatrix} \frac{\partial x'}{\partial x} & \frac{\partial x'}{\partial y} & \frac{\partial x'}{\partial t} \\ \frac{\partial y'}{\partial x} & \frac{\partial y'}{\partial y} & \frac{\partial y'}{\partial t} \\ \frac{\partial t'}{\partial x} & \frac{\partial t'}{\partial y} & \frac{\partial t'}{\partial t} \end{pmatrix}\right] = \det(\mathbf{A}) = 1 \tag{12.63}$$

We can see from Eq. (12.62) that the 3D FT of a 2D pattern in translation is an intersection of an oblique plane having the normal $\mathbf{w} = (\mathbf{v}^T, 1)^T$,

$$k_xv_x + k_yv_y + k_t = \mathbf{w}^T\mathbf{k} = 0 \tag{12.64}$$

and a cylinder,[6]

---

[6] Note that the cylinder is given by the 2D FT of the pattern stacked in the depth, i.e., $\tilde{G} = G$ for all values of k_t.

$$\tilde{G}(k_x, k_y, k_t) = G(k_x, k_y) \tag{12.65}$$

The inclination of the plane, Eq. (12.64), is steered by the velocity, $\mathbf{v}$, in that it controls the normal of the plane, $\mathbf{w}$

$$\mathbf{w} = \begin{pmatrix} \mathbf{v} \\ 1 \end{pmatrix} \tag{12.66}$$

This normal vector, and thereby the velocity, can be estimated from the image data by the least significant eigenvector of the structure tensor, k_3. By construction, the third element of $\mathbf{w}$ in Eq. (12.66) must be 1 for its first two elements to equal to $\mathbf{v}$. Accordingly, we obtain $\mathbf{v}$ from k_3 as follows.

$$k_3 = (k_x, k_y, k_t) \quad \Rightarrow \quad \mathbf{v} = (\frac{k_x}{k_t}, \frac{k_y}{k_t})^T \tag{12.67}$$

Some studies refer to Eq. (12.62) as "a tilting of G". The product $G\delta$ is thus a cut of the cylinder $\tilde{G}$ consisting of the 2D FT of the still pattern by a thin oblique plane, δ, representing the motion plane, Eq. (12.64). Evidently, such a cut is not a true tilting of $G(\mathbf{k})$ in the FT domain because the coordinates of $G(\mathbf{k})$ would have undergone an orthogonal CT, which Eq. (12.62) does not depict. Instead, the transformation is an inverse projection, i.e., the FT of the still image is a projection of the motion plane to the k_x, k_y plane. Even if the still image is a band-limited function, without a limitation on the speed, a translation can tilt the motion plane to reach arbitrary high values for k_t. Accordingly, a translation is a band-enlarging operation for band-limited functions.

The velocity in Eq. (12.67) requires that $k_t \neq 0$. When $\mathbf{k}_t \to 0$ the velocity risks increasing beyond every bound, causing numerical instability in computations. The question for what patterns k_t becomes small, and thereby the unambiguous motion estimation becomes unstable, is discussed in the next paragraph.

The Velocity of Two Nonparallel Lines and Sensitivity Analysis

We assume now that we have an image,

$$g(x_0, y_0) \tag{12.68}$$

defined on E_2, but that this image contains two sets of isocurves with distinct directions,[7] e.g., two sets of nonparallel lines in the image $g(x_0, y_0)$. Accordingly, the 2D vectors $\mathbf{a}_1, \mathbf{a}_2$ represent the respective normal vectors of the assumed line sets,

$$\mathbf{a}_1 = (\cos(\theta_1), \sin(\theta_1))^T, \quad \mathbf{a}_2 = (\cos(\theta_2), \sin(\theta_2))^T, \quad \text{where} \quad \theta_1 \neq \theta_2. \tag{12.69}$$

---

[7] Such an image is not linearly symmetric in 2D since it contains more than one direction. For this reason g is not a function originally defined on 1D originally but instead on 2D. Sums of two linearly symmetric functions with different directions are examples of such images.

When the line sets translate with a common velocity vector $\mathbf{v}$ so that a point at s_0 moves to s:

$$\mathbf{v} = (v_x, v_y)^T \quad \Rightarrow \quad s = (x, y)^T = s_0 + t\mathbf{v} = (x_0 + v_x t, y_0 + v_y t)^T \quad (12.70)$$

We obtain by this CT an image f that is continuous in x, y, t

$$f(x, y, t) = g(x - v_x t, y - v_y t) \tag{12.71}$$

in analogy with the discussion in the previous section. Because $(s - s_0) = t\mathbf{v}$, is what we are interested in, the orthogonal projections of the (common) velocity $\mathbf{v}$ on the normal directions of the lines are the individual speeds that would have been perceived by the observer, if there was only one direction in the image. In that case, the components of the velocity in the directions of $\mathbf{a}_1$, and $\mathbf{a}_2$ and the lines themselves would define two sets of oblique planes containing the lines. The normal vectors of these planes generated by the motion of each line are given by

$$\tilde{\mathbf{k}}_1 = (\cos(\theta_1), \sin(\theta_1), -\mathbf{a}_1^T \mathbf{v})^T \quad \text{and} \quad \tilde{\mathbf{k}}_2 = (\cos(\theta_2), \sin(\theta_2), -\mathbf{a}_2^T \mathbf{v})^T$$

according to the discussion preceeding Eq. (12.53). The cross-product of these 3D normals will then be orthogonal to both, yielding the vector pointing in the direction of the intersection (lines) of the nonparallel plane sets.

$$
\begin{aligned}
\tilde{\mathbf{k}}_3 = \tilde{\mathbf{k}}_1 \times \tilde{\mathbf{k}}_2 &= \begin{pmatrix} \cos(\theta_1) \\ \sin(\theta_1) \\ -v_x \cos(\theta_1) - v_y \sin(\theta_1) \end{pmatrix} \times \begin{pmatrix} \cos(\theta_2) \\ \sin(\theta_2) \\ -v_x \cos(\theta_2) - v_y \sin(\theta_2) \end{pmatrix} \\
&= \begin{pmatrix} \sin(\theta_1)(-v_x \cos(\theta_2) - v_y \sin(\theta_2)) - \sin(\theta_2)(-v_x \cos(\theta_1) - v_y \sin(\theta_1)) \\ -\cos(\theta_1)(-v_x \cos(\theta_2) - v_y \sin(\theta_2)) + \cos(\theta_2)(-v_x \cos(\theta_1) - v_y \sin(\theta_1)) \\ \cos(\theta_1)\sin(\theta_2) - \sin(\theta_1)\cos(\theta_2) \end{pmatrix} \\
&= \begin{pmatrix} -v_x \cos(\theta_2)\sin(\theta_1) + v_x \sin(\theta_2)\cos(\theta_1) \\ v_y \sin(\theta_2)\cos(\theta_1) - v_y \sin(\theta_1)\cos(\theta_2) \\ \sin(\theta_2 - \theta_1) \end{pmatrix} \\
&= \begin{pmatrix} v_x \\ v_y \\ 1 \end{pmatrix} \sin(\theta_2 - \theta_1) = \begin{pmatrix} \mathbf{v} \\ 1 \end{pmatrix} \sin(\theta_2 - \theta_1)
\end{aligned}
$$

The vector $\tilde{\mathbf{k}}_3$ can be estimated by the least significant eigenvector of the structure tensor of the function $f(x, y, t)$, i.e., $\mathbf{k}_3$, up to a scale.

$$\tilde{\mathbf{k}}_3 = \mathbf{k}_3 \quad \Leftrightarrow \quad \begin{pmatrix} \mathbf{v} \\ 1 \end{pmatrix} \sin(\theta_2 - \theta_1) = \begin{pmatrix} k_x \\ k_y \\ k_t \end{pmatrix} \tag{12.72}$$

This reconfirms the velocity estimate computations suggested in the previous paragraph, Eq. (12.67), as

$$\mathbf{v} = (\frac{k_x}{k_t}, \frac{k_y}{k_t})^T \tag{12.73}$$

However, most important, Eq. (12.72) shows that the condition $k_t \neq 0$ will be best fulfilled when the difference between the two directions is maximum, $(\theta_2 - \theta_1) = \pi/2$. Ideally this happens when the 2D image has perfectly balanced directions, i.e., both eigenvalues of the 2D structure tensor of g are large and are equal to each other. Ill-conditioned numerical computations can be avoided by assuring that: (i) there is a nonzero motion, i.e., $\|\frac{\partial f}{\partial t}\| > 0$; (ii) the (2D spatial) gradient in the 2D image is nonzero, i.e., $\|\nabla g\| > 0$; (iii) the 2D image g lacks linear symmetry such that there are at least two distinct directions in it, i.e., the most significant eigenvalue of the 2D structure tensor of g has the multiplicity 2. We summarize these findings in the following lemma.

Lemma 12.6 (Spatial directions constraint). *The lack of linear symmetry is a sufficient and necessary condition for an image $g(x, y)$ to satisfy for a translation of it be computable from the corresponding spatiotemporal image $f(x, y, t)$. The minimum number of directions that must be contained in g to allow computation of an unambiguous translational motion is 2, provided that they are sufficiently distinct.*

$\blacklozenge$

12.7 Discrete Structure Tensor by Tensor Sampling in ND

In discretizing the ND tensor, we will follow an analogous approach to that developed in Sect. 10.11; for this reeason we only state the results without derivations or proofs. The computation and discretization in the r-domain is done by utilizing the Parseval–Plancherell theorem and by assuming that the interpolation function and the multiplicative window defining a local image, when need be, are two Gaussians with σ_p and σ_w, respectively.

The structure tensor for E_3

The structure tensor for the 3D Euclidean space was defined in the spectral domain and for continuous images via Eq. (12.13). In numerous situations, these quantities need to be estimated on a discrete (Cartesian) 3D grid in the spatial domain, (the r-domain). Furthermore, this should, in many applications, be done quite often, e.g., for local images around every point.

Lemma 12.7. *The structure tensor is estimated in the TLS error sense, by averaged tensor (outer) products*

$$\mathbf{S} = \frac{1}{4\pi^2} \sum_l (\nabla f_l)(\nabla f_l)^T m_l. \tag{12.74}$$

where ∇f_l is the gradient of $f(\mathbf{r})$ at the point $\mathbf{r}_l$, which belongs to a regular discrete grid, and m_l is an averaging kernel that consists of a discrete Gaussian. The discrete gradients are estimated by filtering the original image f_l with a kernel consisting of a discrete gradient of a Gaussian.

$\blacklozenge$

Eq. (12.74) is equivalent to a discrete convolution by a Gaussian if $\mathbf{S}$ is computed for every point in the original discrete image. Since the values of m_l, the filter coefficients, decrease rapidly outside of a circle with radius equal to a few standard deviations of the Gaussians used, we can truncate the infinite filter when its coefficients are sufficiently small. The advantages of Gaussians in 3D Euclidean spaces are even more imposing than those in 2D: their separability, compactness, and being fully isotropic. The conclusions, including the lemma, are evidently valid also in ND.

The structure tensor for sampled motion images

Both translational and affine model-based motion parameter estimations developed in the previous sections originally formulated the problem as a direction estimation problem in the continuous domain. The optical flow vectors, formally in 2D, are encoded in the the 3D structure tensor, which in turn can be estimated by by use of lemma 12.7. However, some issues raised in the context of sampling band-enlarging operators in Sect. 8.3, need to be considered.

First, the structure tensor of any dimension is a band-enlarging operator because it requires multiplications of functions having the same bandwidth, due to the tensor product $\nabla f \nabla^T f$. The tensor image $\nabla f \nabla^T f$ is also a band-limited function, but the bandwidth is now larger than that of the bandwidth of the constituent vector image ∇f. To make sure that $\nabla f \nabla^T f$ is sampled without aliasing, the components $D_{r_i} f$, where $r_i = x, y, t$, must be oversampled with a factor 2. Alternatively, the maximum frequencies of F should stay within half the Nyquist cube.[8] This is typically done by incorporating a lowpass filtering directly into the kernel coefficients of the operators D_{r_i}. Thus, in each direction the Gaussian interpolation functions used to reconstruct f from its samples must be at least $\sigma_p \approx 1.3$, as discussed in Sect. 9.3. Furthermore, since the physical dimensions of spatial coordinates x, y are different than that of the temporal coordinate t, the used σ_p are the same for the spatial coordinates but different than that of the spatial coordinate, σ_p. Accordingly, the derivation operator is the sampling of the derived interpolation function:

$$D_{r_i}\mu(x,y,t) = \exp\left(-\frac{x^2+y^2}{2\sigma_{p^s}^2}\right)\exp\left(-\frac{t^2}{2\sigma_{p^t}^2}\right), \qquad \text{with} \qquad r_i = x, y, t.$$

Second, translating a 2D band-limited image, $g(x,y)$, to produce a 3D spatio–temporal image, $f(x,y,t)$ is a band-enlarging operation. This is concluded from Eq. (12.62), which states that the 3D FT, $F(k_x, k_y, k_t)$, is given by the cylinder $G(k_x, k_y)$ cut by the motion plane, i.e., the plane with the normal $(\mathbf{v}, 1)^T$. Accordingly, if $g(x,y)$ is a band-limited function, i.e., $G(k_x, k_y)$ vanishes when $\mathbf{k}_s = (k_x, k_y)^T$ is outside of a compact region,

---

[8] Recall that the Nyquist cube is the cube with surfaces intersecting the frequency coordinate axes at π and surface normals coinciding with the frequency axes. Half the Nyquist cube is similar, but the surface intersections are at $\frac{\pi}{2}$ on the frequency axes.

$$\|\mathbf{k}_s\| > K_s \quad \Rightarrow \quad G(k_x, k_y) = 0 \tag{12.75}$$

Because of this $|k_t|$ will have an upper bound for the meaningful spatial frequencies

$$\|\mathbf{k}_s\| \le K_s \quad \Rightarrow \quad |k_t| \le vK_s = K_t \tag{12.76}$$

Then, if the translation speed v is limited, the spatio–temporal image is band-limited, too. Furthermore, if the temporal axis is sampled with the sampling period $\frac{2\pi}{2K_t}$ or tighter, the speeds of points moving with speeds not greater than v will be recoverable, because the motion planes generated by such translations will have a smaller inclination angle with the k_x, k_y plane. In consequence of this, it is sufficient that $v_{\max}$, K_t, and K_s satisfy

$$v_{\max} \le \frac{K_t}{K_s} \tag{12.77}$$

where $v_{\max}$ is the largest speed of any point in $f(x, y, t)$, to recover the spatio–temporal image, and thereby represent the motion without distortion, via the samples $f(x_j, y_j, t_j)$. We summarize these results in the following lemma:

Lemma 12.8. *Let $g(x, y)$ be a band-limited function with the maximum spatial frequency K_s and f be a translated version of it, $f(x, y, t) = g(x - v_x t, y - v_y t)$, with the velocity $\mathbf{v} = (v_x, v_y)^T$. Then,*

$$K_s |\mathbf{v}| = K_t \tag{12.78}$$

where K_t is the maximal temporal frequency that can occur in F. If the structure tensor is to be sampled, then K_s and K_t must satisfy

$$K_s \le \frac{\pi}{2} \qquad K_t \le \frac{\pi}{2} \tag{12.79}$$

assuming normalized spatial and temporal sampling periods.

♦

If we assume that both $K_s = K_t = \pi$, which normalizes the sampling period to unity to the effect that the sampling points are one unit apart in x, y, and t directions, then the maximal speed of the translation is bounded by

$$v_{\max} \le \frac{K_t}{K_s} = 1 \tag{12.80}$$

To recover larger speeds from sampled images, either K_t must be increased (the temporal axis is sampled more densely) or K_s should be decreased (x, y axes are low-pass filtered or the image is enlarged). When the structure tensor is used to recover the velocity, one must make sure that both K_t and K_s are less than $\pi/2$, assuming normalized distances between the image pixels and image frames.

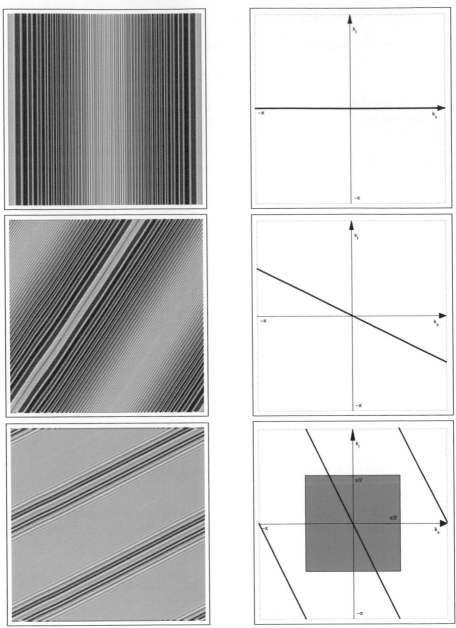

Fig. 12.4. (*Left*) A pattern undergoing a translation of (*top*) 0, (*middle*), 0.67, and (*bottom*) 2 pixels/frame. (*Right*) The schematic FTs are also shown

We illustrate the signal discretization effects of translational motion by Fig. 12.4, where a 1D pattern, containing the full range of the frequencies, is translated with zero, 0.67, and 2 pixels per frame. In the color images, the vertical axis is the time axis. The image is actually a gray image, see Fig. 8.8, but to illustrate the translation effectively, every gray tone has been replaced by a unique color. The corresponding schematic FTs of the signals are shown besides, with the red lines representing the nonzero power. For small motions there is no aliasing wrap-around, because the temporal sampling frequency is sufficient to discretize the motion faithfully. However, when the motion magnitude is increased above 1 pixel/frame, the sampling rate becomes too low for a satisfactory representation. The result is a distortion (aliasing), or frequency wrap-around caused by the repetition of the Nyquist square in the k_t direction. Had the spatio–temporal signal been oversampled with a factor of 2 in the x, as well as in the t directions, the extra red lines, caused by the repetition in the k_t direction, would not appear. This is because the signal would then be confined to the cyan-colored zone shown in the figure. In conclusion, the motion sequences in the spatial as well as temporal domains must be oversampled with a factor 2, to avoid sampling aliasing caused by high spatial frequency contents (quick variations) moving fast. Independently, this is also what is required if the structure tensor is to be used to compute the velocity.

12.8 Affine Motion by the Structure Tensor in 7D

Whereas a translational model is often adequate to describe the motion in a local image, it becomes sometimes insufficient to describe complex motion when the image patch to be analyzed is enlarged and thereby generally the complexity of the motion is increased. A more elaborate model such as an *affine motion* model will generally be better placed to describe the motion of a rigid object, particularly if the field of view is small enough [132, 141]. Here we only treat the case when all image points move according to the same affine motion model, parameters whose need to be identified.

As before, we assume that BCC is valid, meaning that the spatio–temporal image is generated from one frame of the image, i.e., the brightness distribution originating from a certain instant. The generation of the rest of the spatio–temporal image will be done by applying the *affine coordinate transformation* to the spatial coordinates, $\mathbf{s} = (x, y)^T$:

$$\mathbf{s}^* = s + \delta t[\mathbf{A}_0 \mathbf{s} + \mathbf{v}_0] \quad \Rightarrow \quad f(x, y, t) = g(s + \delta t[\mathbf{A}_0 \mathbf{s} + \mathbf{v}_0]) \tag{12.81}$$

Consequently, the affine model is characterized by a velocity field $\mathbf{v}$ expressed in the parametric form:

$$\mathbf{v}(\mathbf{s}) = \mathbf{A}_0 \mathbf{s} + \mathbf{v}_0 \tag{12.82}$$

with $\mathbf{v}_0$, $\mathbf{A}_0$ being a 2D vector and 2×2 invertible matrix, respectively. There are two real parameters in the constant translation $\mathbf{v}_0$, and four in the matrix $\mathbf{A}_0$, representing rotation, scaling, and the (two) shearing deformations, totaling to six degrees of

Fig. 12.5. Affine motion is estimated from 5 images (*right*) [61], using the eigenvectors of (12.92). One of the original frames is shown on the *left*. On the *right*, the *red* car region, and the *cyan-labeled* background region have different affine motion parameters which were found automatically

freedom in the affine model. Naturally, more complex flow fields can be modeled by the affine model than the translation model, which has only two degrees of freedom. The affine matrix $\mathbf{A}_0$ can be decomposed into [58, 64, 132],

$$\mathbf{A}_0 = \mathbf{A}_0^s + \mathbf{A}_0^r + \mathbf{A}_0^{h_1} + \mathbf{A}_0^{h_2} \tag{12.83}$$

where the matrices on the right side control the amounts of rotation, scaling, shearing of the first type, and shearing of the second type, respectively.

$$\mathbf{A}_0^s = \begin{pmatrix} s & 0 \\ 0 & s \end{pmatrix}, \quad \mathbf{A}_0^r = \begin{pmatrix} 0 & -r \\ r & 0 \end{pmatrix}, \quad \mathbf{A}_0^{h_1} = \begin{pmatrix} h_1 & 0 \\ 0 & -h_1 \end{pmatrix}, \quad \mathbf{A}_0^{h_2} = \begin{pmatrix} 0 & h_2 \\ h_2 & 0 \end{pmatrix}$$

The CT in Eq. (12.81) is a differential equation modeling the speed or the optical flow field, because

$$\frac{\mathbf{s}^* - \mathbf{s}}{\delta t} \to \frac{d\mathbf{s}}{dt} = \mathbf{v}(x, y, t) \tag{12.84}$$

when δt approaches to zero. Accordingly, the trajectory of a point in the image plane, $\mathbf{s}(t) = (x(t), y(t))^T$, can be solved analytically. The six parameters represented by $\mathbf{v}_0$ and $\mathbf{A}_0$ steer the tangent curves of $\mathbf{s}$. It can be shown that a small motion along these curves can be achieved by an *affine infinitesimal operator*[9] [58],

$$D_\zeta = k_1 D_{\xi_1} + k_2 D_{\xi_2} + \cdots k_7 D_{\xi_7} = \sum_{j=1}^{7} k_j D_{\xi_j} \tag{12.85}$$

---

[9] The CT generated by the affine model is a Lie group of transformations of one parameter, time.

with

$$
\begin{aligned}
D_{\xi_1} &= D_x \\
D_{\xi_2} &= D_y \\
D_{\xi_3} &= xD_x + yD_y \\
D_{\xi_4} &= -yD_x + xD_y \\
D_{\xi_5} &= xD_x - yD_y \\
D_{\xi_6} &= yD_x + xD_y \\
D_{\xi_7} &= D_t
\end{aligned}
\tag{12.86}
$$

under which a spatio–temporal image generated by the affine coordinate transformation is invariant:

$$
D_\zeta f(x,y,t) = \sum_j k_j D_{\xi_j} f(x,y,t) = 0
\tag{12.87}
$$

From this equation we will attempt to solve the unknown parameter vector:

$$
\mathbf{k} = (k_1, \cdots k_7)^T, \qquad \text{with} \qquad \|\mathbf{k}\| = 1.
\tag{12.88}
$$

The solution is evidently not possible if we only know the 7D measurement vector $\mathbf{D}_\xi f$ defined as

$$
\mathbf{D}_\xi f = (D_{\xi_1} f, D_{\xi_2} f, \cdots D_{\xi_7} f)^T
\tag{12.89}
$$

at a single point $\mathbf{r} = (x, y, t)^t$. Equation (12.87) is effectively an equation of a hyperplane in E_7.

$$
\mathbf{k}^T \mathbf{D}_\xi f(\mathbf{r}) = 0
\tag{12.90}
$$

The equation is satisfied in many points ideally because f is in that case a spatio–temporal image that is truly generated from a static image by use of an *affine coordinate transformation*. In that case F will be concentrated to a 6D plane, see theorem 12.1 and lemma 12.7. Accordingly, estimating the *affine motion parameters* is a linear symmetry problem in E_7. It can be solved in the TLS error sense by minimizing the error

$$
e(\mathbf{k}) = \int \|\mathbf{k}^T \mathbf{D}_\xi f\|^2 dx dy dt = \mathbf{k}^T [\int (\mathbf{D}_\xi f)(\mathbf{D}_\xi^T f) dx dy dt] \mathbf{k} = \mathbf{k}^T \mathbf{S} \mathbf{k}
\tag{12.91}
$$

under the constraint $\|\mathbf{k}\| = 1$. Here, the matrix $\mathbf{S}$ is the structure tensor defined as

$$
\mathbf{S} = \int (\mathbf{D}_\xi f)(\mathbf{D}_\xi^T f) dx dy dt
\tag{12.92}
$$

The TLS estimate of the parameter vector $\mathbf{k}$ is given by the least significant eigenvector of $\mathbf{S}$. The necessary differential operators and the integrals are possible to implement by use of Eqs. (12.86), and (12.89) via separable convolutions with kernels derived from Gaussians [59]. Estimating $\mathbf{S}$ by sampling the structure tensor in ND is discussed in Sect. 12.7. Figure 12.5 illustrates a motion image sequence. The car is moving to the left whereas a video camera tracks the car and keeps it approximately at the center of the camera view. As a result there are two motion regions, the car's and the background's. The eigenvectors of Eq. (12.92) are computed twice, to estimate both the background and the car motion using the affine model which in turn has helped to find the motion boundaries [61]. The boundary estimation aided by motion parameters is far more accurate than only using static image frames (e.g., see the antenna of the car which has also a cable, loosely attached to it), although not perfect (e.g., the dark region behind the car is not well segmented).

12.9 Motion Estimation by Differentials in Two Frames

We begin with writing down the *velocity of a particle*, $\mathbf{v} = (v_x, v_y)^T$, moving in the (x, y)-plane, as defined in mechanics:

$$\mathbf{v} = \frac{d\mathbf{s}(t)}{dt} = (\frac{dx(t)}{dt}, \frac{dy(t)}{dt})^T = (v_x, v_y)^T \tag{12.93}$$

where $\mathbf{s} = (x, y)^T$ is the coordinate of the particle and t is the time coordinate. Let the function $f(x, y, t)$ represent the spatio–temporal image of such moving parti-cles,x where f is the gray intensity. A change of the gray intensity can be expected when changing the coordinates $\mathbf{s} = (x, y)^T$ and/or t. From calculus we can conclude that a small intensity change in f can be achieved by small changes of all three vari-ables of f. To be exact, the change df is controllable by the independent changes dx, dy, dt via:

$$df = dt\frac{\partial f}{\partial t} + dx\frac{\partial f}{\partial x} + dy\frac{\partial f}{\partial y} \tag{12.94}$$

or equivalently,

$$\frac{df}{dt} = \frac{\partial f}{\partial t} + \frac{dx}{dt}\frac{\partial f}{\partial x} + \frac{dy}{dt}\frac{\partial f}{\partial y} \tag{12.95}$$

Here we assumed that f is differentiable, i.e., all of its partial derivatives exist con-tinuously. It is of great interest to find a path $\mathbf{s}(t)$ such that f does not change at all, i.e.,

$$\frac{df}{dt} = 0 \tag{12.96}$$

This path will yield the isocurves of $f(x, y, t)$, which in turn offers an opportunity to track points. Under the condition that the path is unique, this is the same as having tracked a point if the BCC holds [110, 122], see Definition (12.2). This conclusion is reasonable because the image of a moving 3D point is a moving 2D point, which normally changes its gray value only insignificantly, at least during a sufficiently short observation time:

BCC: $$\frac{\partial f}{\partial t} + v_x\frac{\partial f}{\partial x} + v_y\frac{\partial f}{\partial y} = 0 \tag{12.97}$$

Conversely, if an image $f(x, y, t)$ satisfies this equation we can conclude that the motion satisfies the BCC [110, 157]. This is why the equation is frequently quoted as the *BCC equation* in image analysis studies.

If one attempts to solve $\mathbf{v} = (v_x, v_y)^T$ from the BCC when the partial derivatives of f are known at a given space–time x, y, t, one fails because $\mathbf{v}$ contains two real variables, whereas we have only one equation. However, if the equation holds at several image points $\mathbf{s} = (x_k, y_k)^T$ at a given time t_0, then we can write several such equations. This happens typically for a local image pattern $g(x, y)$ wherein all points translate with the same velocity $\mathbf{v}$. This situation is similar to the example in Fig. 12.2 (right), where the coherent translation of points is depicted. Accordingly, the BCC for a discrete 2D neighborhood $f(x_k, y_k, t_0)$ yields[10]

$$
\begin{pmatrix}
\frac{\partial f(x_1,y_1,t_0)}{\partial t} \\
\frac{\partial f(x_2,y_2,t_0)}{\partial t} \\
\vdots \\
\frac{\partial f(x_N,y_N,t_0)}{\partial t}
\end{pmatrix}
= -
\begin{pmatrix}
\frac{\partial f(x_1,y_1,t_0)}{\partial x} & \frac{\partial f(x_1,y_1,t_0)}{\partial y} \\
\frac{\partial f(x_2,y_2,t_0)}{\partial x} & \frac{\partial f(x_2,y_2,t_0)}{\partial y} \\
\vdots & \vdots \\
\frac{\partial f(x_N,y_N,t_0)}{\partial x} & \frac{\partial f(x_N,y_N,t_0)}{\partial y}
\end{pmatrix}
\begin{pmatrix}
v_x \\
v_y
\end{pmatrix}
\qquad (12.98)
$$

Here, $f(x_k, y_k, t_0)$ with $k \in \{1 \cdots N\}$ represents $f(x, y, t)$ evaluated at the kth point of an image neighborhood. This is an overdetermined system of linear equations of the form

$$
\mathbf{d} = -\mathbf{D}\mathbf{v} \qquad (12.99)
$$

with $\mathbf{v}$ being unknown and

$$
\mathbf{d} =
\begin{pmatrix}
\frac{\partial f(x_1,y_1,t_0)}{\partial t} \\
\frac{\partial f(x_2,y_2,t_0)}{\partial t} \\
\vdots \\
\frac{\partial f(x_N,y_N,t_0)}{\partial t}
\end{pmatrix}, \qquad
\mathbf{D} =
\begin{pmatrix}
\frac{\partial f(x_1,y_1,t_0)}{\partial x} & \frac{\partial f(x_1,y_1,t_0)}{\partial y} \\
\frac{\partial f(x_2,y_2,t_0)}{\partial x} & \frac{\partial f(x_2,y_2,t_0)}{\partial y} \\
\vdots & \vdots \\
\frac{\partial f(x_N,y_N,t_0)}{\partial x} & \frac{\partial f(x_N,y_N,t_0)}{\partial y}
\end{pmatrix}
\qquad (12.100)
$$

are to be assumed known. Suggested by Lucas and Kanade [157], this is a linear regression problem for optical flow estimation. The standard solution of such a system of equations is given by the MS estimate, obtained by multiplying the equation with $\mathbf{D}^T$ and solving for the 2×2 system of equations for the unknown $\mathbf{v}$:

$$
\mathbf{D}^T \mathbf{d} = -\mathbf{D}^T \mathbf{D} \mathbf{v} \qquad (12.101)
$$

The solution exists if the matrix

$$
\mathbf{S} = \mathbf{D}^T \mathbf{D} = \sum_k (\nabla_{\mathbf{s}_k} f) \cdot (\nabla_{\mathbf{s}_k}^T f) \qquad (12.102)
$$

where

$$
\nabla_{\mathbf{s}_k} f =
\begin{pmatrix}
\frac{\partial f(x_k,y_k,t_0)}{\partial x} \\
\frac{\partial f(x_k,y_k,t_0)}{\partial y}
\end{pmatrix}
\qquad (12.103)
$$

However, $\mathbf{S}$ is the structure tensor for the 2D discrete image $f(x_k, y_k, t_0)$. A unique solution exists if $\mathbf{S}$ is nonsingular, a situation that occurs if the image lacks linear

---

[10] For a fixed t_0, $f(x, k, y_k, t_0)$ is a 2D function.

symmetry. Conversely, if $\mathbf{S}$ has an eigenvalue that vanishes, there is no unique velocity that can be estimated from the image measurements. Having an overdetermined
system of equations and yet not being able to solve for the velocity may appear counterintuitive, but it is explained by the fact that the 2D pattern $f(x_k, y_l, t_0)$ is linearly
symmetric when an eigenvalue of $\mathbf{S}$ vanishes. Such images have parallel lines as
isocurves, that is, if translated along these lines no difference in gray values will be
noticed. Accordingly, this situation is the same as the one in the example of Fig. 12.2
(left), where a translating line has been depicted. Because lines and edges are common in real images, singular structure tensors are also common in image sequences.
In consequence, an appropriate nonsingularity test before solving Eq. (12.101) must
be applied.

The solution of Eq. (12.101) necessitates the estimation of the structure tensor for
a 2D image neighborhood, which we know how to do from Sect. 10.11. However,
the vector $\mathbf{d}$ is also needed. It is customary to estimate it via a temporal difference

$$\frac{\partial f(x_k, y_k, t_0)}{\partial t} = f(x_k, y_k, t_1) - f(x_k, y_k, t_0) \qquad (12.104)$$

between two successive image frames [157]. Accordingly, the optical flow technique
discussed above is possible to compute from just two frames.

The technique described in this section and the tensor approach discussed in
Sects. 12.5 and 12.6 are related since they are both gradient-based. The main difference is that the tensor approach solves the regression problem in the TLS sense,
whereas here it is solved in the MS sense, meaning that the regression error in the
time direction is assumed to be noise-free, see Sect. 10.10. Because of this, the estimation is not independent of the coordinate system, in contrast to the tensor averaging approach. A second difference lies in that the tensor approach uses 3D gradients in which both types of translations are jointly represented, whereas Lucas and
Kanade's approach uses 2D gradients and frame differences to estimate the translation of points, which decreases its noise tolerance. However, it should be pointed out
that the 2D approach is computationally less demanding and therefore faster.

12.10 Motion Estimation by Spatial Correlation

We assume that the discrete image frame $f(x_k, y_k, t_0)$, which is typically a local
image patch, satisfies the BCC constraint while it undergoes a translation with the
displacement vector $\mathbf{v} = (v_x, v_y)^T$. In other words, the 2D pattern $f(x_k, y_k, t_0)$ and
$f(x_k, y_k, t_1)$ are the same or change insignificantly except for a translational CT in
the spatial coordinates, $\mathbf{s}_k = (x_k, y_k)^T$:

$$\mathbf{s}(t_1) = \mathbf{s}(t_0) + \mathbf{v}(t_1 - t_0) = \mathbf{s}(t_0) - \mathbf{v} \qquad (12.105)$$

where we wrote the position of a point in a 2D frame as a function of the time, $\mathbf{s}(t)$,
and assumed that the temporal sampling period is normalized, $(t_1 - t_0) = 1$. Because
the two image frames satisfy the BCC, one can write

$$\Delta f = f(x_k(t_1), y_k(t_1), t_1) - f(x_k(t_0), y_k(t_0), t_0)) = 0 \qquad (12.106)$$

so that the substitution of Eq. (12.105) yields

$$\textbf{DFD:} \quad \Delta f = f(x_k, y_k, t_1) - f(x_k - v_x, y_k - v_y, t_0) = 0 \qquad (12.107)$$

The spatial coordinates x_k, y_k refer to those at time t_1, i.e., $x_k(t_1)$, $y_k(t_1)$. The difference function Δf is also called the *displaced frame difference* DFD, in the literature because one moves a frame towards another using a displacement. In practice, however, the DFD is not exactly zero because the BCC is satisfied only approximately. The problem is instead reformulated so that the $\mathcal{L}^2$ norm of the DFD is minimized over $\mathbf{v}$:

$$\|\Delta f\|^2 = \sum_k (f(x_k, y_k, t_1) - f(x_k - v_x, y_k - v_y, t_0))^2 = \sum_k (f_k - \tilde{f}_k)^2 \quad (12.108)$$

where f_k and $\tilde{f}_k$ are the sampled image frame at time t_1 and the sampled translated image frame[11] at time t_0. The minimization is achieved by assuming that f is a local patch, typically a 7×7 square, and the displacement is limited, $|\mathbf{v}| < C$, typically $C = 7$. Thus, a typical implementation would search for a point yielding the least $\|\Delta f\|^2$ by varying $\mathbf{v}$ for a local f with the size of 7×7 in a search window of a 15×15. Equation (12.108) measures the distance between two images after one is translated towards the other. Altenatively, one can compare the frame to its translated previous frame by using a similarity measure. Defining the discrete gray values in the 2D image patches as vectors i.e.,

$$\mathbf{f} = (\cdots f_{k-1}, f_k, f_{k+1} \cdots)^T, \qquad \text{and} \qquad \tilde{\mathbf{f}} = (\cdots \tilde{f}_{k-1}, \tilde{f}_k, \tilde{f}_{k+1} \cdots)^T, \qquad (12.109)$$

the Schwartz inequality,

$$\cos(\theta) = \frac{|\langle \mathbf{f}, \tilde{\mathbf{f}} \rangle|}{\|\mathbf{f}\| \|\tilde{\mathbf{f}}\|} = \frac{|\mathbf{f}^T \tilde{\mathbf{f}}|}{\|\mathbf{f}\| \|\tilde{\mathbf{f}}\|} \leq 1 \qquad (12.110)$$

can be used used as a similarity measure. One can search for the pattern $\tilde{\mathbf{f}}$ that is most parallel to $\mathbf{f}$ among patterns in its vicinity by maximizing $\cos(\theta)$, which is insensitive to multiplicative changes of $\tilde{\mathbf{f}}$. Seeking the optimal $\tilde{\mathbf{f}}$ by maximizing $\cos(\theta)$ requires repetitive scalar products, which is also a correlation. When $\|\mathbf{f}\| = \|\tilde{\mathbf{f}}\|$ holds, optimizing the objective functions in Eq. (12.108) and Eq. (12.110) yield identical $\mathbf{v}$ as is seen by the expansion

$$\|\mathbf{f} - \tilde{\mathbf{f}}\|^2 = \langle \mathbf{f} - \tilde{\mathbf{f}}, \mathbf{f} - \tilde{\mathbf{f}} \rangle = \|\mathbf{f}\|^2 + \|\tilde{\mathbf{f}}\|^2 - 2\mathbf{f}^T \tilde{\mathbf{f}} \qquad (12.111)$$

which is minimized at the same time as $\mathbf{f}^T \tilde{\mathbf{f}}$ is maximized. This is why using both variants of the objective functions is known as correlation-based optical flow. A major difference is, however, that the ratio $\cos(\theta)$ is less affected by brightness changes because these are approximated well by multiplicative amplifications of images.

---

[11] This is also referred to as the *warped image*.

12.11 Further Reading

Roughly, the types of motion estimation can be categorized by whether they use *sparse* local image data, called feature points, or *all* local image data. Those using sparse data rely heavily upon existence of discriminatory local images that can be identified from local information and their relative positions to other feature points. Typically, in computer vision and cartography, various correlates of lack of linear symmetry are utilized to identify these points [28, 74, 97]. An alternative approach is to try to identify the presence of a specific symmetry of the local image, e.g., see Sect. 11.11, which, in addition to detection, delivers the geometric orientation of the pattern the target of tracking. By its nature, this approach is two-dimensional [198]. Combined with snakes and energy minimization techniques, it can cope with tracking deformable objects as well as deformable boundaries, e.g. [148]. The problem of correspondence, i.e., which sparse point corresponds to which between two frames, if at all present in both, is, however, still a nontrivial issue along with the occlusion of points by moving objects.

In this presentation, we have mainly dwelled on dense motion estimation techniques, known as optical flow fields. Such motion estimations are used in numerous applications. Probably one of the most widely used application is in image compression, where the displacements fields serve to reduce the redundancy by qualified guesses as to where an image patch will likely go to in the next frame. This is called prediction which of course is never perfect but it is sufficient if the guess is roughly correct, because then the error will have less variations than the original sequence due to almost correct guesseses. Coding a function that has lower variations results in compression of the data necessary to represent the image sequence [120].

A basic assumption in the dense translational motion field estimation is that a 2D image undergoes motion with its pattern basically intact, i.e., the gray levels and their geometic distributions are conserved, leading to the BCC equation. This is, however, not enough to solve for the velocity and must be regularized [15]. We have presented two effective regularizations, tensor averaging in 3D and differential averaging in 2D. However, other approaches to regularization exist, e.g., those, using the membrane differential equation, [110], adding gradient propagation, [167], combining BCC and local feature invariance, [200]. An alternative approach to the partial derivative-based techniques is to estimate the translational motion in the local spectral domain. As has been discussed, the motion of points and lines are concentrated to a plane and a line in the spectrum. The 3D local spectrum can, however, be sampled by means of a Gabor decomposition, which reveals whether or not there is a certain concentration of the spectral energy, and hence the motion information. The study in [101] has proposed an algorithm for detecting optical flow using the magnitude of 3D Gabor filters on the basis of the extracted spectral energy via a set of Gabor filters. A plane is fit to the power spectrum by a full search of the tilt parameters. Using an analogue of the tensor approach discussed in Sect. 10.13, but in 3D, [138] does a similar fit to the power spectrum. The advantage is a full search is avoided. The report of [73] solves the same tilt estimation problem by using the Gabor filter phase, where phase-

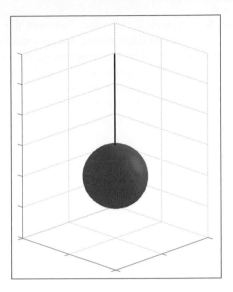

Fig. 12.6. Does the apparent motion of the specular light represent the motion of the sphere?

unwrapping discontinuities are avoided by use of the derivatives of the phase that is continuous in contrast to the phase itself.

The motion estimation using affine model (six parameters) is also dense, although its support region is generally much larger than the translational model (two parameters) to yield a reliable estimate. By contrast, it can describe more complex motion fields than the translational model. However, it is also true that when the image support is large, even the affine model will be quickly unsufficient, typically because the image will cover two or more regions moving independently according to different motion models. The situation with several motions require an estimation of the moving regions, called layers, and the motion parameters, [3, 9, 30, 60] simultaneously or in alternation, where one is improved while the other is kept unchanged. In a layer there is a single motion following a motion model with fixed but unknown parameters. This happens in TV like image sequences where the images will usually contain several independently moving objects and/or regions, e.g., see Fig. 12.5.

12.12 Appendix

The Rotating Sphere Experiment

That optical flow fields may not necessarily identify motion fields faithfully has been known, as is often illustrated by the rotating sphere phenomenon, Fig. 12.6. In this experiment, a hanging sphere with uniform color, illuminated by a light source, is brought to rotation about the vertical axis. An image sequence generated by an observing camera that is fixed in the room will identify this motion as a zero motion because there will not be a statistical difference between the observations of the sphere at different time instances. In other words, nonzero motion field can generate a zero optical flow field as suggested by the experiment. By contrast, if the sphere and the observer are immobile but the light source is brought into motion, the sphere will have different light distributions falling on its surface, which will be observed as a nonzero optical flow field by the observer. Accordingly, a zero motion field can generate a nonzero optical flow field.

13

World Geometry by Direction in N Dimensions

The origins of the concept "perspective" as it came to be used in the field of computer vision can be traced to two key fields, photogrammetry and geometry. Although the problems and the concepts studied in the latter since the early 1800s accounts for a large portion of contemporary computer vision, these fields have a long history that covers several schools of thought and art. Many of these concepts date back at least to the Renaissaince or even earlier. In 1480, Leonardo da Vinci formulated a definition of the perspective images [56] that was very close to our contemporary understanding of it. Scientists continued the work of da Vinci on projections and geometry. Albrecht Duerer created an instrument that could be used to create a true perspective drawing in 1525 [93]. Girard Desargues contributed to the foundation of projective geometry, Traité de la section perspective (1636). Other significant contributions were by Johan Heinrich Lambert, with the treatise "Perspectiva Liber" (The Free Perspective, 1759), and the establishment of the relationship between projective geometry and photogrammetry (1883), by R. Sturms and G. Haick [56]. In the subsequent sections, we will outline the basic principles of measurements of world geometry from photographs, to help us understand the structure of a world scene.

13.1 Camera Coordinates and Intrinsic Parameters

We assume that we have a perspective camera. This can be imagined as a box with a very small hole through which the light rays hit the image plane behind. For this reason the perspective camera is also called the *pinhole camera*. It is an ideal *camera obscura* of the kind that was carried on horses or on the back of the artist in the medieval age. The pinhole camera is illustrated in Fig. 13.1 with the image plane shown in green.

First, we make some notational precisons. Points are represented by capital letters such as P, O, Q. Since the geometry of points and vectors between them in the ordinary Euclidean space E_3 is discussed in this chapter, an alternative notation will be used for vector representation. Consequently, to represent a vector between the

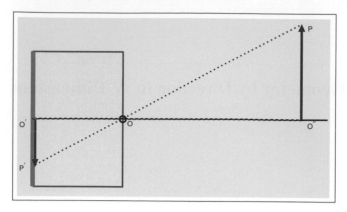

Fig. 13.1. An illustration of the pinhole camera, also known as the perspective camera

world points P and Q the notation $\overrightarrow{PQ}$ will be used. The primed points represent mapped versions of the corresponding points to a plane.

The axis $\overrightarrow{OO'}$ is known as the *optical axis* of the camera. The image is formed in the plane orthogonal to the optical axis, shown in green in Fig. 13.1. Exploiting the congruency between the triangles[1] of the pinhole camera, an equivalent geometry for points mapped through the camera can be obtained (Fig. 13.2). The points that have the same labels correspond logically to each other when compared to those in Fig. 13.1. The image plane is again shown in green. Note that the focal plane or the image plane is now placed between "the pinhole" and the object, whereas in its physical realization it is the pinhole that is between the object and the image plane. Use of this geometry as a reference model for the *projective camera* is mostly adopted in computer vision studies, and it is equivalent to the physical model shown in Fig. 13.1. One advantage is that the image of an object is not turned upside down.

The parameters that depend on the hardware and the design of the camera are called *intrinsic parameters*. Referring to Fig. 13.2, the distance

$$\|\overrightarrow{OO'}\| = f \tag{13.1}$$

called the *focal length*,[2] is the foremost one. Because digital images are used in the practice of computer vision, there are more parameters that are intrinsic. The parameters s_x, s_y, representing the size of 1 pixel in the length unit, e.g., meter, are the most obvious. In a CCD sensor, for example, these are determined by how many pixels there are in horizontal and vertical directions and how large the total width and

---

[1] If two triangles have the same angles, then they are said to be congruent.

[2] The tiny hole in the pinhole camera is effectively replaced by a lens in practice. This has the advantage that the camera can receive more light since the lens is larger than the tiny hole, allowing imaging with less light. Also, a lens effectively captures one plane of the world and focuses it on the image plane, allowing us to map objects that are at a certain distance, only.

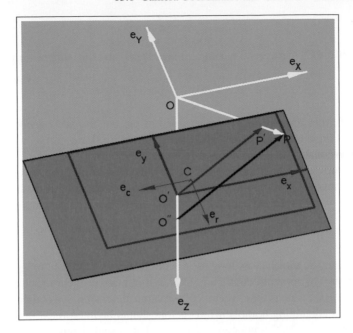

Fig. 13.2. The perspective camera model is illustrated by projection of a point P to its image point P'

the total height of the sensor is in the length unit. Next, is the less obvious parameter set

$$\overrightarrow{O'C} = (c_0, r_0)^T \tag{13.2}$$

which is the translation vector between the digital image center and the analog image center, the image of the optical axis. For technical reasons, such as limitations and constraints of mechanical construction and lens imperfections, the digital image rectangle, shown in magenta in Fig. 13.2 is not centered around the optical axis but around a point that is marked as C in the figure.

We can represent the world point P by the vector

$$\overrightarrow{OP} = (X, Y, Z)^T \tag{13.3}$$

where the numerical values of the vector components are represented relative to the camera basis, e_X, e_Y, and e_Z. Likewise we can represent an image point P' by the vector

$$\overrightarrow{O'P'} = (x, y)^T \tag{13.4}$$

where the vector components are given relative to the (analog) image coordinates, e_x, e_y with the frame origin placed at the image of the optical axis, the point O'. Using the congruency between the triangle $OO'P'$ and the triangle $OO''P$, we can write

$$\frac{f}{Z} = \frac{x}{X} \qquad \frac{f}{Z} = \frac{y}{Y} \tag{13.5}$$

which yields the CT between the analog image frame and the camera frame

$$x = \frac{f}{Z}X, \qquad y = \frac{f}{Z}Y \tag{13.6}$$

Before further discussion, it is necessary to present the *homogeneous coordinates*. Any n dimensional vector represented in a certain frame can be written as an $n + 1$-dimensional vector by adding a redundancy in terms of an extra dimension. The coordinates in the augmented version are called the homogeneous coordinates, including a possible common scale factor λ. These are obtained as follows for the dimension $n = 2$. For dimensions higher than 2 the procedure is analogous.

$$\text{If} \quad \overrightarrow{O'P'} = (x, y)^T \quad \Rightarrow \quad (\overrightarrow{O'P'})_H = \lambda (x, y, 1)^T \tag{13.7}$$

The real scalar λ is arbitrary as long as it is not 0. We will mark homogenized vectors here with $(\cdot)_H$ to avoid confusion, although in matrix algebra such vectors are treated in the same way as other vectors of the same dimension. An equality sign "$=$" in expressions containing homogenized vectors should be interpreted in the sense of equivalent classes, i.e., a homogenized vector that is scaled remains the same homogenized vector. If the homogeneous coordinates of a point in the image plane are known, then the image coordinates are recoverable as

$$(\overrightarrow{O'P'})_H = (X, Y, Z)^T \Rightarrow \overrightarrow{O'P'} = \frac{1}{Z}(X, Y)^T \tag{13.8}$$

We note that $\overrightarrow{OP}$ is not uniquely determined by its image, i.e., $\overrightarrow{OP'}$, or $\overrightarrow{O'P'}$. In fact, not only P, but any point P'' (not shown in Fig. 13.2) on the infinite line represented by the vector $\overrightarrow{OP}$ will have the image P'. Accordingly, the coordinates of a point P in the 3D world are homogeneous coordinates for the point P' if the focal length is assumed to have the unit length. This assumption is not a loss of generality because when length measurements are done in the focal length, or equivalently when an appropriate image scaling is performed, then $f = 1$. However, to do this requires an estimation of f, for reasons as follows. In this sense, a 3D homogeneous vector derived from a point in the 2D image plane is thus the infinite line represented by the vector that joins the projection center, O, to the image point. The line, among others, passes through all 3D world points whose image is given by the intersection between the line and the image plane. These results are summarized by the following lemma:

Lemma 13.1. *Let the analog picture in a projective camera be described by the basis vectors e_x, e_y, placed at O', the image of the projection center. Then a point P is imaged at point P', with the coordinates*

$$(\overrightarrow{O'P'})_{AH} = \mathbf{M}_A \cdot (\overrightarrow{OP})_C, \quad \text{with} \quad \mathbf{M}_A = \begin{pmatrix} f & 0 & 0 \\ 0 & f & 0 \\ 0 & 0 & 1 \end{pmatrix}, \tag{13.9}$$

where $(\overrightarrow{xOP})_C = (X, Y, Z)^T$ and $(\overrightarrow{O'P'})_{AH} = Z \cdot (x, y, 1)^T$ are the coordinates of the point P in the camera frame, and the homogeneous coordinates of the point P' in the analog image frame, respectively.

◆

Not surprisingly, the homogeneous coordinate concept is the result of geometry studies in mathematics, to which A.F. Möbius (1790–1868) contributed greatly. It has turned out to be an important tool of computer vision as well as computer graphics in modern times. Among others, it is used to make affine transformations linear, e.g., a 3D rotation and translation can be implemented as a single 4×4 matrix multiplication, thanks to the relationship between the projective spaces and Euclidean spaces. Besides that, the representation allows a structured treatment of geometric concepts such as points, infinite lines, planes, parallelism, and bundles [70, 91].

Above, the quantities x, and y were in length units, e.g., meters or focal length, whereas one refers to a point in a digital image by its column count in the left–right direction, c, and by its row count in the top–down direction, r. Having this, and the geometric relationship

$$\overrightarrow{O'P'} = \overrightarrow{O'C} + \overrightarrow{CP'} \tag{13.10}$$

in mind, the basis vectors $\mathbf{e}_c, \mathbf{e}_r$ placed at point C define the digital image frame, represented by the column and row basis vectors, $\mathbf{e}_c, \mathbf{e}_r$ placed at the point C, which is, in general, not the same as O'. The quantities c, r are then coordinates in this basis. The coordinates of a point in the digital image frame can be transformed to the analog image frame by the following equations:

$$x = -(c - c_0)s_x \\ y = -(r - r_0)s_y \tag{13.11}$$

The pixel counts $(c, r)^T$ are with reference to the central point, C. The two minus signs in Eq. (13.11) are motivated as follows. For the x-direction one should note that the delivered digital image is the one observed by an observer behind the "image screen" at point O so that an increase in c count results in a decrease in x. For the y-direction one needs a minus sign because the row count r grows in the opposite direction of y. By using Eq. (13.11) one can then obtain

$$\begin{cases} c = -\frac{x}{s_x} + c_0 \\ r = -\frac{y}{s_y} + r_0 \end{cases} \Leftrightarrow \begin{cases} cZ = (-\frac{x}{s_x} + c_0)Z \\ rZ = (-\frac{y}{s_y} + r_0)Z \\ Z = Z \end{cases} \tag{13.12}$$

In matrix form, this result is restated as follows:

Lemma 13.2. *Let the frame representing the digital picture in a projective camera be $\mathcal{D}$ with the basis vectors $\mathbf{e}_c, \mathbf{e}_r$, placed at a point C in the image. Then the coordinates of a point P' represented in the analog picture frame $\mathcal{A}$ are transformed to frame $\mathcal{D}$ as follows:*

$$(\overrightarrow{CP'})_{DH} = \mathbf{M}_D(\overrightarrow{O'P'})_{AH} \tag{13.13}$$

with

$$(\overrightarrow{CP'})_{DH} = Z\begin{pmatrix} c \\ r \\ 1 \end{pmatrix}, \quad \mathbf{M}_D = \begin{pmatrix} -\frac{1}{s_x} & 0 & c_0 \\ 0 & -\frac{1}{s_y} & r_0 \\ 0 & 0 & 1 \end{pmatrix}, \quad (\overrightarrow{O'P'})_{AH} = Z\begin{pmatrix} x \\ y \\ 1 \end{pmatrix}$$

$$(13.14)$$

$\blacklozenge$

In the lemma, $(\overrightarrow{CP'})_{DH}$ and $(\overrightarrow{O'P'})_{AH}$ are the homogeneous coordinates of the point P' in frame $\mathcal{D}$, and the homogeneous coordinates of the point P' in frame $\mathcal{A}$, respectively. By combining lemmas 13.1 and 13.2, one can thus obtain a single linear transformation from frame $\mathcal{C}$ to frame $\mathcal{D}$, without passing through $\mathcal{A}$. This is made explicit in lemma 13.3 .

Lemma 13.3. *Given a point P and its camera frame coordinates, $(\overrightarrow{OP})_C$, its perspective transformation P' has digital image frame coordinates that can be obtained linearly from the former by*

$$(\overrightarrow{CP'})_{DH} = \mathbf{M}_I(\overrightarrow{OP})_C \qquad (13.15)$$

with

$$(\overrightarrow{CP'})_{DH} = \begin{pmatrix} cZ \\ rZ \\ Z \end{pmatrix}, \quad (\overrightarrow{OP})_C = \begin{pmatrix} X \\ Y \\ Z \end{pmatrix} \qquad (13.16)$$

and

$$\mathbf{M}_I = \mathbf{M}_D\mathbf{M}_A = \begin{pmatrix} -\frac{1}{s_x} & 0 & c_0 \\ 0 & -\frac{1}{s_y} & r_0 \\ 0 & 0 & 1 \end{pmatrix}\begin{pmatrix} f & 0 & 0 \\ 0 & f & 0 \\ 0 & 0 & 1 \end{pmatrix} = \begin{pmatrix} -f_x & 0 & c_0 \\ 0 & -f_y & r_0 \\ 0 & 0 & 1 \end{pmatrix} \qquad (13.17)$$

Here M_I is the intrinsic matrix *that encodes the hardware parameters of the camera with $f_x = \frac{f}{s_x}$, and $f_y = \frac{f}{s_y}$ whereas $(\overrightarrow{OP})_C$, and $(\overrightarrow{CP'})_{DH}$ are coordinates of a point in the camera frame, $\mathcal{C}$, and the homogeneous coordinates corresponding to the perspective projected point in the digital image frame, $\mathcal{D}$, respectively.*

$\blacklozenge$

The coordinates of the vector $\overrightarrow{OP}$ are relative the basis $\mathbf{e}_X, \mathbf{e}_Y, \mathbf{e}_Z$, and the vector $\overrightarrow{CP'}$ is represented via its 3D *homogeneous* coordinates. From the concept of homogeneous coordinates, it then follows that

$$\text{if } (\overrightarrow{CP'})_{DH} = (P'_x, P'_y, P'_z)^T \;\Rightarrow\; (\overrightarrow{CP'})_D = \frac{1}{P'_z}(P'_x, P'_y)^T = (c, r)^T$$

$$(13.18)$$

Conversely,

$$\text{if } (\overrightarrow{CP'})_D = \begin{pmatrix} c \\ r \end{pmatrix} \quad\Rightarrow\quad (\overrightarrow{CP'})_{DH} = \lambda\begin{pmatrix} c \\ r \\ 1 \end{pmatrix} \qquad (13.19)$$

with λ being an arbitrary real scalar.

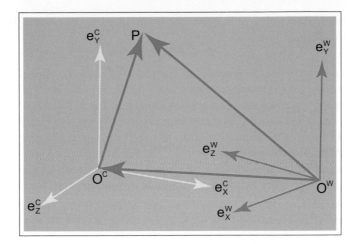

Fig. 13.3. The camera and the world coordinate systems are shown in *yellow* and *green*, respectively

13.2 World Coordinates

In this section we place the camera frame, $\mathcal{C}$, into another frame, that we call the world frame, $\mathcal{W}$ (Fig. 13.3). We will study how points represented in the camera frame can be transferred to the world frame. This is useful because the camera frame and the world frame are often in motion w.r.t. each other while the camera observes the world. The camera and the world frames do not, in general, have axis parallel basis vectors, which means that there is a rotation matrix (a tensor) $\mathbf{R}$ having the elements[3]

$$\mathbf{R} = \begin{pmatrix} R_{11} & R_{12} & R_{13} \\ R_{21} & R_{22} & R_{23} \\ R_{31} & R_{32} & R_{33} \end{pmatrix} \tag{13.20}$$

that can align them. Equivalently, the elements of $\mathbf{R}$ can be used to express the camera basis by a linear combination of the world basis:

$$\begin{cases} \mathbf{e}_X^C = R_{11}\mathbf{e}_X^W + R_{12}\mathbf{e}_Y^W + R_{13}\mathbf{e}_Z^W \\ \mathbf{e}_Y^C = R_{21}\mathbf{e}_X^W + R_{22}\mathbf{e}_Y^W + R_{23}\mathbf{e}_Z^W \\ \mathbf{e}_Z^C = R_{31}\mathbf{e}_X^W + R_{32}\mathbf{e}_Y^W + R_{33}\mathbf{e}_Z^W \end{cases} \Leftrightarrow \left[\mathbf{e}_X^C, \mathbf{e}_Y^C, \mathbf{e}_Z^C\right] = \left[\mathbf{e}_X^W, \mathbf{e}_Y^W, \mathbf{e}_Z^W\right] \cdot \mathbf{R}^T \tag{13.21}$$

Here $\mathbf{e}_X^W$ represents a world basis vector and $\mathbf{e}_X^C$ represents a camera basis vector, respectively. The interpretation of the other symbols is analogous. Note that entities like $\mathbf{e}_X^W$ are abstract vectors that do not require a particular frame to exist. Defining the camera and the world basis sets as

$$\mathbb{C} = \left[\mathbf{e}_X^C, \mathbf{e}_Y^C, \mathbf{e}_Z^C\right], \quad \mathbb{W} = \left[\mathbf{e}_X^W, \mathbf{e}_Y^W, \mathbf{e}_Z^W\right] \tag{13.22}$$

---

[3] A rotation matrix $\mathbf{R}$ is an orthogonal matrix, i.e. $\mathbf{R}^T\mathbf{R} = \mathbf{I}$ and $\mathbf{R}^{-1} = \mathbf{R}^T$.

one can write the basis change between the world and the coordinate frames, Eq. (13.21), as follows.

$$\mathbb{C} = \mathbb{W}\mathbf{R}^T \tag{13.23}$$

The quantities $\mathbb{C}, \mathbb{W}$ appear as row vectors in (13.22) in brackets. However, it is important to note that their elements are not scalars, but basis vectors! To mark that this is a formalism to represent the basis vectors jointly, in Eq. (13.22) we used $\mathbb{C}, \mathbb{W}$, which are in a different "vector" notation than the ordinary vector notation (boldface letters). Accordingly, the multiplication in the basis change equation involves the symbolic quantities such as e_X^C, not scalars. The brackets are used throughout this chapter to mark that we are constructing a matrix or a vector by juxtaposing other entities, which can be other matrices or vectors in coordinate or abstract representation.

A vector $\overrightarrow{O^C P}$ expressed in the camera frame can be written as

$$\overrightarrow{O^C P} = X^C \mathbf{e}_X^C + Y^C \mathbf{e}_Y^C + Z^C \mathbf{e}_Z^C \tag{13.24}$$

Here, X^C, Y^C, Z^C are scalars, i.e., coordinates, that are the projections of $\overrightarrow{O^C P}$ on $\mathbf{e}_X^C, \mathbf{e}_Z^C, \mathbf{e}_Z^C$, respectively. Using the definitions for $\mathbb{C}, \mathbb{W}$ in (13.22) we can write the equation of coordinates, (13.24), as

$$\overrightarrow{O^C P} = \mathbb{C} \cdot (\overrightarrow{O^C P})_C, \quad \text{where} \quad (\overrightarrow{O^C P})_C = (X^C, Y^C, Z^C)^T. \tag{13.25}$$

Note that we used the subscript $(\cdot)_C$ to mark that the vector in question is no longer an abstract vector, but it is expressed in the camera frame by means of a specific triplet of scalars, the coordinates. Accordingly, the same vector will be represented as a *different* triplet of scalars in the world frame. The new coordinates can be obtained by substituting (13.22) in (13.25), yielding:

$$\overrightarrow{O^C P} = \mathbb{C}(\overrightarrow{O^C P})_C = \mathbb{W}\mathbf{R}^T \cdot (\overrightarrow{O^C P})_C \tag{13.26}$$

Equation (13.25) represents $\overrightarrow{O^C P}$ in the camera frame. Using an analogous notation, it can be represented in the world frame as well:

$$\overrightarrow{O^C P} = \mathbb{W}(\overrightarrow{O^C P})_W \tag{13.27}$$

Representing the same vector in the same frame, the right-hand sides of Eqs. (13.26) and (13.27) can be compared to each other, establishing the following identity between the coordinate triplets of the camera and the world frames as:

$$(\overrightarrow{O^C P})_W = \mathbf{R}^T \cdot (\overrightarrow{O^C P})_C \tag{13.28}$$

or

$$(\overrightarrow{O^C P})_C = \mathbf{R} \cdot (\overrightarrow{O^C P})_W \tag{13.29}$$

In the latter, we note that the coordinate transformation uses the matrix $\mathbf{R}$, which is the *inverse* of the matrix used in the basis transformation, Eq. (13.23), $\mathbf{R}^T$. We summarize this result in lemma 13.4.

Lemma 13.4. *If the two bases* $\mathbb{C}, \mathbb{W}$ *are related to each other as* $\mathbb{C} = \mathbb{W}\mathbf{R}^T$, *with* $\mathbf{R}$ *being an invertible matrix and* P *being a point in the space* E_3, *then*

$$(\overrightarrow{O^W P})_C = \mathbf{R} \cdot (\overrightarrow{O^W P})_W \qquad (13.30)$$

where $(\overrightarrow{O^W P})_C$ *and* $(\overrightarrow{O^W P})_W$ *are the coordinates of* $\overrightarrow{O^W P}$ *in the bases* $\mathbb{C}$ *and* $\mathbb{W}$, *respectively.*

♦

To reach this conclusion, we used only fundamental principles of linear algebra. It is worth pointing out that even if the CT matrix $\mathbf{R}$ had not been an orthogonal matrix, it would still relate to the basis change matrix inversely. Also, because a vector is equivalent to all its translated versions, the above result is not only applicable to a point, but to any vector. Also, the two involved frames, $\mathbb{C}, \mathbb{W}$, and the vector space which they describe do not need to be 3D. We summarize this generalization as a lemma,

Lemma 13.5. *If two bases* $\mathbb{C}, \mathbb{W}$ *are related to each other as* $\mathbb{C} = \mathbb{W}\mathbf{R}^{-1}$ *with* $\mathbf{R}$ *being an invertible matrix, and* $\overrightarrow{AB}$ *being a vector in the vector space, then*

$$(\overrightarrow{AB})_C = \mathbf{R} \cdot (\overrightarrow{AB})_W \qquad (13.31)$$

where $(\overrightarrow{AB})_C$ *and* $(\overrightarrow{AB})_W$ *are the coordinates of* $\overrightarrow{AB}$ *in the bases* $\mathbb{C}$ *and* $\mathbb{W}$ *respectively.*

♦

Returning to the 3D vector space containing the world and camera frames, we sum the vectors between the three points O^W, O^C, P (Fig. 13.3) to obtain the equation

$$\overrightarrow{O^W O^C} + \overrightarrow{O^C P} + \overrightarrow{P O^W} = 0 \qquad (13.32)$$

Evidently, the relationship holds in any frame of the vector space, provided that all three vectors are represented in that same frame. Using the world frame, we obtain

$$(\overrightarrow{O^W P})_W = (\overrightarrow{O^W O^C})_W + (\overrightarrow{O^C P})_W \qquad (13.33)$$

In practice, however, the vector $\overrightarrow{O^C P}$ is available in the camera frame. The substitution of Eq. (13.28) in this equation yields

$$(\overrightarrow{O^W P})_W = (\overrightarrow{O^W O^C})_W + \mathbf{R}^T \cdot (\overrightarrow{O^C P})_C \qquad (13.34)$$

or

$$(\overrightarrow{O^C P})_C = \mathbf{R} \cdot (\overrightarrow{O^W P})_W - \mathbf{R} \cdot (\overrightarrow{O^W O^C})_W \qquad (13.35)$$

The latter is an affine relationship between the vectors $(\overrightarrow{O^C P})_C, (\overrightarrow{O^W P})_W$, i.e., the position vectors of the point P expressed in the camera and world frames.

This is because the term $-\mathbf{R} \cdot (\overrightarrow{O^W O^C})_W$ is constant for all points P as soon as $\mathbf{R}, (\overrightarrow{O^W O^C})_W$ are fixed. The parameters $\mathbf{R}, (\overrightarrow{O^W O^C})_W$ are called extrinsic parameters because they encode the direction and the position of the camera frame relative to the world frame. Classically, we can make the relationship between the two position vectors linear by representing $(\overrightarrow{O^W P})_W$ in its homogeneous coordinates. By augmenting the matrix $\mathbf{R}$, we thus obtain

$$(\overrightarrow{O^C P})_C = \mathbf{M}_E (\overrightarrow{O^W P})_{WH} \qquad (13.36)$$

where[4]

$$\mathbf{M}_E = [\mathbf{R}, -\mathbf{R}(\overrightarrow{O^W O^C})_W] = [\mathbf{R}, (\overrightarrow{O^C O^W})_C], \quad (\overrightarrow{O^W P})_{WH} = \begin{bmatrix} (\overrightarrow{O^W P})_W \\ 1 \end{bmatrix} \quad (13.37)$$

represent the *extrinsic matrix* and the homogeneous coordinates of the point P w.r.t. the world frame, respectively. Here we have used lemma 13.4 to obtain $(\overrightarrow{O^C O^W})_C = -\mathbf{R}(\overrightarrow{O^W O^C})_W$, which is the relationship between the two representations of the displacement vector between the camera and the world frames. The extrinsic matrix is accordingly a 3×4 matrix obtained by juxtaposing $\mathbf{R}$ with the column vector of the interframe displacement, which we define as follows for convenience:

$$\mathbf{t} = (t_X, t_Y, t_Z)^T = (\overrightarrow{O^C O^W})_C = -\mathbf{R}(O^W O^C)_W \qquad (13.38)$$

We summarize these results as follows:

Lemma 13.6. *Given the camera and the world frames* $\mathbb{C}, \mathbb{W}$ *at the different origins* O^C, O^W, *and that they are related to each other as* $\mathbb{C} = \mathbb{W}\mathbf{R}^T$, *then the coordinates of a point in one frame can be transformed linearly to those in the other as*

$$(\overrightarrow{O^C P})_C = \mathbf{M}_E (\overrightarrow{O^W P})_{WH} \qquad (13.39)$$

where

$$\mathbf{M}_E = [\mathbf{R}, \mathbf{t}], \quad (\overrightarrow{O^W P})_{WH} = \begin{bmatrix} (\overrightarrow{O^W P})_W \\ 1 \end{bmatrix} \qquad (13.40)$$

with

$$\mathbf{t} = (\overrightarrow{O^C O^W})_C \qquad (13.41)$$

◆

---

[4] Note that we construct matrices from other matrices by use of their symbols and brackets $[\cdot, \cdot]$, and then multiply (!) these following the rules of block matrix operations in linear algebra. A summary is given in the Appendix, Sect. 13.8.

We can now "transport" a point represented in the world coordinates to the perspective image, where we will be able to find its coordinates in the digital image frame. We do that first by transferring the coordinates of the point in the world frame to the camera frame via Eq. (13.36) and then taking these to the digital image frame via (13.15), which is realized by matrix multiplications as stated in lemma 13.7.

Lemma 13.7. *A point P represented in the world basis is projected to a point in the digital image basis via a linear transformation if both coordinates are homogeneous:*

$$(\overrightarrow{CP'})_{DH} = \mathbf{M}(\overrightarrow{O^W P})_{WH} = \mathbf{M}_I \mathbf{M}_E (\overrightarrow{O^W P})_{WH} \qquad (13.42)$$

where $\mathbf{M} = \mathbf{M}_I \mathbf{M}_E$ *is called the* camera matrix.[5]

♦

In conclusion, for every perspective camera there exists a 3×4 camera matrix:

$$\mathbf{M} = \begin{pmatrix} M_{11} & M_{12} & M_{13} & M_{14} \\ M_{21} & M_{22} & M_{23} & M_{24} \\ M_{31} & M_{32} & M_{33} & M_{34} \end{pmatrix} = \mathbf{M}_I \mathbf{M}_E \qquad (13.43)$$

where $\mathbf{M}_I, \mathbf{M}_E$ are defined by Eqs. (13.17) and (13.40), respectively. In the following section, we outline the fundamentals of determining $\mathbf{M}, \mathbf{M}_I, \mathbf{M}_E$.

13.3 Intrinsic and Extrinsic Matrices by Correspondence

In practice, $\mathbf{M}$ is not known to a sufficient degree of accuracy for a variety of reasons, e.g., either or both of $\mathbf{M}_I, \mathbf{M}_E$ are noisy or unavailable. Because of this, $\mathbf{M}$ will be assumed here to be unknown. We will estimate $\mathbf{M}, \mathbf{M}_I, \mathbf{M}_E$ from known correspondences between a set of image and world points. Defining the pair of homogeneous position vectors (with $\lambda \neq 0$) that correspond to the same point in the world frame and in the image frame, respectively, as

$$\mathbf{p} = (X, Y, Z, 1)^T = (\overrightarrow{O^W P})_{WH} \quad \lambda \mathbf{p'} = \lambda(c, r, 1)^T = (\overrightarrow{CP'})_{DH} \qquad (13.44)$$

we note that these must satify Eq. (13.42):

$$\mathbf{M}\mathbf{p} - \lambda \mathbf{p'} = 0 \qquad (13.45)$$

If one uses the definition of $\mathbf{M}$ given by Eq. (13.43), this produces three equations:

$$XM_{11} + YM_{12} + ZM_{13} + M_{14} - c\lambda = 0 \qquad (13.46)$$
$$XM_{21} + YM_{22} + ZM_{23} + M_{24} - r\lambda = 0 \qquad (13.47)$$
$$XM_{31} + YM_{32} + ZM_{33} + M_{34} - \lambda = 0 \qquad (13.48)$$

---

[5] The matrix $\mathbf{M}$ is also known as the *projection matrix*.

However, pulling λ from Eq. (13.48) and substituting it in Eqs. (13.46) and (13.47) reduces the number of equations to 2.

$$X M_{11} + Y M_{12} + Z M_{13} + M_{14} - x(X M_{31} + Y M_{32} + Z M_{33} + M_{34}) = 0 \quad (13.49)$$
$$X M_{21} + Y M_{22} + Z M_{23} + M_{24} - y(X M_{31} + Y M_{32} + Z M_{33} + M_{34}) = 0 \quad (13.50)$$

Now let

$$\mathbf{c}(\mathbf{p}, \mathbf{p}') = (X, Y, Z, 1, 0, 0, 0, 0, -cX, -cY, -cZ, -c)^T \quad (13.51)$$
$$\mathbf{r}(\mathbf{p}, \mathbf{p}') = (0, 0, 0, 0, X, Y, Z, 1, -rX, -rY, -rZ, -r)^T \quad (13.52)$$

where $\mathbf{c}, \mathbf{r}$ are 12–D vectors produced by the pair of vectors $\mathbf{p}, \mathbf{p}'$ representing a point and its image according to Eq. (13.42). Furthermore, let

$$\mathbf{m} = (M_{11}, M_{12}, M_{13}, M_{14}, M_{21}, M_{22}, M_{23}, M_{24}, M_{31}, M_{32}, M_{33}, M_{34})^T, \quad (13.53)$$

which is a vector version of the unknown matrix $\mathbf{M}$. Consequently, we can express Eqs. (13.49) and (13.50) as the equation pair

$$\mathbf{c}^T \mathbf{m} = 0 \quad (13.54)$$
$$\mathbf{r}^T \mathbf{m} = 0 \quad (13.55)$$

where we have omitted displaying the dependency of $\mathbf{c}, \mathbf{r}$ on $\mathbf{p}, \mathbf{p}'$ for convenience. Geometrically, the homogeneous equation[6]

$$\mathbf{s}^T \mathbf{m} = 0 \quad (13.56)$$

represents the equation of a (hyper)plane in E_{12}, where the constant vector $\mathbf{m}$ is the normal of the plane. For a corresponding pair $\mathbf{p}, \mathbf{p}'$, we can obtain two points in this 12D space, i.e., $\mathbf{s}^1 = \mathbf{c}$ and $\mathbf{s}^2 = \mathbf{r}$ given by Eqs. (13.54) and (13.55), that satisfy Eq. (13.56).

Let $\mathcal{S} = \{\mathbf{c}, \mathbf{r}\}$ be a set of correspondence vectors.[7] Because of measurement noise, members of $\mathcal{S}$ will only approximately satisfy Eq. (13.56). Then, a TLS solution is the preferable procedure to find $\mathbf{m}$. In that, one attempts to minimize

$$e(\mathbf{m}) = \int_{\mathcal{S}} |\mathbf{s}^T \mathbf{m}|^2 h(\mathbf{s}) d\mathbf{s} = \mathbf{m}^T \left[\int_{\mathcal{S}} \mathbf{s}\mathbf{s}^T h(\mathbf{s}) d\mathbf{s} \right] \mathbf{m}, \quad \text{with} \quad \|\mathbf{m}\| = 1,$$

$$(13.57)$$

which is the linear symmetry problem in E_{12}, see theorem 12.1. Here h is a function that represents the certainty on the correspondence data $\mathbf{s}$, which is equivalent

---

[6] An equation is homogeneous if it is of the form $\mathbf{A}\mathbf{x} = \mathbf{0}$, with $\mathbf{x}$ being the unknown vector. Only if the null space of $\mathbf{A}$ is nonnil, does another solution than $\mathbf{x} = \mathbf{0}$ exist.

[7] The set can be a dense set, e.g., a jointly observed surface in the world frame, and in the digital image frame. Although the term "digital" implies sampling, strictly speaking, sampling is not necessary for the conclusions of this chapter. This is because even the transformation from $(x, y, 1)^T$ to $(c, r, 1)^T$ is continuous!

to a probability density in statistics, and a mass density in mechanics. The integral reduces to a summation for a discrete set $\mathcal{S} = \{\mathbf{s}^j\}$. When the certainty of a correspondence is available, which can be a computed confidence on the position of the points or the strength of the salient points being matched, it is used. When this is not available, h can be approximated by the constant function 1. Accordingly, for a discrete set of correspondences $\mathcal{S}$

$$\mathcal{S} = \{\mathbf{c}^1, \mathbf{r}^1, \mathbf{c}^2, \mathbf{r}^2, \cdots, \mathbf{c}^N, \mathbf{r}^N\} \tag{13.58}$$

and assuming that no certainty data is available for simplicity, the problem consists in minimizing

$$\mathbf{m}^T \left(\sum_j \mathbf{c}^j \mathbf{c}^{jT} + \sum_j \mathbf{r}^j \mathbf{r}^{jT} \right) \mathbf{m} = \mathbf{m}^T \mathbf{B}^T \mathbf{B} \mathbf{m} = 0, \quad \text{with} \quad \mathbf{B} = \begin{pmatrix} \mathbf{B}_c \\ \mathbf{B}_r \end{pmatrix}. \tag{13.59}$$

where

$$\mathbf{B}_c = \begin{pmatrix} \mathbf{c}^{1T} \\ \mathbf{c}^{2T} \\ \vdots \\ \mathbf{c}^{NT} \end{pmatrix}, \qquad \mathbf{B}_r = \begin{pmatrix} \mathbf{r}^{1T} \\ \mathbf{r}^{2T} \\ \vdots \\ \mathbf{r}^{NT} \end{pmatrix}. \tag{13.60}$$

Therefore,

$$\mathbf{B} = \begin{pmatrix} X^1 & Y^1 & Z^1 & 1 & 0 & 0 & 0 & 0 & -c^1 X^1 & -c^1 Y^1 & -c^1 Z^1 & -c^1 \\ X^2 & Y^2 & Z^2 & 1 & 0 & 0 & 0 & 0 & -c^2 X^2 & -c^2 Y^2 & -c^2 Z^2 & -c^2 \\ \vdots & \vdots & \vdots & \vdots & \vdots & \vdots & \vdots & \vdots & \vdots & \vdots & \vdots & \vdots \\ X^N & Y^N & Z^N & 1 & 0 & 0 & 0 & 0 & -c^N X^N & -c^N Y^N & -c^N Z^N & -c^N \\ 0 & 0 & 0 & 0 & X^1 & Y^1 & Z^1 & 1 & -r^1 X^1 & -r^1 Y^1 & -r^1 Z^1 & -r^1 \\ 0 & 0 & 0 & 0 & X^2 & Y^2 & Z^2 & 1 & -r^2 X^2 & -r^2 Y^2 & -r^2 Z^2 & -r^2 \\ \vdots & \vdots & \vdots & \vdots & \vdots & \vdots & \vdots & \vdots & \vdots & \vdots & \vdots & \vdots \\ 0 & 0 & 0 & 0 & X^N & Y^N & Z^N & 1 & -r^N X^N & -r^N Y^N & -r^N Z^N & -r^N \end{pmatrix} \tag{13.61}$$

Equivalently, the problem is to solve the system of equations [1]:

$$\mathbf{B}\mathbf{m} = 0 \tag{13.62}$$

The solution can be determined only up to a multiplicative constant because of the homogeneity of the underlying equation, i.e., if $\mathbf{m}$ is a solution then so is $\gamma\mathbf{m}$ with $\gamma \neq 0$. The solution is given by the least significant eigenvector[8] of $\mathbf{B}^T\mathbf{B}$. If the solution is to be unique and nontrivial, then the least eigenvalue of $\mathbf{B}^T\mathbf{B}$ should ideally be 0 (close to 0 in practice), and it should have multiplicity 1.

---

[8] Equivalently, the solution can be found by the least singular vector of $\mathbf{B}$ as obtained from a singular value decomposition. We will discuss this decomposition, which is used frequently to solve homogeneous equations (i.e. those like Eq. (13.62), efficiently, in Sect. 15.3.

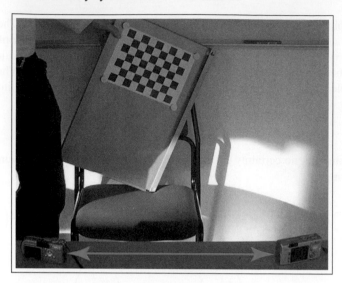

Fig. 13.4. The checkerboard consists of *black and white squares* with side lengths of 28 mm. Such patterns are frequently used to calibrate cameras, i.e., their intrinsic and extrinsic parameters are computed

When **m**, and thereby **M**, is estimated, however, $\mathbf{M}_I$ and $\mathbf{M}_E$ are yet to be determined. To do that, we spell out **M** in terms of its physical parameters.

$$
\mathbf{M} = \begin{pmatrix} -f_x & 0 & c_0 \\ 0 & -f_y & r_0 \\ 0 & 0 & 1 \end{pmatrix} \begin{pmatrix} R_{11} & R_{12} & R_{13} & t_X \\ R_{21} & R_{22} & R_{23} & t_Y \\ R_{31} & R_{32} & R_{33} & t_Z \end{pmatrix} \tag{13.63}
$$

$$
= \begin{pmatrix} -f_x R_{11} + c_0 R_{31}, & -f_x R_{12} + c_0 R_{32}, & -f_x R_{13} + c_0 R_{33}, & -f_x t_x + c_0 t_z \\ -f_y R_{21} + r_0 R_{31}, & -f_y R_{22} + r_0 R32, & -f_y R_{23} + r_0 R_{33}, & -f_y t_y + r_0 t_z \\ R_{31}, & R_{32}, & R_{33}, & t_z \end{pmatrix}
$$

Accordingly, and by observing that the first three columns of the last row in **M** is a row of an orthogonal matrix that should have the norm 1 [70], we can compute first γ

$$
\sqrt{M_{31}^2 + M_{32}^2 + M_{33}^2} = \gamma \sqrt{R_{31}^2 + R_{32}^2 + R_{33}^2} = \gamma \tag{13.64}
$$

and normalize **M** as

$$
\mathbf{M} \leftarrow \frac{1}{\gamma} \mathbf{M} \tag{13.65}
$$

Next we calculate c_0, r_0, f_x, f_y by building pairwise scalar products between the subrows of Eq. (13.63) that contain the elements of **R** to conclude that

Fig. 13.5. A camera to be calibrated takes the picture of a known pattern

$$(M_{11}, M_{12}, M_{13}) \begin{pmatrix} M_{31} \\ M_{32} \\ M_{33} \end{pmatrix} = c_0$$

$$. (M_{11}, M_{12}, M_{13}) \begin{pmatrix} M_{31} \\ M_{32} \\ M_{33} \end{pmatrix} = r_0$$

$$M_{11}^2 + M_{12}^2 + M_{13}^2 = f_x^2 + c_x^2$$
$$.M_{21}^2 + M_{22}^2 + M_{23}^2 = f_y^2 + c_y^2.$$

where the observation that different rows of $\mathbf{R}$ are orthogonal has been utilized. In the equations, M_{ij} are the known parameters, to the effect that the intrinsic parameters can be pulled out without effort. Once the intrinsic parameters are known, they can be substituted into Eq. (13.63) so that $\mathbf{t}$ is identified from the last column, whereas R is identified from the remaining columns, i.e., the first 3×3 block matrix on the left.

Using the above procedure, from every $\mathbf{M}$, one can estimate $\mathbf{M}_I, \mathbf{M}_E$. However, for every registered view of the calibration pattern, one can obtain an $\mathbf{M}$ and thereby a pair of $\mathbf{M}_I, \mathbf{M}_E$. A series of views containing the pattern can thus be shot by moving and/or tilting it randomly, yielding a sequence of matrices $\mathbf{M}_I$ and $\mathbf{M}_E$, where $\mathbf{M}_I$ is the estimate of a constant matrix, the intrinsic parameters. The intrinsic parameters of the camera should remain unchanged so that an average of the matrices $\mathbf{M}_I$ will be a better estimate of the intrinsic matrix as compared to that of an individual calibration. The matrix $\mathbf{M}_E$ will nevertheless be different for each calibration because each estimated $\mathbf{M}_E$ encodes the rotation and the distance of the calibra-

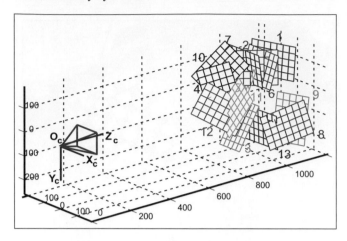

Fig. 13.6. The various placements of the calibration pattern relative to the camera as computed from the extrinsic parameters, $\mathbf{M}_E$

tion pattern, which is the world frame, at different positions but relative to the fixed camera frame. Equivalently, each $\mathbf{M}_E$ expresses the position and the rotation of the camera frame with respect to the world frame, which can be assumed fixed.

Example 13.1. *The correspondences can be established by letting the camera observe a known pattern, the calibration pattern. Figure 13.4 illustrates a digital camera calibration set-up. Two digital cameras of different brands will be calibrated, i.e., the intrinsic and extrinsic parameters of each of the cameras w.r.t. a (world) coordinate frame will be determined. Each camera images calibration pattern whose geometry is fully known, e.g, in the case of the figure this means that the checkerboard consists of black and white squares with side length of 28 mm.*

The world coordinate frame is assumed to be attached to the pattern (Fig. 13.5). Because the sizes of the squares are known and the world coordinate frame is attached to the calibration pattern, the world point coordinates, X^j, Y^j, Z^j, of the corners of the squares are therefore known. Their corresponding image points c^j, r^j can be identified in the camera image either automatically, e.g., by the lack of linear symmetry technique discussed in Sect. 10.9 or manually. Together, this knowledge allows us to determine c^j, r^j corresponding to the rows of the matrix $\mathbf{B}$. For each view imaged by the camera one can thus compute the data matrix $\mathbf{B}$ and thereby the matrix $\mathbf{B}^T\mathbf{B}$, least eigenvector corresponds to the TLS estimate of $\mathbf{m}$.

Figure 13.6 depicts the 3D placements of the calibration pattern from the camera viewpoint. The graphics were drawn by using 13 estimated matrices $\mathbf{M}_E$ corresponding to 13 different placements of the calibration pattern and using the left camera in Fig. 13.4.

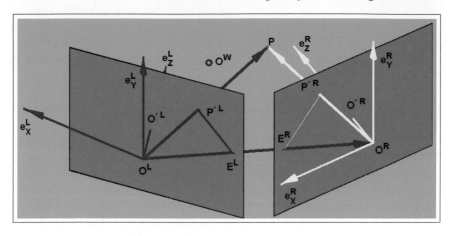

Fig. 13.7. The system graph summarizes a stereo camera set-up with both cameras as well as the corresponding epipoles, E^L, E^R, marked

13.4 Reconstructing 3D by Stereo, Triangulation

In this section we will discuss how to estimate the position of a point in the world by means of two images. The ideas behind stereo processing can also be extended to three or more cameras, although we will not elaborate on them to limit the scope. To be precise, two images taken by two perspective cameras located at different places but observing the point simultaneously will be assumed. The issue of taking pictures simultaneously is evidently an approximation of the reality. When the points in the view fields of the two cameras move much slower than a camera, even one camera taking pictures at two different locations and two different times can be viewed as a reasonable approximation of a stereo system. The points of interest in our system are listed below and are also marked in Fig. 13.7:

P An arbitrary point in the common view field of the cameras
$P\prime^L$ The image of P in the left camera
$P\prime^R$ The image of P in the right camera
O^L The projection center of the left camera
O^R The projection center of the right camera
E^L The image of O^R in the left camera, the left epipole
E^R The image of O^L in the right camera, the right epipole
$O\prime^L$ The image of O^L in the left camera
$O\prime^R$ The image of O^R in the right camera
O^W The projection centre of a fictitious world frame

In Fig. 13.7 we have only shown the left and the right coordinate frames, whereas we have only marked O^W for simplicity because the world frame can be eliminated by transferring the reference to one of the camera frames, here the left. We will elaborate on this further below.

Next we discuss *triangulation*, which is the process of reconstructing the position vector of a 3D point in the reference (here left) coordinate frame. The name triangulation refers to the triangle $O^L P O^R$ because the sought $\overrightarrow{O^L P}$ is obtained by summing up the involved vectors as

$$\overrightarrow{PO^L} + \overrightarrow{O^I O^R} + \overrightarrow{O^R P} = \mathbf{0} \tag{13.66}$$

where coordinates of all vectors are represented in the same frame, the left camera coordinates. Triangulation is evidently not possible when $\|\overrightarrow{OO'}\| = 0$. Joint determination of the position vector corresponding to a world point and the camera parameters is a difficult problem in the presence of measurement noise. Unsurprisingly, there have been numerous research studies [98] on the topic. We describe in the following a basic technique that solves the triangulation problem after that the camera parameters have been determined. We have chosen this technique more for its simplicity and reasonable efficiency than for its superiority.

TLS Triangulation in a Projective Space

Assume that we know the positions of two points P'^L, P'^R in the left and right camera images of the same world point P (Fig. 13.7). Then, according to lemma 13.7, we have

$$\begin{cases} \mathbf{M}^L (\overrightarrow{O^W P})_{WH} = (\overrightarrow{C^L P'^L})_{DHL} \\ \mathbf{M}^R (\overrightarrow{O^W P})_{WH} = (\overrightarrow{C^R P'^R})_{DHR} \end{cases} \tag{13.67}$$

where $\mathbf{M}^L, \mathbf{M}^R$ are the camera parameters obtained from calibrating the left and the right cameras individually against a known world frame placed at the point O^W. Remembering that the cross-products of parallel lines vanish, we rewrite these as

$$\begin{cases} (\overrightarrow{C^L P'^L})_{DHL} \times \mathbf{M}^L (\overrightarrow{O^W P})_{WH} = \mathbf{T}^L \mathbf{M}^L \mathbf{p} = \mathbf{0} \\ (\overrightarrow{C^R P'^R})_{DHR} \times \mathbf{M}^R (\overrightarrow{O^W P})_{WH} = \mathbf{T}^R \mathbf{M}^R \mathbf{p} = \mathbf{0} \end{cases} \tag{13.68}$$

Here $\mathbf{p} = (\overrightarrow{O^W P})_{WH}$ is the unknown, and we expressed the cross-product operation as a matrix multiplication according to:

$$(\overrightarrow{C^L P'^L})_{DHL} \times \cdot = \begin{pmatrix} c^L \\ r^L \\ 1 \end{pmatrix} \times \cdot \quad \Rightarrow \quad \mathbf{T}^L = \begin{pmatrix} 0 & -1 & r^L \\ 1 & 0 & -c^L \\ -r^L & c^L & 0 \end{pmatrix} \cdot \tag{13.69}$$

$$(\overrightarrow{C^R P'^R})_{DHR} \times \cdot = \begin{pmatrix} c^R \\ r^R \\ 1 \end{pmatrix} \times \cdot \quad \Rightarrow \quad \mathbf{T}^R = \begin{pmatrix} 0 & -1 & r^R \\ 1 & 0 & -c^R \\ -r^R & c^R & 0 \end{pmatrix} \cdot \tag{13.70}$$

where the elements of the matrices are derived from known image coordinates of correspondence points in respective cameras. Consequently, we obtain:

$$\begin{cases} \mathbf{T}^L \mathbf{M}^L \mathbf{p} = \mathbf{0} \\ \mathbf{T}^R \mathbf{M}^R \mathbf{p} = \mathbf{0} \end{cases} \tag{13.71}$$

This is equivalent to solving for $\mathbf{p}$ in the homogeneous equation

$$\mathbf{Ap} = \mathbf{0}, \qquad \text{with} \qquad \mathbf{A} = \begin{bmatrix} \mathbf{T}^L \mathbf{M}^L \\ \mathbf{T}^R \mathbf{M}^R \end{bmatrix}, \tag{13.72}$$

which can be done by minimizing $\|\mathbf{p}^T \mathbf{A}^T \mathbf{Ap}\|^2$. The solution is, however, in the world coordinate system, which may not be desirable if there is a series of calibration parameter matrices of a fixed stereo system that amenate from different world coordinates. One might then wish to compute the position of P in the left camera coordinates. We discuss this below.

By using Eq. (13.39) for $(\overrightarrow{O^L P})_{LC}$, the 3D position vector of the point P in the left camera coordinates, one can obtain

$$(\overrightarrow{O^L P})_{LC} = \mathbf{M}_E^L (\overrightarrow{O^W P})_{WH} = \mathbf{M}_E^L \mathbf{p} \tag{13.73}$$

where $\mathbf{R}^L$, $\mathbf{t}^L$ are the rotation and the translation matrices constituting $\mathbf{M}_E^L$, respectively, being the extrinsic matrix of the left camera w.r.t. the world frame. A substitution in the first line of Eq. (13.71) then affords a formulation in the left camera coordinates:

$$\mathbf{T}^L \mathbf{M}^L \mathbf{p} = \mathbf{T}^L \mathbf{M}_I^L \mathbf{M}_E^L \mathbf{p} = \mathbf{T}^L \mathbf{M}_I^L (\overrightarrow{O^L P})_{LC} = \mathbf{0} \tag{13.74}$$

where we have used $\mathbf{M}^L = \mathbf{M}_I^L \mathbf{M}_E^L$. Defing the unknown $\tilde{\mathbf{p}}$ as the homogenized $(\overrightarrow{O^L P})_{LC}$:

$$\tilde{\mathbf{p}} = (\overrightarrow{O^L P})_{LCH} = \begin{bmatrix} (\overrightarrow{O^L P})_{LC} : \\ 1 \end{bmatrix} \tag{13.75}$$

and appending a null column to $\mathbf{M}_I^L$ via $\mathbf{M}_I^L[\mathbf{I}, \mathbf{0}]$, we can thus rewrite Eq. (13.74) and thereby express the left camera equation as

$$\mathbf{T}^L \mathbf{M}_I^L[\mathbf{I}, \mathbf{0}]\tilde{\mathbf{p}} = \mathbf{0} \tag{13.76}$$

The vector $\tilde{\mathbf{p}}$ is, as $\mathbf{p}$ is, homogenized and 4D except that it is in the left camera coordinates.

We now turn to the right camera equation in Eq. (13.71) with the purpose of rewriting the unknown in the left camera coordinates. Remembering from Eq. (13.39) that

$$\mathbf{M}_E^L = [\mathbf{R}^L, \mathbf{t}^L], \qquad \text{and} \qquad \mathbf{p} = (\overrightarrow{O^W P})_{WH} = \begin{bmatrix} (\overrightarrow{O^W P})_W \\ 1 \end{bmatrix}, \tag{13.77}$$

we can rewrite Eq. (13.73):

$$(\overrightarrow{O^L P})_{LC} = \mathbf{M}_E^L \mathbf{p} = \mathbf{R}^L (\overrightarrow{O^W P})_W + \mathbf{t}^L \tag{13.78}$$

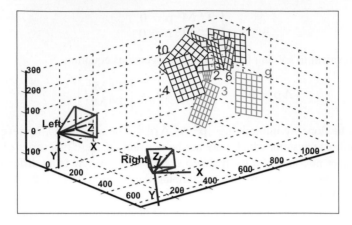

Fig. 13.8. This computed view of camera and checkerboard positions of Fig. 13.4 requires knowledge of $\mathbf{M}_E, \mathbf{M}_I^L, \mathbf{M}_I^R$, which are in turn computed by using known correspondence points

to pull out

$$(\overrightarrow{O^W P})_W = \mathbf{R}^{L^T}(\overrightarrow{O^L P})_{LC} - \mathbf{R}^{L^T}\mathbf{t}^L \tag{13.79}$$

and to obtain $\mathbf{p}$ by homogenization as

$$\mathbf{p} = (\overrightarrow{O^W P})_{WH} = \begin{bmatrix} (\overrightarrow{O^W P})_W \\ 1 \end{bmatrix} = \begin{bmatrix} \mathbf{R}^{L^T}(\overrightarrow{O^L P})_{LC} - \mathbf{R}^{L^T}\mathbf{t}^L \\ 1 \end{bmatrix} \tag{13.80}$$

By substituting this in Eq. (13.71) and using the decomposition of $\mathbf{M}_E^R$ in analogy with Eq. (13.77), we obtain:

$$\mathbf{T}^R\mathbf{M}^R\mathbf{p} = \mathbf{T}^R\mathbf{M}_I^R\mathbf{M}_E^R\mathbf{p} = \mathbf{T}^R\mathbf{M}_I^R[\mathbf{R}^R, \mathbf{t}^R] \begin{bmatrix} \mathbf{R}^{L^T}(\overrightarrow{O^L P})_{LC} - \mathbf{R}^{L^T}\mathbf{t}^L \\ 1 \end{bmatrix}$$

$$= \mathbf{T}^R\mathbf{M}_I^R(\mathbf{R}^R(\mathbf{R}^{L^T}(\overrightarrow{O^L P})_{LC} - \mathbf{R}^{L^T}\mathbf{t}^L) + \mathbf{t}^R)$$

$$= \mathbf{T}^R\mathbf{M}_I^R(\mathbf{R}^R\mathbf{R}^{L^T}(\overrightarrow{O^L P})_{LC} - \mathbf{R}^R\mathbf{R}^{L^T}\mathbf{t}^L + \mathbf{t}^R) \tag{13.81}$$

We now define,

$$\mathbf{R} = \mathbf{R}^R\mathbf{R}^{L^T}, \quad \mathbf{t} = \mathbf{t}^R - \mathbf{R}\mathbf{t}^L, \quad \text{and} \quad \mathbf{M}_E = [\mathbf{R}, \mathbf{t}], \tag{13.82}$$

and call the matrix $\mathbf{M}_E$ the *stereo extrinsic matrix* because it represents the relative rotation and displacement of the two cameras. Using Eq. (13.81) we then obtain the equation of the right camera system as

$$\mathbf{T}^R\mathbf{M}^R\mathbf{p} = \mathbf{T}^R\mathbf{M}_I^R(\mathbf{R}(\overrightarrow{O^L P})_{LC} + \mathbf{t})$$

$$= \mathbf{T}^R\mathbf{M}_I^R[\mathbf{R}, \mathbf{t}]\tilde{\mathbf{p}} = \mathbf{T}^R\mathbf{M}_I^R\mathbf{M}_E\tilde{\mathbf{p}} \tag{13.83}$$

Fig. 13.9. The images registered by the *left* and *right* camera system shown in Fig. 13.4

Finally, we put together (13.76) and (13.83):

$$\begin{cases} \mathbf{T}^L \mathbf{M}_I^L [\mathbf{I}, \mathbf{0}] \tilde{\mathbf{p}} = \mathbf{0} \\ \mathbf{T}^R \mathbf{M}_I^R \mathbf{M}_E \tilde{\mathbf{p}} = \mathbf{0} \end{cases} \tag{13.84}$$

to obtain:

$$\tilde{\mathbf{A}} \tilde{\mathbf{p}} = \mathbf{0}, \qquad \text{with} \qquad \tilde{\mathbf{A}} = \begin{bmatrix} \mathbf{T}^L \mathbf{M}_I^L [\mathbf{I}, \mathbf{0}] \\ \mathbf{T}^R \mathbf{M}_I^R \mathbf{M}_E \end{bmatrix}. \tag{13.85}$$

Equation (13.85) can be solved for $\tilde{\mathbf{p}}$ by minimizing

$$\|\tilde{\mathbf{A}} \tilde{\mathbf{p}}\|^2 = \tilde{\mathbf{p}}^T \tilde{\mathbf{A}}^T \tilde{\mathbf{A}} \tilde{\mathbf{p}} \tag{13.86}$$

The found $\tilde{\mathbf{p}}$ is now in the left camera coordinate system and solves a 4D linear symmetry direction estimation problem. Accordingly, $\tilde{\mathbf{p}}$ is given by the least eigenvector of $\tilde{\mathbf{A}}^T \tilde{\mathbf{A}}$. An analogous solution and interpretation in terms of direction in a 4D space can evidently also be made for Eq. (13.72). We summarize our findings as the following two lemmas.

Lemma 13.8. *Given the intrinsic matrices* $\mathbf{M}_I^L, \mathbf{M}_I^R$ *and the extrinsic matrices,*

$$\mathbf{M}_E^L = [\mathbf{R}^L, \mathbf{t}^L], \qquad \mathbf{M}_E^R = [\mathbf{R}^R, \mathbf{t}^R] \tag{13.87}$$

of two cameras, the stereo extrinsic matrix yields

$$\mathbf{M}_E = [\mathbf{R}, \mathbf{t}] \tag{13.88}$$

where

$$\mathbf{R} = \mathbf{R}^R \mathbf{R}^{L^T}, \qquad and \qquad \mathbf{t} = \mathbf{t}^R - \mathbf{R}\mathbf{t}^L. \tag{13.89}$$

The coordinates of vectors marked with L and R are in the left and right camera coordinates, respectively, whereas the L and R marked matrices rotate the world coordinates to the respective camera coordinates.

♦

Compared with Sect. 13.2, we note that the left camera frame (the reference) here takes the role of the world frame there, (not to be confused with the world frame attached to the calibration pattern, at O^W in Fig. 13.7). Consistently with this, $\mathbf{t}$ represents the position of the left camera center (the world center) in the right camera coordinates, whereas the matrix $\mathbf{R}$ transfers the coordinates of a vector in the left camera (the world) to coordinates in the Right camera, see Figs. 13.3, 13.7 in connection with lemmas 13.6 and 13.8.

Lemma 13.9. *Let the intrinsic matrices $\mathbf{M}_I^L, \mathbf{M}_I^R$ of two cameras and the extrinsic matrix of the corresponding stereo system $\mathbf{M}_E = [\mathbf{R}, \mathbf{t}]$, having the left camera as reference, be given. Then a TLS estimate of a homogenized world point position, $\tilde{\mathbf{p}}$, in the reference frame is given by the solution of a 4D linear symmetry direction determination problem,*

$$\min_{\|\tilde{\mathbf{p}}\|=1} \tilde{\mathbf{p}}^T \tilde{\mathbf{A}}^T \tilde{\mathbf{A}} \tilde{\mathbf{p}}, \qquad with \qquad \tilde{\mathbf{A}} = \begin{bmatrix} \mathbf{T}^L \mathbf{M}_I^L [\mathbf{I}, \mathbf{0}] \\ \mathbf{T}^R \mathbf{M}_I^R \mathbf{M}_E \end{bmatrix}, \tag{13.90}$$

where the image coordinates of a point P are encoded as

$$\mathbf{T}^L = \begin{pmatrix} 0 & -1 & r^L \\ 1 & 0 & -c^L \\ -r^L & c^L & 0 \end{pmatrix}, \qquad \mathbf{T}^R = \begin{pmatrix} 0 & -1 & r^R \\ 1 & 0 & -c^R \\ -r^R & c^R & 0 \end{pmatrix} \tag{13.91}$$

for the left and in the right camera views, respectively.

♦

That $\mathbf{M}_E$ should be independent of our (reasonable but yet) arbitrary placement of the calibration pattern (the world center at O^W) is supported by lemma 13.8. The extrinsic parameters of a stereo system can be obtained by combining the individual calibrations of the two cameras using the lemma. Based on individual camera rotation and translation estimations w.r.t. to the current world coordinates, a series of such measurements can thus be computed from displaced and rotated calibration patterns via Eq. (13.82). The obtained $\mathbf{M}_E$, as well as the intrinsic matrices $\mathbf{M}_I^L, \mathbf{M}_I^R$, can subsequently be averaged to reduce the random measurement errors.

Example 13.2. *Using known patterns, such as a checkerboard pattern, and correspondence between points, we can estimate the extrinsic and intrinsic matrices of a stereo camera system which allows us to determine the positions of the cameras in*

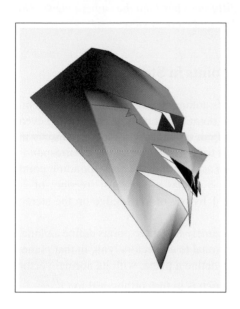
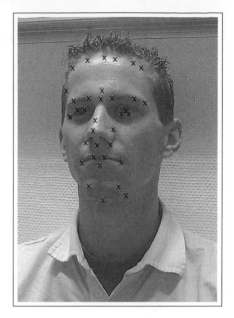

Fig. 13.10. (*Left*) The result of 3D position estimation for a sparse set of points. The planes interpolate linearly between the estimated 3D points. (*Right*) The set of points in the right camera view, for which the 3D positions have been estimated

the world. Fig. 13.8 illustrates a result of the camera calibration for the stereo cameras shown in Fig. 13.4. The positions of the checkerboards as well as the positions of the two cameras observing them are marked. The cameras were static all the time, whereas the world coordinates attached to the checkerboard were displaced.

As in the linear symmetry direction estimate case discussed in Chap. 12, the eigenvalues of the matrix $\tilde{\mathbf{A}}^T \tilde{\mathbf{A}}$ can be sorted and used in various combinations to provide an estimate for the quality of the fitted plane, normal whose now represents the sought position in the triangulation problem. Examples include λ_4, $\lambda_3 - \lambda_4$, or the dimensionless certainty measure:

$$C_t = \frac{\lambda_3 - \lambda_4}{\lambda_3 + \lambda_4} \tag{13.92}$$

where λ_4 is the least significant eigenvalue.

Example 13.3. *Fig. 13.9 shows the images registered by the left and right cameras shown in Fig. 13.4. The cameras were calibrated, i.e., the extrinsic as well as intrinsic parameters are known. In Fig. 13.10 the triangulation is illustrated by using the stereo image pair of Fig. 13.9. The result of triangulation for a set of points is shown on the left. The surfaces are planes that interpolate between the 3D points, whereas color modulates height. Because their 3D positions are known, the object defined by the* *planes could be rotated and imaged from a different view than the camera views. On the right we show the used set of points in the right camera view.*

13.5 Searching for Corresponding Points in Stereo

Finding correspondence points in stereo images automatically is a difficult problem. In this section we discuss how the search for a point can at least be constrained even if the problem remains a difficult one. Simplified, the underlying logics consists in translating the knowledge on the stereo system to limit the freedom of a corresponding point, as to where it can be. Without stereo assumption, the corresponding point can be anywhere in the other camera view, i.e., it has a degree of freedom of 2, whereas we expect to reduce this freedom to 1 by use of knowledge on the stereo set-up.

We will utilize an axiom of the 3D world, namely that three points define a plane. We observe that the normal of a plane is orthogonal to all vectors lying in that plane. The points O^L, P, O^R in a stereo system then define a plane, with its normal vector given by the cross-product $\overrightarrow{O^R O^L} \times \overrightarrow{O^L P}$, which is in turn orthogonal to $\overrightarrow{O^R P}$:

$$(\overrightarrow{O^R P})^T (\overrightarrow{O^R O^L} \times \overrightarrow{O^L P}) = 0 \tag{13.93}$$

All vectors must be represented in the same frame for this equation to hold. We keep the same notation as in previous sections to characterize the stereo system, meaning that

- the relative displacement vector $\mathbf{t} = (\overrightarrow{O^R O^L})_R$ is in the right camera frame, whereas
- the relative rotation matrix $\mathbf{R}$ transfers the coordinates of a vector represented in the left camera to the right camera.

Accordingly, in Eq. (13.93) there is one vector, $\overrightarrow{O^L P}$, that can be expressed most naturally in the left (reference) camera frame, whereas the remaining two are normally expressed in the right camera coordinates. We assume then that the coordinates of all vectors are w.r.t. the right camera basis

$$(\overrightarrow{O^R P})_R^T ((\overrightarrow{O^R O^L})_R \times (\overrightarrow{O^L P})_R) = \mathbf{0} \tag{13.94}$$

so that by the replacement $(\overrightarrow{O^L P})_R = \mathbf{R}(\overrightarrow{O^L P})_L$, we obtain:

$$(\overrightarrow{O^R P})^T (\overrightarrow{O^R O^L} \times (\mathbf{R}(\overrightarrow{O^L P})_L)) = \mathbf{0} \tag{13.95}$$

where, and below, we have dropped marking the right camera frame because all vectors are in the right camera coordinates unless otherwise mentioned. Note however, that this does not mean a change of the world frame. The left camera still plays the role of the world frame in analogy with Sect. 13.2, and consequently, the interpretation of $\mathbf{t}, \mathbf{R}$ remains as listed above. We express the cross-product operation $\overrightarrow{O^R O^L} \times \mathbf{q}$ as a matrix multiplication with a vector $\mathbf{q}$, that is:

$$\overrightarrow{O^R O^L} \times \mathbf{q} = \mathbf{T} \cdot \mathbf{q} \tag{13.96}$$

where, given that

$$\overrightarrow{O^R O^L} = (X, Y, Z)^T \tag{13.97}$$

the matrix $\mathbf{T}$ is determined by the elements of $\overrightarrow{O^R O^L}$ as

$$\mathbf{T} = \begin{pmatrix} 0 & -Z & Y \\ Z & 0 & -X \\ -Y & X & 0 \end{pmatrix} \tag{13.98}$$

The coplanarity equation, (13.95), for points $O^L P O^R$ then yields:

$$(\overrightarrow{O^R P})^T \mathbf{TR}(\overrightarrow{O^L P})_L = 0 \tag{13.99}$$

Consequently, the coplanarity equation can be expressed as

$$(\overrightarrow{O^R P})^T \mathbf{E}(\overrightarrow{O^L P})_L = 0, \tag{13.100}$$

where

$$\mathbf{E} = \mathbf{TR} \tag{13.101}$$

Calling the third components of $(\overrightarrow{O^L P})_L$, $\overrightarrow{O^R P}$ as Z^L, Z^R, and the focus length of the two camera frames as f^L, f^R, respectively, we can, from Eq. (13.100) obtain

$$(\frac{f^R}{Z^R} \overrightarrow{O^R P})^T \mathbf{E}(\frac{f^L}{Z^L} \overrightarrow{O^L P})_L = 0 \tag{13.102}$$

so that

$$(\overrightarrow{O'^R P'^R})^T \mathbf{E} (\overrightarrow{O'^L P'^L})_L = 0 \tag{13.103}$$

Note that both $\overrightarrow{O'^R P'^R}$ and $(\overrightarrow{O'^L P'^L})_L$ are now image points expressed in homogeneous coordinates in the *right* and the *left image planes*, respectively. In what follows we will not mark the homogenized position vectors in the two image planes, because all such vectors are homogenized.

The matrix $\mathbf{E}$, often named the *essential matrix* for the stereo system,[9] is not affected when the point P moves in the space. It changes only when the relative position and gaze of the cameras change by a translation (via $\mathbf{T}$), and/or by a rotation (via $\mathbf{R}$), respectively. The equation can be utilized when determining correspondence

---

[9] Note that $\mathbf{E} = \mathbf{TR}$ and the stereo extrinsic matrix, $\mathbf{M}_E = [\mathbf{R}, \mathbf{t}]$, discussed in Sect. 13.4 are obtained from the same information.

because its validity is required if two points in stereo view correspond to the same 3D scene point, P.

We show next that the identity Eq. (13.103) can also be interpreted as two equations of two lines lying in the left and the right (analog) image frames, respectively. First, we assume that we have a world point P that moves along the line represented by $\overrightarrow{O^R P'^R}$. The right image of such a a point is always P'^R, to the effect that $(O'^R P'^R)^T$ remains unchanged, and one obtains

$$0 = (\overrightarrow{O'^R P'^R})^T \mathbf{E} \, (\overrightarrow{O'^L P'^L})_L = (a, b, c) \begin{pmatrix} x \\ y \\ 1 \end{pmatrix} \qquad (13.104)$$

$$= ax + by + c = 0$$

where $\overrightarrow{O'^L P'^L} = (x, y, 1)^T$ has been assumed. Naturally, $a, b,$ and c are the scalars that must be obtained via

$$(a, b, c) = (\overrightarrow{O'^R P'^R})^T \mathbf{E} \qquad (13.105)$$

which is a vector that has a constant direction,[10] as long as P is a point along the infinite line represented by $\overrightarrow{O^R P'^R}$. Accordingly, the direction of $(a, b, c)^T$ remains unchanged, even for the point O^R. In conclusion, E^L must lie on the line $ax + by + c = 0$ which passes through the point P'^L. This line is sometimes called the (left) *epipolar line* (of point P). An important byproduct of this reasoning is that E^L must lie on all (left) epipolar lines (of all points P in the 3D space) as long as the extrinsic parameters, $\mathbf{E}$, encoding the relative position and gaze of the stereo cameras are unchanged. In other words, all (left) epipolar lines intersect at the (left) epipole, E^L.

Second, assuming that the world point P now moves along the line that corresponds to $\overrightarrow{O'^L P'^L}$, we can find analogous results for the right image frame. Accordingly, it can be shown that the line

$$a'x' + b'y' + c' = 0 \qquad (13.106)$$

with

$$(a', b', c')^T = \mathbf{E}(\overrightarrow{O'^L P'^L})_L \text{ and } \overrightarrow{O'^R P'^R} = (x', y', 1)^T \qquad (13.107)$$

is the right epipolar line with the Right epipole E_R, both defined in analogy with their left counterparts.

The vector $(\overrightarrow{O^L E^L})_L$ is not in the image plane of the left camera, and it is not homogenized, i.e., it is an ordinary 3D vector. Yet, it is in the null space of the matrix $\mathbf{E}$:

---

[10] The vector $\overrightarrow{O'^R P'^R}$ is represented in homogeneous coordinates, i.e., it can be known only up to a scale factor. In turn this causes $(a, b, c)^T$ to change only up to a scale factor when $\overrightarrow{O'^R P'^R}$ is scaled.

$$\mathbf{E}(\overrightarrow{O^L E^L})_L = \mathbf{TR}(\overrightarrow{O^L E^L})_L = \overrightarrow{O^R O^L} \times \overrightarrow{O^L E^L} = \mathbf{0} \qquad (13.108)$$

because it is parallel to $\overrightarrow{O^R O^L}$, and multiplication by $\mathbf{T}$ is a cross-product that is fully determined by $\overrightarrow{O^R O^L}$, Eq. (13.98). Yet, $(\overrightarrow{O^L E^L})_L$ is a homogenized version of $(\overrightarrow{O'^L E^L})_L$, so that the latter is also in the null space of $\mathbf{E}$. Similarly, the homogenized $\overrightarrow{O'^R E^R}$ must be in the null space of $\mathbf{E}^T$ because

$$\mathbf{E}^T \overrightarrow{O^R E^R} = \mathbf{R}^T \mathbf{T}^T \overrightarrow{O^R E^R} \qquad (13.109)$$

which, noting that $\mathbf{T}^T = -\mathbf{T}$, translates to:

$$\mathbf{E}^T \overrightarrow{O^R E^R} = -\mathbf{R}^T \mathbf{T} \overrightarrow{O^R E^R} = -\mathbf{R}^T \overrightarrow{O^R O^L} \times \overrightarrow{O^R E^R} = \mathbf{0} \qquad (13.110)$$

Evidently, the 3×3 matrices $\mathbf{E}, \mathbf{E}^T$ are rank-deficient. We summarize these results in the following lemma.

Lemma 13.10. *Let a stereo system be characterized by a rotation matrix $\mathbf{R}$ such that $\mathbf{R}(\overrightarrow{O^L P})_L = (\overrightarrow{O^L P})_R$, and by a displacement vector $\mathbf{t} = (X, Y, Z)^T = (\overrightarrow{O^R O^L})_R$, such that the matrix $\mathbf{T}$ is determined by the elements of $\mathbf{t}$ as*

$$\mathbf{T} = \begin{pmatrix} 0 & -Z & Y \\ Z & 0 & -X \\ -Y & X & 0 \end{pmatrix} \qquad (13.111)$$

Then, two points P'^L and P'^R in the left and the right (analog) image coordinates are images of the same world point P if and only if

$$(\overrightarrow{O'^R P'^R})_R^T \, \mathbf{E} \, (\overrightarrow{O'^L P'^L})_L = 0, \qquad with \qquad \mathbf{E} = \mathbf{TR}. \qquad (13.112)$$

Furthermore, the homogenized (analog) image coordinates of the left and the right epipoles, E^L, E^R, are in the null spaces of the essential matrix, $\mathbf{E}$, and its transpose, $\mathbf{E}^T$, respectively,

$$\mathbf{E} \, \overrightarrow{(O'^L E^L)}_L = \mathbf{0}, \qquad \mathbf{E}^T (\overrightarrow{O'^R E^R})_R = \mathbf{0} \qquad (13.113)$$

◆

Now we show that the relative position and gaze of the cameras are not needed to find the epipoles, and that the latter implicitly encode the extrinsic parameters of the stereo system. We note first that $(\overrightarrow{O'^L P'^L})_L$ is represented in the analog image coordinates of the left frame. It can also be expressed in the digital image coordinates, i.e., in column and row counts via

$$\overrightarrow{(O'^L P'^L)}_{LD} = \mathbf{M}_I^L \overrightarrow{(O'^L P'^L)}_L \tag{13.114}$$

where $\mathbf{M}_I^L$ is the matrix encoding the intrinsic parameters of the left camera. Similarly, we obtain

$$\overrightarrow{(O'^R P'^R)}_{RD} = \mathbf{M}_I^R \overrightarrow{(O'^R P'^R)}_R \tag{13.115}$$

The epipolar equation (13.112) can then be denoted as

$$\overrightarrow{(O'^R P'^R)}_{RD}^T \mathbf{F} \overrightarrow{(O'^L P'^L)}_{LD} = 0, \tag{13.116}$$

where

$$\mathbf{F} = (\mathbf{M}_I^R)^{-T} \mathbf{E} \, (\mathbf{M}_I^L)^{-1} \tag{13.117}$$

The matrix $\mathbf{F}$ is often called the *fundamental matrix*, which can be obtained via a set of correspondences in digital image coordinates [156].

The coordinates of a vector $\mathbf{q}$ represented in the left analog image coordinates can be transformed back and forth to left digital image coordinates. Evidently, the same can be done for the left camera image. Because of this, identical reasoning can be followed to reach analogous conclusions as those in lemma 13.10, but for the fundamental matrix. This is given precision in the following lemma.

Lemma 13.11. *Let a stereo system be characterized by a rotation matrix $\mathbf{R}$ such that $\mathbf{R}\overrightarrow{(O^L P)}_L = \overrightarrow{(O^L P)}_R$, and by a displacement vector $\mathbf{t} = (X, Y, Z)^T = \overrightarrow{(O^R O^L)}_R$, such that the matrix $\mathbf{T}$ is determined by the elements of $\mathbf{t}$ as*

$$\mathbf{T} = \begin{pmatrix} 0 & -Z & Y \\ Z & 0 & -X \\ -Y & X & 0 \end{pmatrix} \tag{13.118}$$

Then, two points P'^L and P'^R in the left and the right (digital) image coordinates are images of the same world point P if and only if

$$\overrightarrow{(C^R P'^R)}_{RD}^T \mathbf{F} \overrightarrow{(C^L P'^L)}_{LD} = 0, \qquad with \qquad \mathbf{F} = (\mathbf{M}_I^R)^{-T} \mathbf{E} (\mathbf{M}_I^L)^{-1}. \tag{13.119}$$

Furthermore, the homogenized (digital) image coordinates of the left and right epipoles, E^L, E^R, are in the null spaces of the fundamental matrix, $\mathbf{F}$, *and its transpose, $\mathbf{F}^T$, respectively:*

$$\mathbf{F} \overrightarrow{(C^L E^L)}_{LD} = \mathbf{0}, \qquad \mathbf{F}^T \overrightarrow{(C^R F^R)}_{RD} = \mathbf{0} \tag{13.120}$$

$\blacklozenge$

The main points of interest with epipoles and the epipolar lines include the following:

- The corresponding points can be searched along lines that can be effectuated as a 1D search instead of a search in two dimensions. For example, if the point $(\overrightarrow{O'^L P'^L})_{LD}$ is known, then by substituting it in Eq. (13.116) one obtains the search line on which the corresponding unknown point $(\overrightarrow{O'^R P'^R})$ must lie.
- The establishment of corresponding points in images is a crucial problem for 3D geometry reconstruction from stereo.
- The epipolar lines can be utilized to implement triangulation where the rays $\overrightarrow{O^L P'^L}$ and $\overrightarrow{O^R P'^R}$ are guaranteed to intersect.

13.6 The Fundamental Matrix by Correspondence

Assume that the correspondences of (at least) eight points in the digital image coordinates are known. Then Eq. (13.116) can be written for a known pair of points, expressed in homogeneous coordinates, as

$$
\begin{aligned}
0 = (\mathbf{p}^R)^T \mathbf{F}\, \mathbf{p}^L \\
= c^R c^L F_{11} + c^R r^L F_{12} + c^R F_{13} + \\
+ r^R c^L F_{21} + r^R r^L F_{22} + r^R F_{23} + \\
+ c^L F_{31} + r^L F_{32} + F_{33} = 0
\end{aligned}
\tag{13.121}
$$

where the unknown matrix elements,

$$
\mathbf{F} = \begin{pmatrix} F_{11} & F_{12} & F_{13} \\ F_{21} & F_{22} & F_{23} \\ F_{31} & F_{32} & F_{33} \end{pmatrix}
\tag{13.122}
$$

and the known pair of points,

$$
\mathbf{p}^L = (c^L, r^L, 1)^T \quad \text{and} \quad \mathbf{p}^R = (c^R, r^R, 1)^T,
\tag{13.123}
$$

have been utilized. Equation (13.121) is a scalar product between two 9D vectors that reduces to the homogeneous equation

$$
\mathbf{q}^T \mathbf{f} = 0
\tag{13.124}
$$

with

$$
\begin{aligned}
\mathbf{q} &= (\; c^R c^L,\; c^R r^L,\; c^R,\; r^R c^L,\; r^R r^L,\; r^R,\; c^L,\; r^L,\; 1\;)^T \\
\mathbf{f} &= (\; F_{11},\; F_{12},\; F_{13},\; F_{21},\; F_{22},\; F_{23}, F_{31}, F_{32}, F_{33}\;)^T
\end{aligned}
\tag{13.125}
$$

where the vectors $\mathbf{q}$ and $\mathbf{f}$ encode the known data and the unknown parameters, respectively. Geometrically, Eq. (13.124) represents the equation of a hyperplane in E_9, and we are interested once again in finding the direction of this hyperplane, $\mathbf{f}$, when we know a set of correspondences, $\mathcal{Q} = \{\mathbf{q}\}$, of image points in a pair of stereo images.[11] The equation also represents the distance of a point to the hyperplane so that the problem is a linear symmetry direction–fitting problem, see theorem 12.1, in E_9 where one attempts to minimize

$$e(\mathbf{f}) = \int_Q |\mathbf{q}^T\mathbf{f}|^2 d\mathbf{q} = \mathbf{f}^T(\int_Q \mathbf{q}\mathbf{q}^T h(\mathbf{q}))d\mathbf{q}\mathbf{f}, \quad \text{with} \quad \|\mathbf{f}\| = 1 \qquad (13.126)$$

Here h is the strength of the correspondence $\mathbf{q}$, which is equivalent to a probability density in statistics, and a mass density in mechanics. The integral reduces to a summation for a discrete set $Q = \{\mathbf{q}^j\}$, and h reduces to a discrete certainty on correspondence, or a constant if it is not available. Note that the constraint $\|\mathbf{f}\| = 1$ is deduced from the fact that we can determine $\mathbf{f}$ only up to a scale constant. This is because (13.124) is homogeneous, i.e., it is satisfied by $\lambda\mathbf{f}$, where λ is a scalar, if it is satisfied by $\mathbf{f}$. For a discrete Q and no certainty data:

$$Q = \{\mathbf{q}^1, \mathbf{q}^2, \cdots, \mathbf{q}^N\} \qquad (13.127)$$

the problem is confined to minimization of

$$\mathbf{f}^T(\sum_j \mathbf{q}^j\mathbf{q}^{jT})\mathbf{f} = \mathbf{f}^T\mathbf{Q}^T\mathbf{Q}\mathbf{f} = 0, \quad \text{with} \quad \mathbf{Q} = \begin{pmatrix} \mathbf{q}^{1T} \\ \mathbf{q}^{2T} \\ \vdots \\ \mathbf{q}^{NT} \end{pmatrix}. \qquad (13.128)$$

that is, the elements of $\mathbf{Q}$ are given by

$$\mathbf{Q} = \begin{pmatrix} c^{R^1}c^{L^1}, & c^{R^1}r^{L^1}, & c^{R^1}, & r^{R^1}c^{L^1}, & r^{R^1}r^{L^1}, & r^{R^1}, & c^{L^1}, & r^{L^1}, & 1 \\ c^{R^2}c^{L^2}, & c^{R^2}r^{L^2}, & c^{R^2}, & r^{R^2}c^{L^2}, & r^{R^2}r^{L^2}, & r^{R^2}, & c^{L^2}, & r^{L^2}, & 1 \\ \vdots & \vdots & \vdots & \vdots & \vdots & \vdots & \vdots & \vdots & \vdots \\ c^{R^N}c^{L^N}, & c^{R^N}r^{L^N}, & c^{R^N}, & r^{R^N}c^{L^N}, & r^{R^N}r^{L^N}, & r^{R^N}, & c^{L^N}, & r^{L^N}, & 1 \end{pmatrix} \qquad (13.129)$$

Designating the least significant eigenvalue of the 9×9 matrix $\mathbf{Q}^T\mathbf{Q}$ as λ_9, and its corresponding eigenvector as $\mathbf{v}_9$, the solution of (13.128) is given by $\mathbf{f} = \mathbf{v}_9$. Assuming that the thus obtained vector $\mathbf{f}$ has the elements

$$\mathbf{f} = (f_1, f_2, f_3, f_4, f_5, f_6, f_7, f_8, f_9)^T \qquad (13.130)$$

and using (13.122) and (13.125), one can obtain

$$\mathbf{F} = \begin{pmatrix} f_1 & f_2 & f_3 \\ f_4 & f_5 & f_6 \\ f_7 & f_8 & f_9 \end{pmatrix} \qquad (13.131)$$

We recall from lemma 13.11 that if this $\mathbf{F}$ is to be a useful solution to our problem, $\mathbf{F}$ must be rank-deficient, i.e., its least eigenvalue must equal to zero. Ideally, if the

---

[11] In principle, the correspondence set Q could be a dense set in this formalism.

measurement noise of the data is zero, then $\mathbf{F}$ has an eigenvalue that is zero. However, there is no guarantee for this to happen automatically in practice because the measurements on which one bases the calculation of $\mathbf{f}$ are not noise-free. Accordingly, a numerical correction method, guaranteeing the singularity of $\mathbf{F}$, is applied.[12]

Naturally, the epipolar line represented by $(\overrightarrow{O'^{L}E^{L}})_{LD}$ is given by the last row of $\mathbf{V}$, whereas $(\overrightarrow{O'^{R}E^{R}})_{RD}$ is given by the last row of $\mathbf{U}$. The epipoles in (homogeneous) analog image coordinates $x, y, 1$ are given by:

$$(\overrightarrow{O'^{L}E^{L}})_{L} = \mathbf{M}_{I}^{L}{}^{-1}(\overrightarrow{O'^{L}E^{L}})_{LD}$$
$$(\overrightarrow{O'^{R}E^{R}})_{R} = \mathbf{M}_{I}^{R}{}^{-1}(\overrightarrow{O'^{R}E^{R}})_{RD}$$

when the intrinsic matrices $\mathbf{M}_{I}^{L}$ and $\mathbf{M}_{I}^{R}$ are known for both cameras. In that case, $\mathbf{E}$ can be found too (up to a scale factor):

$$\mathbf{E} = \mathbf{M}_{I}^{R}{}^{T}\mathbf{F}\mathbf{M}_{I}^{L} = \mathbf{T}\mathbf{R} \qquad (13.133)$$

Here, the matrix $\mathbf{T}$ encodes the displacement between the stereo cameras, whereas $\mathbf{R}$ represents their relative rotation, lemma 13.11.

13.7 Further Reading

An alternative introduction to camera calibration and stereo vision of world geometry can be found [219]. In [70] a detailed discussion on projective geometry tools and geometry reconstruction by images can be found, including reconstruction by use of more than two views. Modern developments in the theory and practice of scene reconstruction are described in detail and comprehensive background material is provided in the monograph of [98]. In [102] the essentials of camera calibration, including an extended pinhole camera model with a nonlinear lens correction, is discussed. Epipolar constraints are covered in detail in [8]. Studies where wide-baseline stereo issues are discussed include [211, 215]. Even without camera calibration, scene understanding is possible from image views for numerous applications [14, 106, 170]. Robots controlled by active vision [50] or stereo from motion [194] are significant application fields of geometry studies in computer vision. Another significant domain is architectural or similar scenes, where there is little or no motion [55, 86].

---

[12] This can be achieved by the singular value decomposition of $\mathbf{F}$ that we will discuss in Sect. 15.3. The decomposition will yield:

$$\mathbf{F} = \mathbf{U}\mathbf{\Sigma}\mathbf{V}^{T} \quad \Rightarrow \quad \tilde{\mathbf{F}} = \mathbf{U}\mathbf{\Sigma}'\mathbf{V}^{T} \quad \Rightarrow \quad \mathbf{F} \leftarrow \tilde{\mathbf{F}} \qquad (13.132)$$

where $\mathbf{\Sigma}'$ is the same (diagonal) matrix as $\mathbf{\Sigma}$ except that the least significant (diagonal) element, $\Sigma(3, 3)$ is now replaced by 0.

13.8 Appendix

Below we illustrate matrix augmentation and multiplication by use of block matrices:

$$\begin{pmatrix} \overbrace{\begin{pmatrix} 1\ 2 \\ 4\ 5 \end{pmatrix}}^{\mathbf{A}} & \overbrace{\begin{pmatrix} 3 \\ 6 \end{pmatrix}}^{\mathbf{B}} \\ \underbrace{\begin{pmatrix} 7\ 8 \end{pmatrix}}_{\mathbf{C}} & \underbrace{\begin{pmatrix} 9 \end{pmatrix}}_{\mathbf{D}} \end{pmatrix} \begin{pmatrix} \overbrace{\begin{pmatrix} 9\ 8 \\ 6\ 5 \end{pmatrix}}^{\mathbf{A}'} & \overbrace{\begin{pmatrix} 7 \\ 4 \end{pmatrix}}^{\mathbf{B}'} \\ \underbrace{\begin{pmatrix} 3\ 2 \end{pmatrix}}_{\mathbf{C}'} & \underbrace{\begin{pmatrix} 1 \end{pmatrix}}_{\mathbf{D}'} \end{pmatrix} =$$

$$\begin{bmatrix} \mathbf{A} \ \mathbf{B} \\ \mathbf{C} \ \mathbf{D} \end{bmatrix} \begin{bmatrix} \mathbf{A}' \ \mathbf{B}' \\ \mathbf{C}' \ \mathbf{D}' \end{bmatrix} = \begin{bmatrix} \mathbf{AA}' + \mathbf{BC}' & \mathbf{AB}' + \mathbf{BD}' \\ \mathbf{CA}' + \mathbf{DC}' & \mathbf{CB}' + \mathbf{DD}' \end{bmatrix} \qquad (13.134)$$

Part IV

Vision in Multiple Directions

Why do you increase your bonds?
Take hold of your life before your light grows
dark and you seek help and do not find it.

St. Isaac (died 7th century)

14

Group Direction and N-Folded Symmetry

Direction of a line or a set of lines displaced in a parallel fashion were discussed in Chap. 10. Such images, characterized by linearly symmetric isocurves, contain a single direction that is the common direction of the parallel lines. The linearly symmetric image model was helpful to explain many local image phenomena, including edges, lines, simple textures, and a rich variety of corners. In its various extensions, the single direction model could even be useful for describing motion, simple texture, and even geometry. Here, we wish to go one step further and increase the number of directions that can be allowed to be contained in an image. We do this mainly because the linear model, which assumes a single direction in some coordinate system, is not sufficiently powerful to explain certain image phenomena. Texture patterns consisting of repetitive line constallations having multiple directions, even in the local scale, constitute an important example.

14.1 Group Direction of Repeating Line Patterns

We describe the concept of *group direction* by illustrating it through line patterns. Two linearly symmetric images that differ with $\frac{\pi}{4}$ or $\frac{5\pi}{4}$ in their directions are shown in the first column of Fig. 14.1. The relative rotation between the two images can be obtained from the relative directions between the constituent line directions. Using the double-angle representation, the direction of the lines can be made unique,

$$\theta = 2\varphi \tag{14.1}$$

where φ is the normal of a line direction. Accordingly, θ for the images on the top and on the bottom would be $\theta = \pi$ and $\theta' = \frac{3\pi}{2}$, respectively. The angle θ of the content can thus serve as the direction of the image itself, provided that it is known that the image is linearly symmetric. The relative rotation between the two images becomes $\Delta\theta = \theta' - \theta = \frac{\pi}{2}$, which needs to be divided by 2 to yield $\Delta\varphi$ as $\frac{\pi}{4}$ or $\frac{5\pi}{4}$.

In the second column of Fig. 14.1, the rotated versions of an image containing two sets of parallel lines are shown. The line sets differ directionwise maximally

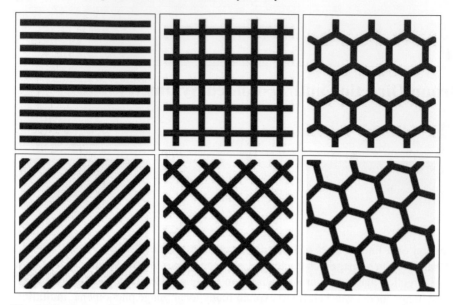

Fig. 14.1. *Top row* illustrates images containing 1, 2, and 3 directions. *Bottom row* shows the same, but with different group directions

within the same image, i.e., the constituent line directions differ by $\frac{\pi}{2}$. The amount of relative rotation between the images is $\frac{\pi}{4}$, $\frac{3\pi}{4}$, or $\frac{5\pi}{4}$. Can we deduce the relative image rotation from the contents? Can we represent the direction of the line constellation in each of the two images separately and uniquely as we did in the linearly symmetric images? Yes, we can use a "quadruple-angle" representation:

$$\theta = 4\varphi \qquad\qquad (14.2)$$

where φ is any of the four possible normal directions, to achieve this. We will call θ the *group direction*. The mapping is continuous w.r.t. any rotation and preserves uniqueness of the rotation in that it continuously maps four different angles to one and the same angle, which in turn represents an *equivalence class* of four angles continuously. Employing the normal directions in the second column of images, we obtain $\theta = 0$ and $\theta = \pi$ as the group directions for the top and the bottom images, respectively. While representing the "direction" of the image, the notion of "group direction" also represents the direction of a constellation of lines that is invariant to specific (group) rotations, justifying its name. In the case of linear symmetries, it was not urgent to "invent" the notion of group direction because images were assumed to contain only one direction. In this case, it was fairly simple to use this as a reference direction of the image itself, allowing us to deduce the relative rotation between two linearly symmetric images.

When there are many directions, the correspondence between the normal direction angles of individual lines and the rotation angle of a line constellation is more

Fig. 14.2. *The first four images* in reading direction are used to construct the *last two test charts*, FMTEST2 and FMTEST3

complex. Therefore a specific notion of group direction θ is needed to differentiate the image direction from individual line normal directions.

The third column shows a more elaborate line configuration. It consists of three distinct directions, none of which is more dominant than the other two. The images are rotated w.r.t. each other with the angle $\frac{\pi}{8} + l\frac{2\pi}{6}$, with l being an integer. In analogy with the above, we can obtain the group direction, uniquely and continuously determining the image and the line constellation directions, by *hexa-angle representation*,

$$\theta = 6\varphi \tag{14.3}$$

The group directions of the images at the top and the bottom then become 0 and $\frac{6\pi}{8}$, respectively.

14.2 Test Images by Logarithmic Spirals

In case of single direction occuring in a local image, we previously discussed the accuracy of direction estimation methods by using a frequency-modulated test chart, see FMTEST in Fig. 10.15. The idea was to observe the output of a method to local image inputs having all possible directions at all possible frequencies. Similarly, we present here two test charts to evaluate methods estimating group directions: one for group directions of local images containing two maximally distinct directions, e.g., the second column of Fig. 14.1, and one for the same, but with third maximally distinct directions, e.g., the 3rd column of Fig. 14.1.

To construct test charts with two distinct directions, we add wave patterns that are orthogonal to each other, namely the first two images of Fig. 14.2 in the reading direction. Each constituent image is locally a sinusiod with isocurves that are approximately parallel. All directions and a wide range of frequencies are represented. When added, the local result contains two wave patterns that are mutually orthogonal with a unique group direction and granularity (frequency), see FMTEST2 shown as the fifth image of Fig. 14.2. The group direction and the granularity change independently in the angular and in the radial directions, respectively.

Similarly, to construct test charts with three distinct directions we add wave patterns that intersect with the angle $\frac{\pi}{3}$ with each other, namely the first, the third, and the fourth images of Fig. 14.2 in reading direction. Each of these spiral[1] images are sinusiod patterns in log-polar coordinates with isocurves that are approximately parallel, locally. The result contains three sinusoid patterns with maximally distinct wave front directions, i.e. the spirals intersect one another at $\frac{\pi}{3}$. There is a continuous change in the group direction (angularly) and the granularity (radially), see FMTEST3 shown as the sixth image in Fig. 14.2.

In the test charts of FMTEST2 and FMTEST3, multiple wave patterns are added. By contrast, in the test chart of FMTEST there is only one sinusoidal wave pattern, locally. Although there are two gradient vectors with opposing directions present in all neighborhoods of FMTEST, there is a unique group direction for each local

---

[1] A circle is a special case of a spiral.

Fig. 14.3. The gradient directions (*left*) in the local images of FMTEST and the corresponding group directions (*right*). The two gradient direction types that can occur in local images, e.g., the *colored region*, are drawn at the same point, to avoid clutter

image. This is illustrated in Fig. 14.3, where the gradient directions and the group directions are overlayed on the original pattern. The gradient and the group direction vectors are drawn for neighborhoods at the same distance from the center and represent the various types of directions that can occur in the respective neighborhoods. Whereas the gradient directions are periodic with 2π, the group directions are periodic with the period π. Per neighborhood, there are two gradient vector types, differing in their directions with π, whereas there is a unique group direction. Because the partial derivatives D_x and D_y are linear, the normal vector at a point, the gradient, is the vectorial summation of the constituent normal vectors. Consequently there are four and six types of gradient vector directions that can locally occur in FMTEST2 and FMTEST3, respectively, but the local group directions are unique for all local images in both test charts. We illustrate this only for the latter, in Fig. 14.4, where we note that the group directions are unique and periodic with the period $\frac{\pi}{3}$, whereas there are six gradient direction types that can occur in a neighborhood, e.g., the green region. As one walks around a circle, the local images rotate continuously, and the group directions reflect this change by rotating continuously.

14.3 Group Direction Tensor by Complex Moments

In Chapter 10 we discussed a technique that allows detection of patterns having one direction, to be precise, linearly symmetric images, while it could quantitate the direction. Studying the FT of images that are linearly symmetric reveals that the power spectrum is concentrated to a line passing through the spectral origin. Here we will be interested in the same except that the number of directions that are jointly present

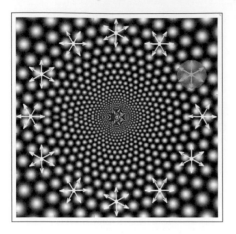 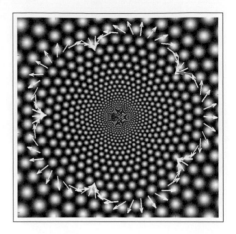

Fig. 14.4. The gradient directions (*left*) in the local images of FMTEST3 and the corresponding group directions (*right*). To avoid clutter, the six gradient direction types that can occur in local images, e.g., the *colored region*, are drawn at the same point

in the image is higher than one. We will call such images *n-folded symmetric*, as the following definition describes:

Definition 14.1. *Let F be the Fourier transform of an image f. If F is zero except on a set of lines, and any of these lines can be obtained from another by an integer multiple of $\frac{2\pi}{n}$ rotations, then f is n-folded symmetric.*

Because the FT is linear, the n-folded symmetric images can be viewed as the superposition of linearly symmetric images that differ directionwise maximally. The isocurves of linearly symmetric images are parallel lines. Accordingly, the definition suggests an extension of the linear symmetry concept to high-order symmetries by isocurve directions of ordinary lines that are jointly present in the image. Note that this extension is different than the generalized structure tensor discussed in Chap. 11, which measures the amount of linear symmetry in non-Cartesian coordinates, i.e. it still represents the presence of a single direction, albeit in a different coordinate system.

Because $|F|$ is invariant to translations of f, a quantity that represents fitting of multiple lines to it will also be invariant to translations. Consequently, fitting multiple lines to $|F|^2$, which is easier to do than to $|F|$, has useful consequences to image analysis. If the fitting error is reasonably low, the obtained directions of individual lines are the gradient directions of the individual isocurves of f. Using the mappings discussed in Sect. 14.1, it will then be straightforward to assign a meaningful and unique group direction to f.

We will first discuss fitting a cross to F in the TLS sense. The error function associated with this problem is not as straightforward to formulate as in Chapter 10. The approach taken here is first to map F to another function through a two-to-one transformation, and then to view it as a line-fitting problem, which we know how to

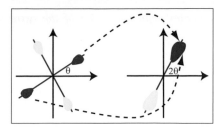

Fig. 14.5. The two-to-one mapping of the spectrum, Eq. (14.5), creating an equivalence of directions differing with $\frac{\pi}{2}$

solve. After presenting the solution to this problem, we will generalize these results by means of a theorem which includes the cross-fitting (and the line-fitting) process as a special case. The proof of the theorem will not explicitly be carried out since it is a straightforward extension of the method described here. We will show that the complex moment I_{pq} of $|F|^2$, with $p - q = 4$, fits a cross (4-folded symmetry) to the function $|F|^2$, and thereby to F in the TLS sense.

Assuming that the spectrum is expressed in polar coordinates, $F(r, \varphi)$, the integral defining the complex moment I_{40} of the power spectrum yields:

$$I_{40} = \int_0^{2\pi}\int_0^{\infty} r^4 \exp(i4\varphi)|F(r, \varphi)|^2 r \mathrm{d}r\mathrm{d}\varphi \tag{14.4}$$

Using the transformation

$$\varphi_1 = 2\varphi \tag{14.5}$$

the integral can be transformed once more, so that it can be identified as I_{20} of a remapped spectrum. The latter mapping is a two-to-one mapping, as illustrated in Fig. 14.5. Consequently, points of the spectrum that have angular coordinates differing by $\frac{\pi}{2}$ are treated as equivalent, i.e. the spectral power $|F|^2$ originating from such points is added, to produce the folded spectral power $|F_1|^2$. The complex moments I_{20}, I_{11} of $|F_1|^2$ are such that they equal I_{40}, I_{22} of $|F|^2$, respectively. Using theorem 10.2, one can then conclude that I_{40}, I_{22} fit a cross to the power spectrum in the TLS sense. The pairs $I_{4,0}, I_{2,2}$ also constitute a tensor because they together represent the structure tensor of a (remapped) power spectrum. The cross-fitting process, a detailed discussion of which is found in [26], leads naturally to the following generalization:

Theorem 14.1 (Group direction tensor I). *A pair of complex moments $I_{n,0}$ and $I_{\frac{n}{2},\frac{n}{2}}$ of $|F|^2$, with $n \neq 0$, determines an optimal fit of a set of lines possessing n-folded symmetry to a function F in the TLS sense. The real quantities $|I_{n0}|, I_{\frac{n}{2},\frac{n}{2}}$ depend on the minimum and the maximum errors of the fit, $e(k_{\max}^n)$, $e(k_{\min}^n)$, respectively,*

$$|I_{n0}| = e(k_{\max}^n) - e(k_{\min}^n) \tag{14.6}$$

$$I_{\frac{n}{2},\frac{n}{2}} = e(k_{\max}^n) + e(k_{\min}^n) \tag{14.7}$$

where $k^n_{\min}, k^n_{\max}$ are complex representations of the extremal group directions, $\theta_{\min}, \theta_{\max}$,

$$k^n_{\min} = \frac{I_{n,0}}{|I_{\frac{n}{2}\frac{n}{2}}|} = \exp(i\theta_{\min}), \qquad and \qquad k^n_{\max} = -k^n_{\min}. \tag{14.8}$$

♦

The quantities $I_{\frac{n}{2},\frac{n}{2}}, I_{n,0}$ represent a tensor because they measure a physical property, unbiased by the observation coordinate system. For odd n, the complex moments $I_{n,0}$ vanish when f is real. This is because the power spectrum of real images is even. Accordingly, to be useful, n must be even when f is real.

14.4 Group Direction and the Power Spectrum

In the theorem discussed above, F was the spectrum of an image without being specific about whether it is local or global, as it holds for both interpretations. However, the case when f is a local image is more interesting for texture analysis applications, therefore we assume this henceforth. The local image can be readily obtained by multiplying the original image with a window concentrated to a small support. Accordingly, $|F|^2$ will be the local power spectrum. This yields a useful interpretation in texture analysis, because the local power spectrum measurements, e.g., $I_{n,0}$, will then reflect the properties of the local texture, which should not change as long as one moves the window within the same texture.

The view supported by psychology experiments suggests that [131] the gray value of a local image in comparison with that of another point in the same neighborhood is more significant than the absolute gray values for human texture discrimination. This is sometimes called the *second-order statistics* because the gray image correlations of *two* points are measured. Some studies limit the definition of texture to images consisting of regions having the same second-order statistics, which depends on the distance and the direction between two points [81,96]. However, more generally a *texture* can be defined as an image consisting of translation-invariant local image properties. The local power spectrum, which is translation-invariant, is therefore rich in texture features. Within homogeneous regions, local phase information which allows us to discriminate between textures consisting of differently shaped lines and edges is neglected, e.g. [57]. In other words, only the directional information of the geometrical structures are taken into account, irrespective of whether, for example, the component lines are caused by crests or valleys. Therefore, when there is a texture with one distinct orientation (linear or 2-folded symmetry) around an inspected point, the power spectrum will be concentrated to a line. When there is a texture with two mutually orthogonal directions (rectangular or 4-folded symmetry) the power spectrum will be concentrated to a cross. It is similar for hexagonal/triangular structures (6-folded symmetry) and octagonal ones (8-folded symmetry). The arguments of the complex moments give the orientation of the estimated n-folded symmetry, whereas the magnitudes give measures of the estimation quality, that is, certainties.

At this point we note that if the complex moments I_{40} and I_{60} of a local power spectrum with an energy concentration having only 2-folded symmetry are computed, the magnitude responses will be high. However, these magnitude responses will be lower relative to the cases in which the spectrum shows 4- and 6-folded symmetries because 2 is a factor of 4 and 6; hence, such a concentration "leaks" to 4- and 6-folded symmetries. The converse is not true, which allows complex moments to discriminate between these symmetries. In the ideal case, a 6-folded symmetric energy concentration gives a zero response for complex moments with $p - q = 4$ and 2. Likewise, a 4-folded symmetric concentration yields a zero response for $p - q = 2$ and 6.

One might think that the arguments of the complex moments which represent $k^2, k^4, \ldots$ according to theorem 14.1 need to be divided by 2, 4, etc., for a straightforward representation of the directions of the component lines or edges. Likewise, the moduli of these moments do not provide for the minimum error directly. Although it is simple to compute this by solving the minimum error in Eqs. (14.6) and (14.7), we will argue that for pattern recognition purposes, the complex number representation of I_{pq} has advantages that allow us to circumvent the following three problems.

First, the *continuity* of the $I_{n,0}$ w.r.t. group directions is possible to achieve in the sense that two such directions that differ a small amount also differ a small amount in the numerical representations afforded by the real and imaginary parts of $I_{n,0}$. That is, a number arbitrarily close to 0 is not arbitrarily close to 2π, whereas the corresponding physical angles (as well as $I_{n,0}$) are [90]. In the Cartesian representation, the complex numbers are continuous with respect to changes in their arguments, except at the origin. Second, the factor n makes the representation of the direction of the symmetry unique. This is because this factor makes the argument of $I_{n,0}$ a *group direction*, as discussed in Sect. 14.1. Eliminating the factor n in the argument would thus discard the equivalence of certain rotation angles which are the same because they have exactly the same physical effect on n-folded symmetric images. Third, knowing the value of $e(k^n_{\min})$ alone is not sufficient to judge the quality of the estimate; one must know whether this error is large or small (the *range* problem). The comparison with the worst case, i.e., $e(k^n_{\max}) - e(k^n_{\min})$, provides a means to assess the quality. The complex quantity I_{20} already represents the difference of the errors through its magnitude, which is real. Alternative quality measures can be found easily, e.g.,

$$\frac{e(k^n_{\max}) - e(k^n_{\min})}{e(k^n_{\max}) + e(k^n_{\min})} = \left| I_{n,0} \right|_{\frac{n}{2}, \frac{n}{2}} \tag{14.9}$$

which is always in the interval $[0,1]$ and attains the end points 0 and 1 if the quality of the fit is totally uncertain and totally certain, respectively.

The continuity, group direction representation, and range problems might easily imperil the performance of e.g., a clustering method that could follow the extraction of orientation and certainty features, if an adequate representation of these features is lacking. By using complex moments in the Cartesian representation one can avoid these three problems.

14.5 Discrete Group Direction Tensor by Tensor Sampling

Theorem 14.1 suggests computing complex moments of the power spectrum $|F|^2$ to estimate the group direction. These computations can also be carried out in the spatial domain by direct tensor sampling, in analogy with the discussion in Sect. 10.11. We can write the integral of the complex moment $I_{n,0}$ as follows, if we assume that the spectrum is represented in its Cartesian coordinates, $F(\omega_x, \omega_y)$, and n is nonzero and even:

$$
\begin{aligned}
I_{n,0} &= \int_{E_2} (\omega_x + i\omega_y)^n (\omega_x - i\omega_y)^0 |F(\omega_x, \omega_y)|^2 d\omega_x d\omega_y \\
&= \int_{E_2} [(\omega_x - i\omega_y)^{\frac{n}{2}} F]^* [(\omega_x + i\omega_y)^{\frac{n}{2}} F(\omega_x, \omega_y)] d\omega_x d\omega_y \quad (14.10)
\end{aligned}
$$

By use of theorem 7.2, due to Parseval–Plancherel, the complex moments integral can be computed in the spatial domain:

$$
\begin{aligned}
I_{n,0} &= (\frac{1}{2\pi})^2 \int_{E_2} [(D_x - iD_y)^{\frac{n}{2}} f(x,y)]^* [(D_x + iD_y)^{\frac{n}{2}} f(x,y)] dx dy, \\
&= (\frac{1}{2\pi})^2 \int_{E_2} [(D_x + iD_y)^{\frac{n}{2}} f(x,y)]^2 dx dy, \quad (14.11)
\end{aligned}
$$

where f is the inverse FT of F. Further, we can identify $(D_x + iD_y)^{\frac{n}{2}} f$ as a *symmetry derivative* of f, introduced in Section 11.9. Then, assuming that f is band-limited, there exists an interpolation function (μ_1 below) by which we can reconstruct f via its discrete samples

$$
f(\mathbf{r}) = \sum_j f_j \mu_1(\mathbf{r} - \mathbf{r}_j) \quad (14.12)
$$

where $\mathbf{r} = (x, y)^T$ and $\mathbf{r}_j = (x_j, y_j)^T$. However, the function $(D_x + iD_y)^{\frac{n}{2}} f$ is band-limited too, and can be reconstructed from the samples of f:

$$
(D_x + iD_y)^{\frac{n}{2}} f(\mathbf{r}) = \sum_j f_j (D_x + iD_y)^{\frac{n}{2}} \mu_1(\mathbf{r} - \mathbf{r}_j) \quad (14.13)
$$

by linear filtering. Evidently, this function can also be sampled without loss of information on the same grid as the samples f_j are defined, so that

$$
(D_x + iD_y)^{\frac{n}{2}} f(\mathbf{r}_k) = \sum_j f_j (D_x + iD_y)^{\frac{n}{2}} \mu_1(\mathbf{r}_k - \mathbf{r}_j) \quad (14.14)
$$

Assuming that μ_1 is a Gaussian with the variance σ_p^2, the discrete symmetry derivatives of f can be computed by an ordinary discrete linear filtering with the filter

$$
\Gamma^{\{\frac{n}{2}, \sigma_p^2\}}(\mathbf{r}_k - \mathbf{r}_j) = (D_x + iD_y)^{\frac{n}{2}} \frac{1}{2\pi\sigma_p^2} \exp\left(\frac{\|\mathbf{r}_k - \mathbf{r}_j\|^2}{2\sigma_p^2}\right) \quad (14.15)
$$

Ordinarily, one uses the same grid as the f_j to compute its discrete symmetry derivatives, so that the discrete symmetry derivative of f is obtained by convolving the discrete f_k with the following filter, theorem 11.2.

$$\Gamma^{\{\frac{n}{2},\sigma_p^2\}}(\mathbf{r}_j) = (D_x + iD_y)^{\frac{n}{2}} \frac{1}{2\pi\sigma_p^2} \exp(\frac{\|\mathbf{r}_j\|^2}{2\sigma_p^2})$$

$$= (\frac{-1}{\sigma_p^2})^{\frac{n}{2}} \frac{1}{2\pi\sigma_p^2} (x + iy)^{\frac{n}{2}} \exp(\frac{\|\mathbf{r}_j\|^2}{2\sigma_p^2}) \qquad (14.16)$$

In analogy with Chapter 10, where we computed the elements of the structure tensor (the group direction tensor for 2-folded symmetry), we conclude that even the group direction tensor $[(D_x + iD_y)^{\frac{n}{2}} f]^2$ is band-limited because $(D_x + iD_y)^{\frac{n}{2}} f$ is band-limited. Accordingly, there is an interpolation function μ_2 that can reconstruct the former from its discrete elements as

$$[(D_x + iD_y)^{\frac{n}{2}} f]^2(\mathbf{r}) = \sum_j [(D_x + iD_y)^{\frac{n}{2}} f]^2(\mathbf{r}_j)\mu_2(\mathbf{r} - \mathbf{r}_j) \qquad (14.17)$$

Then, assuming a Gaussian as an interpolator, an element of the group direction tensor for n-folded symmetry:

$$I_{n,0} = \int_{E_2} \sum_j [(D_x + iD_y)^{\frac{n}{2}} f]^2(\mathbf{r}_j)\mu_2(\mathbf{r} - \mathbf{r}_j)dxdy$$

$$= \sum_j [(D_x + iD_y)^{\frac{n}{2}} f]^2(\mathbf{r}_j) \int_{E_2} \mu_2(\mathbf{r} - \mathbf{r}_j)dxdy$$

$$= \sum_j [(D_x + iD_y)^{\frac{n}{2}} f]^2(\mathbf{r}_j)\mu_2(\mathbf{r}_j)dxdy \qquad (14.18)$$

is obtained by an ordinary Gaussian smoothing of the (pixelwise) square of the complex image delivered by the computation scheme given in Eq. (14.14).

Following a similar reasoning, we can obtain the remaining tensor element, $I_{\frac{n}{2},\frac{n}{2}}$, which is always real and nonnegative. We summarize our findings as a theorem.

Theorem 14.2 (Group direction tensor II). *The complex group direction number, $k_{min}^n = \exp(i\theta_{min})$, associated with an n-folded symmetric line set fitted to the power spectrum in the TLS error sense and the extremal TLS errors of the fit are given by*

$$I_{n,0} = (e(k_{max}^n) - e(k_{min}^n)) \exp^{i\theta_{min}} = \frac{1}{4\pi^2} \int [(D_x + iD_y)^{\frac{n}{2}} f(\mathbf{r})]^2 d\mathbf{r}$$

$$= \langle [\Gamma^{\{\frac{n}{2},\sigma_p^2\}} * f]^2 \rangle \qquad (14.19)$$

$$I_{\frac{n}{2},\frac{n}{2}} = (e(k_{max}^n) + e(k_{min}^n)) = \frac{1}{4\pi^2} \int |(D_x + iD_y)^{\frac{n}{2}} f(\mathbf{r})|^2 d\mathbf{r}$$

$$= \langle |\Gamma^{\{\frac{n}{2},\sigma_p^2\}} * f|^2 \rangle \qquad (14.20)$$

where $\Gamma^{\{\frac{n}{2},\sigma_p^2\}}$ is defined according to (11.97).

♦

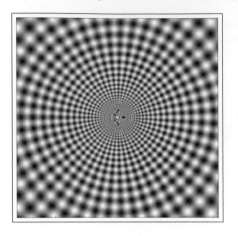

Fig. 14.6. The test image FMTEST2 and I_{40} computed for all local images. The intensity and the hue are modulated by the magnitude and the argument of I_{40}

The structure tensor in theorems 10.2, and 10.4 can be obtained as special cases of this theorem with $n = 2$. Furthermore, we ended up with the same basic computational scheme consisting of three consecutive operations—two linear-filtering steps, linked by an intermediary, nonlinear-mapping step:

First Linear-Filtering: Apply a symmetry derivative filtering,
Nonlinear-Mapping: Apply a pointwise squaring,
Second Linear-Filtering: Apply a symmetry derivative filtering.

The symmetry derivative filter in the first step, $(D_x + iD_y)^{\frac{n}{2}}$, is more elaborate as compared to $D_x + iD_y$, which is used in linear symmetry detection. The last step uses the zero-order symmetry derivative filter in both cases, average-filtering.

Surprisingly, even the GST scheme, developed to detect nontexture patterns, e.g., parabolic patterns, log-spirals, crosses, etc., see lemma 11.6, shares the same computational steps as above. Additionally, the filters used in the first step here are similar to those derived for the last step of the GST scheme. By contrast, in the first step of the GST, the simple, first-order symmetry derivative filter, $D_x + iD_y$, is mandatory whereas more elaborate filters are reserved for the last step. As will be discussed below, and in Section 14.6, to allow high-order symmetries at the first step ensures that multiple directions can be seen as equivalent by the detection process. This will in turn provide translation-invariant *texture* features.

It is worth noting that linearly symmetric patterns can conceptually belong to both object-, and texture-type patterns. In isolation, they can on one hand be viewed as being harmonic monomial objects with the trivial harmonic mapping (the identity transformation in GST) whereas on the other hand, they are the simplest textures containing line patterns, those having a *single direction* when repeated.

To visualise the effect of computing complex moments in the local power spectrum, we show I_{40} applied to FMTEST2 in Fig. 14.6. The hue and intensity represent

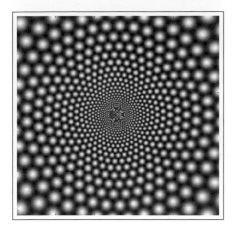 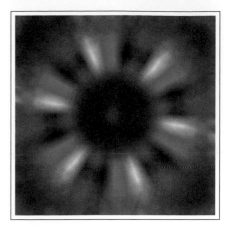

Fig. 14.7. The test image FMTEST3 and I_{60} computed for all local images. In the *color image* the intensity and the hue represent the magnitude and the argument of I_{60}, respectively

the group direction θ and the certainty of the group direction, respectively. They are obtained from the argument and the magnitude of I_{40} at each local image via the three-step complex filtering suggested by theorem 14.2. In the convolution of the first step, delivering symmetry derivatives of local images, the filter $\Gamma\{2, 0.64\}$ was used. After squaring, the image was filtered through $\Gamma\{0, 5.76\}$, which is an ordinary Gaussian with $\sigma = 2.4$ that defines the extent of a local image. The orginal image is preprocessed by a bandpass filter to facilitate the specialization of the scheme to a certain frequency range only. This means that the scheme with the above filters is specialized to detect joint occurence of two directions (4-folded symmetric images) at a certain range around a tune-on frequency, which in the result is observed as nonblack. In the figure the original is placed beside the result to facilitate visual identification of the effective frequency range of the scheme. We also observe that the hue of the colors change continuously and uniformly in the angular direction with the period $\frac{\pi}{2}$. This is manifested by the fact that the same color appears regularly four times circularly. Accordingly, the group direction has been extracted faithfully by I_{40}. A detailed discussion on this can be found in [134].

A similar processing of FMTEST3 to obtain I_{60} yields Fig. 14.7. The hue and intensity represent the group direction θ and the certainty as before. The convolution of the first step utilizes the complex filter $\Gamma\{3, 0.64\}$, whereas the last step uses the same Gaussian filter as above, $\Gamma\{0, 5.76\}$. The orginal image is subjected to a bandpass-filtered prior I_{60} computation to allow the frequency range specialization of the scheme to a certain frequency range only. The filtering scheme with the above filters detects joint occurrence of three directions (6-folded symmetry). As evidence of accuracy of the group direction computation, the hue of the colors changes continuously and uniformly in the angular direction with the period $\frac{\pi}{3}$. The latter is observed by noting that the colors repeat regularly six times circularly.

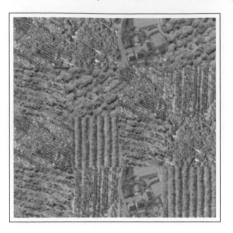

Fig. 14.8. The P1 texture patch composition consisting of real aerial images (*left*) and its automatic segmentation using group direction tensors for 2, 4, and 6-folded symmetries

14.6 Group Direction Tensors as Texture Features

In the previous sections, we discussed group direction features that were shown to be invariant to translation. This is a property one wishes to have in texture features because textures are precisely defined as images in which the same geometric structure repeats itself. The texture properties measured around any point should not vary as long as the point is within the same texture. How powerful the group direction features are and how they can be used as texture features will be studied in further detail next.

The group direction of a texture for n-folded symmetry is determined by the two complex moments,

$$I_{n,0}, \qquad \text{and} \qquad I_{\frac{n}{2},\frac{n}{2}}, \qquad (14.21)$$

of the power spectrum. This s pair represents three real scalars, the first being a complex scalar (two real scalars), the second being a (nonnegative) real scalar. They can be applied to the original image directly or to a bandpass-filtered version of the original, forcing the quantities to be measured for only a certain frequency range (scale) of the image. Assuming bandpass filtering, the result of which is a number of images that differ in their absolute frequency contents, then one can extract the group direction tensor elements for every image, increasing the description power. Bandpass decompositions can be efficiently implemented by a variety of ways, e.g., the Laplace pyramid discussed in Sect. 9.5. Therefore the two complex moments provide evidence for the joint presence of $\frac{n}{2}$ directions contained in a narrow frequency ring and within a local support around a point in an image. We discuss the practical issues by way of two examples below.

Example 14.1. *In Fig. 14.8 we show the P1 texture patch that has to be segmented based on local image features. The texture patches in the composition are cut from* aerial images. Before being put together they are also mean and variance normalized photometrically in that all patches have the same mean and the same variance. This is done to attempt that segmentation be done on elaborate texture measurements rather than based on simple (nontexture) features such as gray level and variance, which can be influenced by illumination. Since the boundaries are known and the patches are photometrically normalized, an advantage of using such images is that one can evaluate and compare the results with other segmentation techniques by sharing the test image and its results only. The ground truth of the boundaries is shown in Fig. 16.8

The segmentation is to be done by some grouping, in unsupervised manner. How to do this grouping and estimating of region boundaries (without training a neural network or a classifier) is suggested in Chapter 16. To stay within the scope of this chapter, however, we only study the texture features. First, five fully circular (isotropic) subbands are obtained by convolution. The (absolute) frequency centers of the subbands are in (octavelike) geometric progression with the factor 1.2 between successive subbands, see Fig. 9.14. The following real measurement for each subband has been computed:

	I_{11}	I_{20}	I_{40}	I_{60}	Total
#Real scalars	1	2	2	2	7

so that effectively 35 measurements per image point are available. The elements I_{22}, I_{33} are excluded because the respective measurements are well approximated through I_{11} when the input image is a subband that contains only a narrow ring of frequency range. In the latter case, $I_{\frac{n}{2},\frac{n}{2}}$ for all n, estimate the variance within a narrow ring of frequencies. The mentioned grouping followed by a boundary estimation using the above features have been applied. The result is shown as a color image in the Fig. 14.8 where color represents the identity label of the found textures. Accordingly, the same color represents the same texture. All seven texture patches are correctly identified, and the boundaries are reasonably well drawn without training.

Example 14.2. In Fig. 14.9 we show a more difficult texture patch composition consisting of seven different textures in a 4×4 arrangement as before. Again, the individual patches are photometrically normalized, and the same texture measures detailed above have been utilized in the automatic grouping and boundary estimation process. The seven texture patches are correctly identified, and the boundaries are reasonably well drawn, although the boundary quality is somewhat worse.

Exercise 14.1. We have studied only $I_{n,0}, I_{\frac{n}{2},\frac{n}{2}}$ above. What do other complex moments of the power spectrum with $p \geq q$, e.g., $I_{4,2}, I_{5,1}$, can you estimate? Does I_{qp} contain new information compared to I_{pq}?
HINT: $(\omega_x + i\omega_y)^p(\omega_x - i\omega_y)^q = (\omega_x + i\omega_y)^{p-q}|\boldsymbol{\omega}|^{2q}$.

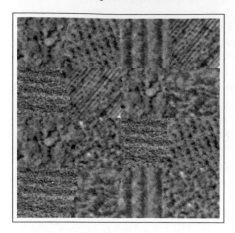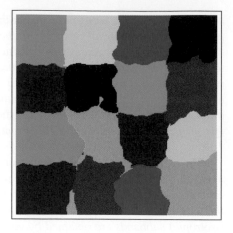

Fig. 14.9. The P2 texture patch composition consisting of real aerial images (*left*) and its automatic segmentation using group direction tensors for 2, 4, and 6-folded symmetries.

14.7 Further Reading

Texture as a vision problem has been covered by a large body of publications in psychology. Texture recognition in its various forms appears in many applications, e.g., as diagnostic tools, multi-spectral image segmentation, and tracking in image sequences [78,82,233]. The prevailing view is that the second order statistics of images is the only property that humans can discriminate in textures. The study in [178] suggests certain Lie operators, which are differential operators, to define textures. While the complex moments provide for optimal solutions to the line-fitting, cross-fitting, etc., problems in the power spectrum, they correspond precisely to Lie operators that can identify the direction(s) along which the image is translation invariant. As has been discussed here, these complex moments can be computed by means of symmetry derivatives applied to subbands of the original in the spatial domain. The scheme was possible to implement via separable filtering thanks to direct tensor sampling and the theoretical results studied in Sect. 11.9. However, the group direction tensor can also be obtained by spectrum sampling in analogy with Sect. 10.13. The theoretical details of this approach to implement the group direction tensor are studied in [26], whereas experimental results can be found in [25]. Even if a texture contains symmetries of high order when no subband decomposition is applied, it can still be discriminated against another texture by use of only structure tensor features applied at the subbands level [134], e.g., the P3 patch shown in Fig. 16.8. By contrast, some textures cannot be discriminated by applying the structure tensor to subbands, but need the descriptive power of higher order symmetry features, e.g., Figs. 14.9 and 16.8. Accordingly, as the number of textures involved in the discrimination increases, and/or the complexity of the individual textures increases, the structure tensor features need to be completed. This can be done by the group direction tensors.

Part V

Grouping, Segmentation, and Region Description

Who has ever seen, that a breach became as a mirror?
Two parties looked thereinto;
it served for those without and those within.
They saw therein as with eyes,
the Power that breaks down and builds up:
They saw Him who made the breach
and again repaired it.
Those without saw His might;
they departed and tarried not till evening:
those within saw His help;
they gave thanks yet sufficed not.

The Nisibene hymns,
St. Ephrem (A.D. 303–373)

Reducing the Dimension of Features

For many applications, a dimensionality reduction results in improved signal separation in the presence of noise. We will discuss the underlying concept in the principal components analysis section below. However, dimension reduction is a general problem that has been the subject of intensive study in different disciplines. While principal components analysis is the earliest and simplest technique, and is widely used for its efficiency, there are other approaches, such as independent component analysis, neural networks, and self-organizing maps [29, 117, 142], that have proven to be more powerful in many applications. For reason of limited scope, we must, however, restrict our discussion to principal components analysis. An example of its use in face recognition is discussed in Sect. 15.2, whereas another example for which reduced dimension is a prerequisite, texture analysis, will be illustrated in Sect. 16.7 when discussing clustering and boundary estimation in textures.

15.1 Principal Component Analysis (PCA)

There are many names to principal component analysis (PCA), as it has been used in numerous disciplines and applications. Examples include color representation, texture segmentation, multispectral image classification, face recognition, source separation, visualization, and image database queries [100, 172, 179, 202, 218, 220]. Synonyms of PCA include *Karhunen–Loéve (KL) transform*, [133], Hotelling transform, eigenvalue analysis, eigenvector decomposition, and spectral decomposition. In image analysis it is used to reduce dimensions, and to find subspaces in which recognition works better than taking the full space. Not only does PCA reduce the size of the data for the sake of efficient storage, transmission, and processing advantages.

Assume that we have observed a set of K vectors

$$\mathbb{O} = \{\mathbf{f}_k\} \tag{15.1}$$

i.e., the *observation set*, in an M-dimensional vector space.

The coordinates of the observation set can be represented by M scalars in some basis

$$\mathbb{B}_M = \{\boldsymbol{\psi}_1, \boldsymbol{\psi}_2, \cdots, \boldsymbol{\psi}_M\} \tag{15.2}$$

yet to be specified, as

$$\mathbf{f}_1 = \begin{pmatrix} \mathbf{f}_1(1) \\ \mathbf{f}_1(2) \\ \mathbf{f}_1(3) \\ \vdots \\ \mathbf{f}_1(M) \end{pmatrix}, \ \mathbf{f}_2 = \begin{pmatrix} \mathbf{f}_2(1) \\ \mathbf{f}_2(2) \\ \mathbf{f}_2(3) \\ \vdots \\ \mathbf{f}_2(M) \end{pmatrix}, \cdots, \mathbf{f}_k = \begin{pmatrix} \mathbf{f}_k(1) \\ \mathbf{f}_k(2) \\ \mathbf{f}_k(3) \\ \vdots \\ \mathbf{f}_k(M) \end{pmatrix}, \cdots, \mathbf{f}_K = \begin{pmatrix} \mathbf{f}_K(1) \\ \mathbf{f}_K(2) \\ \mathbf{f}_K(3) \\ \vdots \\ \mathbf{f}_K(M) \end{pmatrix}, \tag{15.3}$$

where $\mathbf{f}_k(m)$ is the mth component of the vector $\mathbf{f}_k$. Each vector $\mathbf{f}_k$ can then be written as

$$\mathbf{f}_k = \sum_{m=1}^{M} \mathbf{f}_k(m)\boldsymbol{\psi}_m \tag{15.4}$$

By using all M basis vectors, we can thus represent any of the observed $\mathbf{f}_k$ without error. This remains true even if we choose another basis set containing M orthogonal vectors, as long as we include all M basis vectors in the expansion in Eq. (15.4).

Does the basis we choose really matter? Yes, it does, because in applications we cannot always afford to choose complete bases of M vectors for a variety of reasons, including that M can be too large. One must then expand each of the observed $\mathbf{f}_k$ by using fewer vectors:

$$\tilde{\mathbf{f}}_k = \sum_{m=1}^{N} \mathbf{f}_k(m)\boldsymbol{\psi}_m, \qquad \text{where} \qquad N < M \tag{15.5}$$

Note that the only difference between Eq. (15.5) and Eq. (15.4) is in the number of the terms in the summation, N and M, respectively. All terms in Eq. (15.5) exist in Eq. (15.4), but not vice versa. The vectors

$$\tilde{\mathbf{f}}_1, \tilde{\mathbf{f}}_2 \cdots, \tilde{\mathbf{f}}_k, \cdots, \tilde{\mathbf{f}}_K \tag{15.6}$$

only approximate the corresponding observation, because the approximation error

$$\|\mathbf{f}_k - \tilde{\mathbf{f}}_k\| \tag{15.7}$$

is usually not zero.

Here, we are interested in finding an orthonormal (ON) basis, $\mathbb{B}_N$:

$$\mathbb{B}_N = \{\boldsymbol{\psi}_1, \cdots, \boldsymbol{\psi}_N\}, \qquad \text{with} \qquad \langle \boldsymbol{\psi}_i, \boldsymbol{\psi}_j \rangle = \delta_{ij}, \tag{15.8}$$

that is "most economical" among all possible ON basis sets. Note that $\mathbb{B}_N$ is a "truncated" $\mathbb{B}_M$ in that it has fewer basis vectors. Economical means that, despite the fact that the basis $\mathbb{B}_N$ has fewer basis vectors than the full set, it should still represent $\mathbb{O}$ with a smaller *basis truncation error*,

$$\frac{1}{K} \sum_{k}^{K} \|\mathbf{f}_k - \tilde{\mathbf{f}}_k\|^2 \tag{15.9}$$

than all alternative bases. This should be true on equal footing, i.e., when no more than N basis vectors are used in each of the expansions $\tilde{\mathbf{f}}_k$. Accordingly, we are interested in minimizing the error, by varying $\mathbb{B}_N$ for the same observation set $\mathbb{O}$:

$$
\begin{aligned}
\sum_k^K \|\mathbf{f}_k - \tilde{\mathbf{f}}_k\|^2 &= \sum_k^K \langle \mathbf{f}_k - \tilde{\mathbf{f}}_k, \mathbf{f}_k - \tilde{\mathbf{f}}_k \rangle \\
&= \sum_k^K \langle \mathbf{f}_k, \mathbf{f}_k \rangle + \langle \tilde{\mathbf{f}}_k, \tilde{\mathbf{f}}_k \rangle - 2\langle \mathbf{f}_k, \tilde{\mathbf{f}}_k \rangle \\
&= \sum_k^K \|\mathbf{f}_k\|^2 + \|\tilde{\mathbf{f}}_k\|^2 - 2\langle \mathbf{f}_k, \tilde{\mathbf{f}}_k \rangle
\end{aligned}
\tag{15.10}
$$

By substituting Eqs. (15.4) and (15.5) in the last term of this equation, we then obtain:

$$
\begin{aligned}
\sum_k^K \|\mathbf{f}_k - \tilde{\mathbf{f}}_k\|^2 &= \sum_k \left(\|\mathbf{f}_k\|^2 + \|\tilde{\mathbf{f}}_k\|^2 - 2\langle \sum_{m=1}^M \mathbf{f}_k(m)\boldsymbol{\psi}_m, \sum_{m=1}^N \mathbf{f}_k(m)\boldsymbol{\psi}_m \rangle \right) \\
&= T + \sum_k \left(\|\tilde{\mathbf{f}}_k\|^2 - 2\langle \sum_{m=1}^N \mathbf{f}_k(m)\boldsymbol{\psi}_m + \sum_{m=N+1}^M \mathbf{f}_k(m)\boldsymbol{\psi}_m, \sum_{m=1}^N \mathbf{f}_k(m)\boldsymbol{\psi}_m \rangle \right) \\
&= T + \sum_k \left(\|\tilde{\mathbf{f}}_k\|^2 - 2\langle \tilde{\mathbf{f}}_k, \tilde{\mathbf{f}}_k \rangle - 2\langle \sum_{m=N+1}^M \mathbf{f}_k(m)\boldsymbol{\psi}_m, \sum_{m=1}^N \mathbf{f}_k(m)\boldsymbol{\psi}_m \rangle \right)
\end{aligned}
\tag{15.11}
$$

where

$$
T = \sum_k \|\mathbf{f}_k\|^2
\tag{15.12}
$$

is constant w.r.t. the changes of $\mathbb{B}_M$ because the length of $\mathbf{f}_k$ is the same (a zero-order tensor) in every basis. In Eq. (15.11), the first scalar product term $\langle \tilde{\mathbf{f}}_k, \tilde{\mathbf{f}}_k \rangle$, can be identified as $\|\tilde{\mathbf{f}}_k\|^2$, whereas the second scalar product term vanishes because of orthogonality of the involved $\boldsymbol{\psi}_m$, to yield

$$
\frac{1}{K} \sum_k^K \|\mathbf{f}_k - \tilde{\mathbf{f}}_k\|^2 = \frac{T}{K} - \frac{1}{K} \sum_k^K \|\tilde{\mathbf{f}}_k\|^2
\tag{15.13}
$$

Since T is a constant, minimizing this expression is equivalent to maximizing

$$
\begin{aligned}
\sum_k^K \|\tilde{\mathbf{f}}_k\|^2 &= \sum_{k=1}^K \langle \tilde{\mathbf{f}}_k, \tilde{\mathbf{f}}_k \rangle = \sum_{k=1}^K \sum_{m=1}^N |\mathbf{f}_k(m)|^2 \\
&= \sum_{k=1}^K \sum_{m=0}^N |\langle \boldsymbol{\psi}_m, \mathbf{f}_k \rangle|^2 = \sum_{k=1}^K \sum_{m=0}^N \langle \boldsymbol{\psi}_m, \mathbf{f}_k \rangle \langle \mathbf{f}_k, \boldsymbol{\psi}_m \rangle
\end{aligned}
\tag{15.14}
$$

The scalar product of the vector space, the Euclidean F_M, to which both the vectors of $\mathbb{O}$ and $\mathbb{B}_M$ belong, is given by $\langle \boldsymbol{\psi}_m, \mathbf{f}_m \rangle = \boldsymbol{\psi}_m^T \mathbf{f}_k$, so that the highest bound of

$$\sum_{k}^{K} \|\tilde{\mathbf{f}}_k\|^2 = \sum_{k=1}^{K} \sum_{m=0}^{N} \langle \boldsymbol{\psi}_m, \mathbf{f}_k \rangle \langle \mathbf{f}_k, \boldsymbol{\psi}_m \rangle = \sum_{k=1}^{K} \sum_{m=0}^{N} \boldsymbol{\psi}_m^T \mathbf{f}_k \mathbf{f}_k^T \boldsymbol{\psi}_m$$

$$= \sum_{m=0}^{N} \boldsymbol{\psi}_m^T \left(\sum_{k=1}^{K} \mathbf{f}_k \mathbf{f}_k^T \right) \boldsymbol{\psi}_m = K \sum_{m=0}^{N} \boldsymbol{\psi}_m^T \mathbf{S} \boldsymbol{\psi}_m \leq K \sum_{m=1}^{N} \lambda_{(m)} \quad (15.15)$$

is reached when the $\boldsymbol{\psi}_m$ are the N most significant[1] eigenvectors of

$$\mathbf{S} = \frac{1}{K} \sum_{k=1}^{K} \mathbf{f}_k \mathbf{f}_k^T \quad (15.16)$$

which is the scatter matrix for the observation vectors $\mathbb{O}$. This is because $\mathbf{S}$ is, by construction, a positive semi-definite matrix, meaning that $0 \leq \mathbf{g}^T \mathbf{S} \mathbf{g}$ for all $\mathbf{g} \in E_M$, including for $\mathbf{g} = \boldsymbol{\psi}_m$. Accordingly, substituting Eq. (15.15) in Eq. (15.13), yields

$$\frac{1}{K} \sum_{k}^{K} \|\mathbf{f}_k - \tilde{\mathbf{f}}_k\|^2 = \frac{T}{K} - \frac{1}{K} \sum_{k}^{K} \|\tilde{\mathbf{f}}_k\|^2 \geq \frac{T}{K} - \sum_{m=1}^{N} \lambda_{(m)} = \sum_{m=N+1}^{M} \lambda_{(m)} \quad (15.17)$$

and one can similarly conclude that the lowest bound of the approximation error is reached if $\mathbb{B}_N$ is chosen as the N most significant eigenvectors of $\mathbf{S}$. Here, we have used

$$\frac{T}{K} = \frac{1}{K} \sum_{k=1}^{K} \|\mathbf{f}_k\|^2 = \lambda_{(1)} + \cdots + \lambda_{(M)}, \quad (15.18)$$

which can be obtained by remembering that the approximation is error-free, i.e. $\tilde{\mathbf{f}}_k = \mathbf{f}_k$, if one uses all basis vectors, i.e. $N = M$, in Eqs. (15.15) and (15.12). Expression (15.17) tells that if we decide to use N eigenvectors (and not the full set of M) then the lowest possible error of approximation is the sum of the $N - M$ least significant eigenvalues of $\mathbf{S}$, and that this error is achievable if one chooses the N most significant eigenvectors of $\mathbf{S}$. An estimation of $\mathbf{S}$ can be obtained from the observed vectors.

For computational convenience, the scatter matrix computation can be achieved as a pure matrix multiplication, since

$$\mathbf{S} = \frac{1}{K} \sum_{k} \mathbf{f}_k \mathbf{f}_k^T = \frac{1}{K} [\mathbf{f}_1, \cdots, \mathbf{f}_K] \begin{bmatrix} \mathbf{f}_1^T \\ \vdots \\ \mathbf{f}_K^T \end{bmatrix} = \frac{1}{K} \mathbf{O} \mathbf{O}^T \quad (15.19)$$

Because $\mathbf{S}$ is symmetric positive semidefinite, its eigenvectors are orthogonal, guaranteeing the optimal basis to be orthogonal. Once the new basis is found, the observation data can be projected onto the basis vectors as

$$\tilde{\mathbf{O}}^T = \mathbf{O}^T \mathbf{B}_N \quad (15.20)$$

---

[1] The notation $\lambda_{(m)}$ represents sorted eigenvalues, $\lambda_{(M)} \leq \cdots \leq \lambda_{(1)}$.

where

$$\tilde{\mathbf{O}} = [\tilde{\mathbf{f}}_1, \cdots, \tilde{\mathbf{f}}_N], \quad \mathbf{O} = [\mathbf{f}_1, \cdots, \mathbf{f}_K], \quad \mathbf{B}_N = [\psi_1, \cdots, \psi_N] \tag{15.21}$$

are the juxtaposed vector coordinates corresponding to the data, a subset of the basis, and the new data, respectively. The resulting data vectors have N coordinates for $\tilde{\mathbf{f}}_k$, down from the original M for $\mathbf{f}_k$, and yet achieve an approximation of the original data with the minimal error of Eq. (15.9). Since $\mathbf{B}_M$ has full rank with orthogonal columns, using the full set in Eq. (15.20) would rotate the original coordinates. Consequently, Eq. (15.20) is a truncated version of the rotated coordinates, where the corresponding basis vectors are numbered according to their ability to represent the observed data.

Before the rotation, it is often wise to translate the data to the mass center, especially in equally quantized feature spaces (e.g., images are usually quantized uniformly to yield 256 gray values), to avoid numerical problems stemming from the limited dynamic range. Data translation can also be preferable from a pattern discrimination viewpoint too, because the basis rotated at the mass center to yield the minimal error of representation is not the same as a basis attached elsewhere and rotated there. The rotation at the mass center will rank the directions according to the maximal distance of the data to the mass center and the directional variation of the data (which we discuss at the end of this section). If two classes are to be discriminated in a certain direction in the new basis, it is better that there is a great variation in that direction than no variation, since the latter suggests that the classes leave the same footprint, and hence cannot be separated in that particular direction.

Translating the observed vectors to the centroid:

$$\mathbf{f}_k \leftarrow (\mathbf{f}_k - \mathbf{f}_c), \qquad \text{where} \qquad \mathbf{f}_c = \frac{1}{K} \sum_{k=1}^{K} \mathbf{f}_k, \tag{15.22}$$

and searching for a new basis minimizing the approximation error

$$\frac{1}{K} \sum_{k=1}^{K} \|\mathbf{f}_k - \tilde{\mathbf{f}}_k\|^2 \tag{15.23}$$

among the bases attached to the mass center can be conveniently achieved by preprocessing. The scatter matrix obtained after shifting the data to the mass center is also known as the *covariance matrix*. Alternatively, the covariance matrix can be computed as

$$\mathbf{C} = \frac{1}{K} \sum_{k=1}^{K} (\mathbf{f}_k - \mathbf{f}_c)(\mathbf{f}_k - \mathbf{f}_c)^T \tag{15.24}$$

where $\mathbf{f}_k$ is the unshifted observation data. The eigenvectors of $\mathbf{C}$ are frequently called the *principal components*, whereas the rotated coordinates $\tilde{\mathbf{O}}$, relative to the mass center, are called the Karhunen–Loéve (KL) transform.

We could use the same arguments as above to find analogous results for the dimension reduction of the complex vector space, C_M, allowing us to formulate the following lemma, which summarizes the conclusions so far.

Lemma 15.1. *Let* $\mathbf{O} = [\mathbf{f}_1, \cdots, \mathbf{f}_K]$, *where* $\mathbf{f}_k$ *is a vector in* C_M. *The ON basis* $\mathbf{B}_N = [\boldsymbol{\psi}_1, \cdots, \boldsymbol{\psi}_N]$ *that minimizes*

$$\frac{1}{K}\sum_k^K \|\mathbf{f}_k - \tilde{\mathbf{f}}_k\|^2, \qquad with \qquad \tilde{\mathbf{f}}_k = \sum_m^N \mathbf{f}_k(m)\boldsymbol{\psi}_m, \tag{15.25}$$

for every integer $N : N < M$, *is given by the first* N *eigenvectors of the scatter matrix:*

$$\mathbf{S} = \frac{1}{K}\sum_{k=1}^K \mathbf{f}_k\mathbf{f}_k^H = \frac{1}{K}\mathbf{OO}^H \tag{15.26}$$

where the eigenvalues are sorted as $\lambda_{(1)} \geq \cdots \geq \lambda_{(M)}$. *The new coordinates are given by*

$$\tilde{\mathbf{O}}^T = \mathbf{O}^T\mathbf{B}_N, \qquad with \qquad \tilde{\mathbf{O}} = [\tilde{\mathbf{f}}_1, \cdots, \tilde{\mathbf{f}}_N]. \tag{15.27}$$

$\blacklozenge$

Now we consider the following problem. Assuming that the data has its mass center in the origin, we wish to search for $N = M - 1$ basis vectors to approximate the observations $\mathbf{O}$ in the TLS sense. The problem can be solved by application of the lemma, evidently. Alternatively, we can conceive it as a hyperplane-fitting problem:

$$\mathbf{f}^T\boldsymbol{\psi} = 0 \tag{15.28}$$

where $\boldsymbol{\psi}$ is the normal of an unknown $(M - 1)$-dimensional hyperplane passing through the origin in E_m. Given the observations $\mathbf{O} = [\mathbf{f}_1, \cdots, \mathbf{f}_K]$, which contain noise, the equation will not be satisfied exactly but can be solved in the TLS sense, by searching for a $\boldsymbol{\psi}$ minimizing

$$\frac{1}{K}\|\mathbf{O}^T\boldsymbol{\psi}\|^2 = \frac{1}{K}\boldsymbol{\psi}^T\mathbf{OO}^T\boldsymbol{\psi}, \qquad where \qquad \|\boldsymbol{\psi}\| = 1, \tag{15.29}$$

which is solved by the least significant eigenvector of $\mathbf{OO}^T$. When the data are projected onto this plane by

$$\mathbf{f} - \langle \boldsymbol{\psi}, \mathbf{f} \rangle \boldsymbol{\psi} \tag{15.30}$$

we have thus reduced its dimension by 1. The error in the hyperplane approximation is λ_M, the least eigenvalue of $\mathbf{S} = \frac{1}{K}\mathbf{OO}^T$. Naturally, one could fit planes recursively to eliminate more and more dimensions until reaching any desired dimension N. Accordingly, we conclude that although conceptually different, the dimension reduction and the direction estimation are equivalent mathematically, because both are solved by an eigen-analysis of the same scatter matrix. We summarize this as a lemma.

Lemma 15.2 (Direction and PCA). *A solution* $\boldsymbol{\psi}$ *of a homogeneous equation*

$$\mathbf{O}^T\boldsymbol{\psi} = \mathbf{0}, \tag{15.31}$$

where $\|\psi\| = 1$ and $\mathbf{O}$ is an $M \times K$ matrix, that minimizes the TLS error $\|\mathbf{O}^T\psi\|^2$ is given by the least significant eigenvector of $\mathbf{O}\mathbf{O}^T$. The solution coincides with the normal direction of the hyperplane having the least orthogonal distance to the points given by the columns of $\mathbf{O}$. If the multiplicity of the least eigenvalue is ν, the dimension of the fitted hyperplane is $M - \nu$.

♦

15.2 PCA for Rare Observations in Large Dimensions

If the dimension of the observation vector space is very large, it is generally very likely that $\mathbf{S}$ will be singular. The scatter matrix is guaranteed to be singular when the number of observed vectors is less than the dimensions, $K < M$. This happens often in image analysis.

A well-known example of large dimensionality in features is obtained when the observed feature vectors consist of entire images, e.g., face images that we will also illustrate in an example below. The image then becomes an observation vector by scanning it in a certain fashion, for example in the reading direction. For a 1000×1000 image, M is equal to 1 million, for a 100×100 image the dimensionality is 10000, etc. To obtain a reliable estimate of the scatter matrix one has to collect tens of thousands of observations even for small, low-resolution images, whereas for high-resolution images this is simply not practicable. The result is that one has to use the available observations, despite the fact that they are almost never sufficient, to yield a good estimate of the scatter matrix. Another difficulty when dealing with large-dimensional PCA is that building the scatter matrix grows from being a trivial task in few dimensions, to an impractical task in large dimensions. This is because every outer (tensor) product in the sum of Eq. (15.26) needs $M(M + 1)/2$ multiplications, and there are K such matrices in the summation. In the ideal case, K should furthermore be at least as many as the scatter matrix coefficients, $M(M + 1)/2$. It follows then that the number of multiplications is $KM(M+1)/2$, which approaches $O(M^4)$.

In this section we will discuss how to obtain the KL coefficients when $K \ll M$ without performing large numbers of arithmetic operations. The symmetric matrix $\mathbf{O}\mathbf{O}^T$, and thereby $\mathbf{S}$, has the size $M \times M$, whereas the matrix $\mathbf{O}^T\mathbf{O}$, also symmetric, has the size $K \times K$. Although different, these two matrices share eigenvalues because

$$\mathbf{O}\mathbf{O}^T\psi_m = \lambda_m\psi_m, \qquad \Rightarrow \qquad \mathbf{O}^T\mathbf{O}(\mathbf{O}^T\psi_m) = \lambda_m(\mathbf{O}^T\psi_m), \qquad (15.32)$$

where $\lambda_m > 0$. Also, inspecting (15.32), we see that if ψ_m is an eigenvector of $\mathbf{O}\mathbf{O}^T$, then

$$\psi'_m = \mathbf{O}^T\psi_m \qquad (15.33)$$

is an eigenvector of $\mathbf{O}^T\mathbf{O}$. Thus the two matrices have eigenvectors that differ only by a fixed projection. Multiplying this equation with $\mathbf{O}$

$$\mathbf{O}\psi'_m = \mathbf{O}\mathbf{O}^T\psi_m = \lambda_m\psi_m \qquad (15.34)$$

shows that the eigenvectors of $\mathbf{OO}^T$ are related to those of $\mathbf{O}^T\mathbf{O}$ through a projection. Accordingly, given $\mathbf{O} = [\mathbf{f}_1, \cdots, \mathbf{f}_K]$, with $\mathbf{f}_k \in E_M$, the eigenvectors of $\mathbf{OO}^T$ can be obtained from those of $\mathbf{O}^T\mathbf{O}$ as follows:

1. Compute the sorted eigenvalues and eigenvectors of $\mathbf{O}^T\mathbf{O}$ as λ_m, ψ'_m. The eigenvalues are also the eigenvalues of $\mathbf{OO}^T$.
2. Obtain the (unnormalized) eigenvectors ψ_m as $\mathbf{O}\psi'$, (15.34).

Exercise 15.1. *How is it that the $M \times M$ matrix $\mathbf{OO}^T$ has M eigenvalues that are the same as the K eigenvalues of the $K \times K$ matrix $\mathbf{O}^T\mathbf{O}$, and yet $K \ll M$?*

Example 15.1. *In some face recognition techniques, recognition is done by comparing the gray images, interpreted as (very) high dimensional vectors in a Hilbert space with the scalar product, Eq. (3.27). A similarity or dissimilarity measure between two face images can then be used, e.g. the directional difference, (3.56), or the Euclidean distance, to decide whether or not the images represent the same person. Alternatively, a trained classifier, such as a neural network [29, 49], or the Mahalanobis distance [62], can be used in this decision making. One assumes then the reference images of the clients, $\mathbf{O} = [\mathbf{f}_1, \cdots, \mathbf{f}_K]$ with $\mathbf{f}_k \in E_M$. In order for this to work, the images must be scale- and position-normalized, so that little or no background is present and the eyes are essentially in the same position in all images. It has been shown, see [202, 220] and others, that this recognition is improved if the dimension of the face images is first reduced to N by PCA, where typically $N < K \ll M$. The resulting subspace is also called the face space. The similarity of two face images are then measured in this subspace, spanned by the N most significant eigenvectors, also called eigenfaces.*

In the top row of Fig. 15.1, which shows six images of two persons, some samples of a face training set $\mathbf{O}$ are illustrated. Representing 64×96 images, the corresponding face vectors have $M = 6144$ dimensions. The training set $\mathbf{O}$ consists of $K = 738$ such face vectors. The mean of the training set is subtracted from the training set, and also later from the test set, (15.22). The actual scatter matrix $\mathbf{OO}^T$ is 6144×6144, whereas the alternative scatter matrix $\mathbf{O}^T\mathbf{O}$, by which the eigenvectors can be computed, is 738×738. The 24 most significant eigenvectors are shown as pseudo colored images in the same figure, in reading order. Using a test set containing 369 images (different than those in $\mathbf{O}$) and retaining the $N = 30$ most significant eigenvectors, one could obtain 89% recognition in the senspe that the person to be recognized was assigned the top rank. The recognition was 96% if the correct image was found among the top 3 ranks [154].

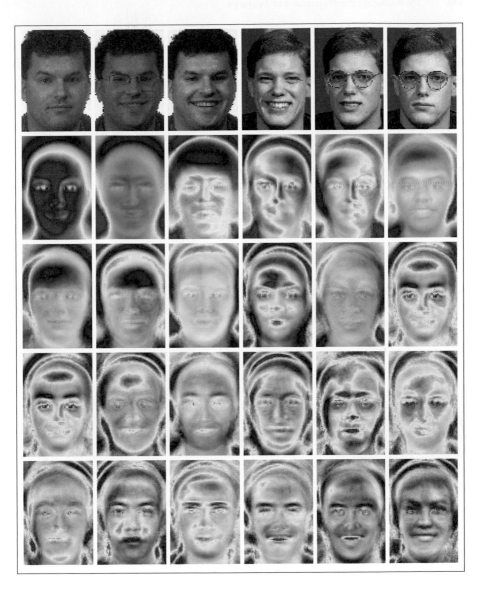

Fig. 15.1. *Top row* shows samples of 64×96 face images of two persons [154]. The *color images* show the 24 most significant basis faces that are found by PCA

15.3 Singular Value Decomposition (SVD)

Here we discuss a powerful tool, the singular value decomposition (SVD) to solve homogeneous linear equations of the type

$$\mathbf{O}^T \psi = \mathbf{0}, \qquad \text{where} \qquad \mathbf{O} \text{ is } M \times K, \tag{15.35}$$

in the TLS error sense, which appear in numerous image analysis problems. This is an important class of problems for computer vision for many reasons, including the following.

First, in the direction estimation formulation, the error $\|\mathbf{O}\psi\|^2$ is to be made as small as possible, possibly zero, by a certain solution ψ. The minimum achievable error is $\lambda_{(M)}$, which can be zero in the ideal case or small in a successful hyperplane fitting, if and only if the solution satisfies

$$\mathbf{O}\mathbf{O}^T \psi_{(M)} = \lambda_{(M)} \psi_{(M)} \tag{15.36}$$

where $\lambda_{(M)}$ is the least (eigenvalue) of the scatter matrix $\mathbf{O}\mathbf{O}^T$. With this construction, we can thus transfer the quadratic minimization problem to the homogeneous equation problem:

$$\min_{\psi} \psi \mathbf{O}\mathbf{O}^T \psi \qquad \Leftrightarrow \qquad \mathbf{O}^T \psi = \mathbf{0} \tag{15.37}$$

where we search for the TLS error solution with $\|\psi\| = 1$.

Second, studying lemmas 15.2 and 15.1 shows that the quadratic minimization $\psi^T \mathbf{O}\mathbf{O}^T \psi$ problem which we converged at, first in the linear symmetry direction problem, then in feature extraction including corner features, group direction features, motion estimation, world geometry estimation, and dimension reduction problems, is a very fundamental problem for vision. One can therefore conclude that the problems of vision can be effectively modeled¿ as a *direction estimation* problem, which is in turn equivalent to a homogeneous equation problem. Third, because a linear equation $\mathbf{A}\psi = \mathbf{b}$ can be rewritten as

$$\mathbf{O}^T \psi = \mathbf{b} \qquad \Rightarrow \qquad [\mathbf{O}^T, -\mathbf{b}] \begin{bmatrix} \psi \\ 1 \end{bmatrix} = \mathbf{0} \tag{15.38}$$

even non homogeneous linear equations can be easily treated within homogeneous equation formalism. There are straightforward extensions of this idea employing polynomials as well as other nonlinear transformations as the elements of $\mathbf{O}$, affording nonlinear modeling too. Typically, $\mathbf{O}, \mathbf{b}$ represent the *explanatory variables* and the *response variables*, respectively. Other names for these variables are *input* and *output*, respectively. One searches for ψ, representing the *model variables*, which are also known as the *regression coefficients*. The advantage with a homogeneous equation resides in that it can model noise in both $\mathbf{O}$ and $\mathbf{b}$ in the TLS sense, making the solution independent of measurement coordinates, that is, a tensor solution, see Sect. 10.10. In computer vision, homogenization is in itself an important tool to transform large affine and projective mappings, which are nonlinear, to linear mappings, see Sect. 13.2, in addition to the tensor solution it affords.

To stay within the limits of our scope, the following lemma, which is extensively discussed elsewhere, e.g. [84], is given without proof.

Lemma 15.3 (SVD). *An $M \times K$ matrix* $\mathbf{O}$ *can be decomposed as*

$$\mathbf{O} = \mathbf{U}\mathbf{\Sigma}\mathbf{V}^T \tag{15.39}$$

where $\mathbf{\Sigma}$ *is an $M \times K$ diagonal matrix,[2] and the matrices* $\mathbf{U}, \mathbf{V}$ *are quadratic and orthogonal.*

Note that the matrix $\mathbf{O}$ can be, and in practice is, nonquadratic. If we have $K > M$, we have the following decomposition format for $\mathbf{O}$

$$\begin{pmatrix} \mathbf{O} \\ M\times K \end{pmatrix} = \begin{pmatrix} \mathbf{U} \\ M\times M \end{pmatrix}\begin{pmatrix} \mathbf{\Sigma} \\ M\times K \end{pmatrix}\begin{pmatrix} \mathbf{V}^T \\ K\times K \end{pmatrix} \tag{15.40}$$

whereas if we have $K < M$ the format of the decomposition yields

$$\begin{pmatrix} \mathbf{O} \\ M\times K \end{pmatrix} = \begin{pmatrix} \mathbf{U} \\ M\times M \end{pmatrix}\begin{pmatrix} \mathbf{\Sigma} \\ M\times K \end{pmatrix}\begin{pmatrix} \mathbf{V}^T \\ K\times K \end{pmatrix} \tag{15.41}$$

In both cases, the matrices $\mathbf{O}, \mathbf{\Sigma}$ have the same form, i.e. either they are both "sleeping" as in Eq. (15.40), or they are both "standing" as in Eq. (15.41). Below, we will qualify a matrix as *standing* if it has more rows than columns, and *sleeping* conversely.

Assuming that the diagonal elements of $\mathbf{\Sigma}$ are sorted in descending order, $\sigma_{(11)} \geq \cdots \geq \sigma_{(\kappa\kappa)}$, where $\kappa = \min(M, K)$, and the columns of the matrices $\mathbf{O}, \mathbf{U}, \mathbf{V}$ are sorted accordingly, one can see that the symmetric, semipositive definite matrix $\mathbf{O}\mathbf{O}^T$ can be diagonalized by means of SVD as

$$\mathbf{O}\mathbf{O}^T = \mathbf{U}\mathbf{\Sigma}\mathbf{V}^T\mathbf{V}\mathbf{\Sigma}^T\mathbf{U}^T = \mathbf{U}\mathbf{\Sigma}\mathbf{\Sigma}^T\mathbf{U}^T$$
$$= \mathbf{U}\mathbf{\Lambda}\mathbf{U}^T \tag{15.42}$$

Because the matrix $\mathbf{\Sigma}\mathbf{\Sigma}^T$ is diagonal and has a square form, the eigenvectors of $\mathbf{O}\mathbf{O}^T$ are to be found in the columns of $\mathbf{U}$, whereas its eigenvalues are the diagonal elements $\mathbf{\Sigma}\mathbf{\Sigma}^T$. The latter elegantly shows that $M - K$ eigenvalues are automatically zero when $\mathbf{O}$ is a standing matrix because the corresponding $\mathbf{\Sigma}\mathbf{\Sigma}^T$ will have zeros

---

[2] A diagonal matrix is one that has zeros as offdiagonal elements, i.e., the elements of the diagonal matrix $\mathbf{\Sigma}$ are $\sigma_{ij}\delta(i - j)$.

in its last $M - K$ diagonal elements. By contrast, if $\mathbf{O}$ were a sleeping matrix, the quadratic matrix $\mathbf{OO}^T$ would not be singular in general since no zero values would be automatically generated in the diagonal elements of $\mathbf{\Sigma\Sigma}^T$. Evidently, the number of such null eigenvalues is a consequence of whether or not the equation $\mathbf{O}^T\psi = \mathbf{0}$ is overdetermined, which is paraphrased by the term sleeping $\mathbf{O}$. Similarly, one can diagonalize $\mathbf{O}^T\mathbf{O}$ as

$$
\begin{aligned}
\mathbf{O}^T\mathbf{O} &= \mathbf{V\Sigma}^T\mathbf{UU\Sigma V}^T = \mathbf{V\Sigma}^T\mathbf{\Sigma V}^T \\
&= \mathbf{V\tilde{\Lambda}V}^T
\end{aligned}
\tag{15.43}
$$

to the effect that the eigenvectors of $\mathbf{O}^T\mathbf{O}$ are to be found in the columns of $\mathbf{V}$, whereas its eigenvalues are in the diagonal elements $\mathbf{\Sigma}^T\mathbf{\Sigma}$. The two diagonal matrices $\Sigma^T\mathbf{\Sigma}$ and $\Sigma\mathbf{\Sigma}^T$ contain the same nonzero diagonal elements.

There exist effective software implementations of SVD, and even some where only the first K columns of U are computed in case one has $K \ll M$. This is valuable in image analysis, where one often fits hyperplanes with low dimensions to data in high-dimensional (Hilbert) vector spaces. The SVD is a ubiquous tool that can be generously applied to numerous vision problems.

Grouping and Unsupervised Region Segregation

In this chapter we present the elements of unsupervised texture boundary estimation to address automatic grouping of regions in images. The grouping is based on the assumption that a set of feature vectors is densely available for the image. These can be viewed as layers of images, each representing a property of the image. A simple example is an HSV-type color image, where the first layer represents the hue, the second layer saturation, and the third layer the brightness. However, the feature images need not come from imaging sensors. They could be local image properties, densely computed from ordinary gray value images, such as the texture properties: mean, variance, and direction. A set of N dense feature layers is equivalent to a 2D grid on which the image takes N-dimensional vector values.

The presentation in this chapter focuses on group formation and boundary estimation in 2D images having tensors as pixel values, but it does not presuppose or prescribe a specific set of features, neither imaged nor computed, as this is application dependent. The value of a pixel is then represented by an array, each element of which over the entire image grid can be considered as a separate image, also called a *feature image*, with real pixel values. First an issue critical to the grouping process of images is discussed, the *uncertainty principle*. The presence of noise in the features causes a class overlap that can be reduced in a multiresolution pyramid. The uncertainties in boundaries can be reduced by means of butterfly-shaped smoothing filters, adaptive to boundary directions. We give further precision to these matters in the subsequent sections, where the discussions follow the same principles as the studies in [195], [207], and [227].

16.1 The Uncertainty Principle and Segmentation

Uncertainty in computed image features is similar to the uncertainty principle in physics. A wave describing a particle cannot be highly concentrated simultaneously in position and frequency. In unsupervised image segmentation, both position (class boundaries) and prototype features (local spectral properties) have to be determined. Finding boundaries is dependent on prototypes and vice versa because:

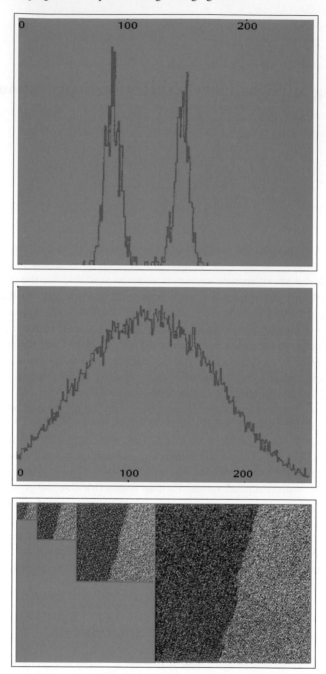

Fig. 16.1. A Gaussian pyramid is shown for a gray image having two classes (*bottom*). The graph on the *top* shows the histogram of the lowest resolution image whereas the graph in the *middle* is ditto for the highest image resolution

- Accurate prototypes are needed to yield accurate boundaries, whereas
- Accurate boundaries are precisely what is needed to compute accurate prototypes.

Therefore, the prototypes and the boundaries cannot be simultaneously determined [227] to yield high accuracy. The presence of noise in the feature space, mainly produced by modeling errors (inappropriate features), is another source of noise which contributes to a class overlap.

The variance of a signal decreases when it is lowpass filtered, i.e., averaged. Because of this, multiple resolutions are crucial in reducing the uncertainty. At lower resolutions the class prototype estimates are better defined than high resolutions. A Gaussian pyramid can be built, up to a predefined level in which each level is obtained by smoothing the preceding finer level. Accordingly, each element of the pixel value tensor, a *feature image*, will have its own pyramid. At the coarsest level, the amount of noise in the features therefore decreases significantly, allowing the feature prototypes to be determined more accurately, but at the expense of the region boundary resolution.

Class uncertainty reduction via pyramids is illustrated in Fig. 16.1 [195, 228]. At the finest level the histogram is unimodal, even if it is possible to see that distinct classes are present. At a lower resolution the noise has been smoothed out significantly, making possible the detection of the two classes (the histogram is bimodal). It is at a low resolution level that a good opportunity to partition the feature space into its constituent N_c classes appears. Partitioning can be seen as finding N_c subsets of feature vectors gathered around their respective prototypes (or class centers) automatically. This can be obtained by applying a clustering algorithm in the *smoothed feature space* (obtained at the coarsest level). For most of these algorithms to work, the number of classes N_c, or an equivalent information, has to be available a priori. Image segmentation experiments using automatic clustering indicate that the results do not critically depend on the choice of a clustering algorithm, provided that the classes are separable by means of the provided features.

Next, isolated points as well as isolated and scattered small classes are eliminated by reassigning them to a class in the spatial vicinity to obtain a spatial connectivity. It is, however, now, when a segmentation at the coarsest level is available, that it becomes evident that the cost of good class separation is bad boundaries.

The last but not the least step is a boundary estimation procedure that gradually improves the class boundaries by traversing down the pyramid. First, at the resolution level where the clustering is performed, the crude boundaries are identified. The children of the *boundary points* define a *boundary region* at the next higher resolution because every pixel at a certain level represents several pixels at the lower resolution. The nonboundary nodes at the children level are given the same labels and properties as their parents. The class uncertainty within the boundary region is high and has to be reduced before reassignment of the boundary vectors. Now that the approximate boundaries and thereby the boundary directions are available, orientation-adaptive filters are used to smooth the boundary regions. For each dominant local direction, a butterflylike averaging filter (see Fig. 16.2 and Sect. 16.6) reduces the influence of feature-vectors across the boundary while decreasing the within-class variance along the boundary. The two halves of the filter (applied separately) produce two responses. The distances between the resulting vectors and the two prototypes, associated to the classes defining the boundary, are computed and a reclassification is performed. The procedure, which will be discussed further in the subsequent sections, can be summarized as follows:

1. Build a multiresolution pyramid up to a certain level. The noise in the feature space is reduced, increasing the separation between the classes at the expense of the spatial resolution.
2. Cluster the data in the smoothed feature space by using algorithms at suitable levels of the pyramid. Reassign isolated pixels as well as small and scattered classes to enforce spatial continuity.
3. Gradually improve the spatial resolution by projecting down the labels and refine the boundaries using orientation-adaptive filters.

16.2 Pyramid Building

Noise reduction can be done by means of smoothing, which can efficiently be implemented in image pyramids [43, 92]. Lower resolution levels are obtained by taking the (weighted) average of small neighborhoods to the next coarser level, as discussed in Sect. 9.5. In the discussion here, an *octave pyramid*, i.e., one in which the image width and heights are halved between subsequent levels, is assumed. Let $\mathbf{f}(\mathbf{r}_k, l)$ be the feature vector having N dimensions (layers) at image location $\mathbf{r}_k$ and level l of a feature pyramid constructed from the highest resolution (lowest level) as follows:

$$\mathbf{f}(\mathbf{r}_k, l+1) = \sum_{\mathbf{r}_p} g(\mathbf{r}_p)\mathbf{f}(2\mathbf{r}_k - \mathbf{r}_p, l) \qquad (16.1)$$

The value of a parent node is the weighted mean of its children within the support of a filter g, such as a Gaussian. The different levels are computed using Eq. (16.1) in a bottom–up manner, starting from level $l = 0$ up to a predefined level $l = L$. The height and width of the image decrease with a factor of 2^l at level l. A pyramid is constructed for each feature separately. Therefore, the data structure can be seen as a set of pyramids or as a single one composed of $M \times 1$ vectors at each image point. The choice of the number of levels is important, as this determines the minimum connected region size in the grouping. If L is too small, the uncertainty is not reduced sufficiently, whereas if L is too high, small regions will disappear. Pyramids also reduce the computational cost by progressively reducing the number of feature vectors on which a clustering algorithm will be applied. Assume, for instance, that the feature images have the size 256×256 and that $L = 3$. Then the number of features is reduced from $65, 536$ to 256 at the level L, a reduction with factor $4^L = 64$. The level L should, however, be chosen as a function of the noise rather than a function of the size. If the number of feature vectors remains prohibitively large for a clustering algorithm, it still can be reduced by taking them from random image sites.

16.3 Clustering Image Features—Perceptual Grouping

At the coarsest level of the pyramid, a *clustering* algorithm is used to find the different classes and their prototypes. It is reasonable to assume that objects with similar properties belong to the same group or cluster, i.e., they are gathered around the same prototype. The problem is to find a partition of the feature space into N_c homogeneous subsets. To solve this problem, a suitable clustering criterion and a similarity (or dissimilarity) measure have to be defined. Being a statistical tool for grouping, clustering is also known as *vector quantization*. Advantages include:

- No training is needed.
- No statistical distribution of the data is to be estimated.

For the clustering to succeed, the features must be powerful enough to group the data into *compact volumes* in the feature space, where compactness depends on the choice of the used metrics in the feature space.

The choice of the number of classes N_c, required by most of the clustering algorithms, is considered to be one of the most fundamental problems in cluster analysis [67]. Usually this information is not known, and partitions of the feature space for different values of N_c are computed.[1] An overview of cluster validity and studies of this nontrivial problem can be found elsewhere [16]. In the following, we will assume that clustering algorithms in which the number of classes N_c (or its equivalent) has to be given. We will, however, discuss reducing the problem of an exact estimation of the class number by overestimating N_c and reassigning small and scattered classes to a class in the spatial neighborhood.

Using a schematic example, we illustrate the *clustering flow chart* in the feature space next. For completeness, in the next section we will summarize the fuzzy C-means clustering algorithm proposed by Bezdek [16], which is an extension of the idea discussed in the example [11], also known as the ISODATA algorithm. This algorithm yields reasonable clustering results while being simple.

Example 16.1. *We wish to cluster the data shown in the left graph of Fig. 16.3. We follow the processing flow illustrated by the chart on the right. The number of classes, N_c, is assumed to be known. Here $N_c = 3$, and the class labels are,* circle, rectangle, *and* square, *for convenience.*

1. **Start:** *At random choose N_c objects, e.g., $2, 3, 8$. These will be the initial class centers (prototypes); assign them a label each, e.g.,* circle, rectangle, *and* square.
2. **Update partitions:** *Given the current class centers, assign all objects to one of the* circle, rectangle, *or* star *classes by use of the least distance criterion. This will yield the current partitioning of the data.*
3. **Update centers:** *Given the current partitioning, find the class centers by computing the class means. This will yield the current class centers.*

---

[1] Hierarchical clustering techniques do not require the number of classes to be known explicitly [28, 191], but they require it implicitly, e.g., they need an input threshold for maximum within class variance.

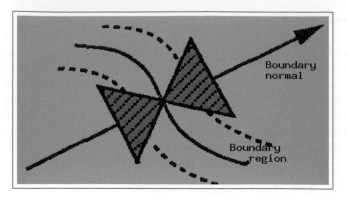

Fig. 16.2. Illustration of an oriented butterfly filter adapting to the normal direction of a boundary point

4. **Exit?** *If the current partitioning or the class centers have changed significantly compared to their previous values, go to step 2; else exit.*

The trace of partitioning would be given by the following listing:

Circle	Rectangle	Star
$\{\underline{2}, 10, 12, 7\}$	$\{\underline{3}, 1, 9, 6, 13, 15\}$	$\{\underline{8}, 5, 14, 4, 11\}$
$\{2, 10, 12, 7, 15\}$	$\{1, 3, 6, 9, 4, 13, 11\}$	$\{5, 8, 14\}$
$\{2, 10, 12, 7, 15\}$	$\{6, 9, 4, 13, 11\}$	$\{5, 8, 14, 1, 3\}$
$\{2, 10, 12, 7, 15\}$	$\{6, 9, 4, 13, 11\}$	$\{5, 8, 14, 1, 3\}$

$$(16.2)$$

Had we called the objects, $2, 3, 8$, as star, rectangle, circle, respectively, the labels of the result would be permuted too. Accordingly, the labels may change up to a permutation, unless they are "learned" by the system separately.

If the clusters are not "ball"-shaped, e.g., elongated or engulfed, there is evidently a risk of erroneous partitioning. This is also related to the distance metrics used in the clustering. Assuming that there is a choice of metrics that can "cluster" the features and the features themselves are powerful, then the clustering method will be able to group the feature data meaningfully. The clustering then will implicitly or explicitly compute a distance matrix $\mathbf{D}$ that describes the distances between all feature vectors $[\mathbf{f}_1, \cdots, \mathbf{f}_K]$:

$$\mathbf{D} = \begin{pmatrix} d_{1,1} & d_{1,2} & \cdots & d_{1,K} \\ d_{2,1} & d_{2,2} & \cdots & d_{2,K} \\ \vdots & & \ddots & \\ d_{K,1} & d_{K,2} & \cdots & d_{K,K} \end{pmatrix} \tag{16.3}$$

Here and elsewhere in this section, we have used the notation $\mathbf{f}_k$ to mean $\mathbf{f}(\mathbf{r}_k, L)$ for notational simplicity, i.e., the clustering is assumed to be done on feature vectors at the highest level of the pyramid. Thus, the distance does not need to be Euclidean (but very well can be), e.g., $\mathcal{L}^1, \mathcal{L}^\infty$. The number of distances to compute grows quadratically. For 1000 feature vectors, one would need to compute over half a million distances in a multidimensional space. This is one of the limitations making clustering algorithms not so practical for large number of features in high-dimensional spaces. The other and more serious limitation is that the convergence problem of the partitioning, which once achieved, is usually not as meaningful as a partitioning obtained in a (unkown) lower-dimensional subspace. This provides yet another motivation to dimension reduction followed by image feature smoothing in a pyramid, before attempting to group image features automatically.

16.4 Fuzzy C-Means Clustering Algorithm

Let $\mathbf{F} = \mathbf{f}_1, ..., \mathbf{f}_K$ be a finite set of feature vectors with $\mathbf{f}_k \in E^M$, and N_c be an integer[2] representing the number of classes to which the feature vectors can belong. The N_c partitioning of the feature space is represented by the $N_c \times K$ matrix $\mathbf{U} = [u_{nk}]$, consisting of the elements u_{nk}, which represent the membership of feature vector $vecf_k$ to the class n. There are K feature vectors and N_c classes, so that k and n are integer indices in the ranges $1, \cdots, K$ and $1, \cdots, N_c$, respectively. In the hard (or crisp) case, the degree of "belongingness" of feature vector $\mathbf{f}_k$ to a class n is either 1 or 0, i.e.,

$$u_{nk} = \begin{cases} 1, \mathbf{f}_k \in \text{class } n, \\ 0, \text{otherwise.} \end{cases} \tag{16.4}$$

In the fuzzy case, u_{nk} gives the "strength" of the membership of feature vector $\mathbf{f}_k$ to class n ($u_{nk} \in [0, 1]$). The *degree of belongingness* u_{nk} can be intuitively seen as a distance from feature vector k to the class n normalized by the sum of distances to all class centers. This representation is in many cases closer to the physical reality in the sense that feature vectors almost never fully belong to one class i.e., there is always a suspicion that an object could have been classified to a different class, albeit with a lower certainty. The two following conditions have to be respected:

$$\sum_{n=1}^{N_c} u_{nk} = 1, \qquad \text{for all } \mathbf{f}_k, \tag{16.5}$$

$$0 < \sum_{k=1}^{K} u_{nk} < K, \qquad \text{for all classes } n. \tag{16.6}$$

In the crisp case, Eq. (16.5) simply means that feature vector k belongs to one class only. The second condition Eq. (16.6) means that no class is empty and no class is all of $\mathbf{F}$. The fuzzy C-means algorithm belongs to the class of objective function methods. Such methods minimize a clustering criterion which is, in this case, the total within-group sum of squared error (WGSS). The fuzzy C-means is the fuzzy extension of the hard C-mean. It minimizes the objective function

$$e_\kappa(\mathbf{U}, \mathbf{V}) = \sum_{k=1}^{K} \sum_{n=1}^{N_c} (u_{nk})^\kappa (d_{nk}(\mathbf{v}_n, \mathbf{f}_k))^2 \tag{16.7}$$

---

[2] Evidently we assume that $N_c \in 2..K - 1$ because if $N_c = 1$ or $N_c = K$, then one would obtain trivial partitionings.

where κ is an empirically chosen weighting exponent controlling the amount of fuzziness $\kappa \in [1, \infty[$. The scalars $u_{nk} \in [0, 1]$ are the elements of the matrix $\mathbf{U} = [\mathbf{u}_1, \cdots, \mathbf{u}_k, \cdots, \mathbf{u}_K]$, which represent a fuzzy partition of the feature vectors such that a column $\mathbf{u}_k$ represents the degree of belongingness of the feature vector $\mathbf{f}_k$ to each of the N_c different classes. In such a partitioning, the class prototypes are represented by the columns of the matrix $\mathbf{V} = [\mathbf{v}_1, \cdots, \mathbf{v}_n, \cdots \mathbf{v}_{N_c},]$. Thus $\mathbf{U}, \mathbf{V}$ are updated by the partitioning process, whereas the feature set $\mathbf{F} = [\mathbf{f}_1, \cdots, \mathbf{f}_k, \cdots \mathbf{f}_K]$, which is the set to be categorized into N_c classes, is fixed. The elements $d_{nk} = \|\mathbf{f}_k - \mathbf{v}_n\|$ where $\| \cdot \|$ is an inner product induced norm on the M-dimensional Euclidean space, E^M, of the features, represents the squared distance between two vectors in the feature space. Restated, the problem is to find the best pair $(\mathbf{U}, \mathbf{V})$ that minimizes the error function e_κ. It can be shown that [16] $e(\mathbf{U}, \mathbf{V})$ may be globally minimal for $(\mathbf{U}, \mathbf{V})$ only if

$$u_{nk} = \left(\sum_{j=1}^{N_c} \left(\frac{d_{nk}}{d_{jk}} \right)^{2/(\kappa-1)} \right)^{-1}, \tag{16.8}$$

$$\mathbf{v}_n = \left(\sum_{k=1}^{K} (u_{nk})^\kappa \mathbf{f}_k \right) / \sum_{k=1}^{K} (u_{nk})^\kappa, \quad \text{for} \quad n \in 1, \cdots, N_c. \tag{16.9}$$

The fuzzy C-means algorithm, which approximates a solution of the minimization problem, can be stated as follows:

1. **Start:** Fix N_c, choose a norm (based on an inner product) $\| \cdot \|$ for E^M, fix κ, where $1 < \kappa \le \infty$, and initialize $\mathbf{U}^{(0)}$.
2. **Update centers:** Calculate the N_c fuzzy cluster centers $\mathbf{V} = [\mathbf{v}_1, \cdots, \mathbf{v}_{N_c}]$ with Eq. (16.9) and the current fuzzy partitions, $\mathbf{U}$.
3. **Update partitions:** Update $\mathbf{U}$ using Eq. (16.8) and the current class centers, $\mathbf{V}$.
4. **Exit ?** If $\|U^{l+1} - U^l\| \le \epsilon$, then stop; otherwise go to step 2.

As $\kappa \to 1$, the fuzzy C-means converge to a "generalized" hard C-means solution (ISODATA) [11], discussed in example 16.1 and Eq. (16.7). This algorithm always reaches a strict local minimum for different initializations of U.

16.5 Establishing the Spatial Continuity

First, we consider the spatially *isolated points* in the resulting image obtained from clustering. Such points are likely to be obtained when the clustering terminates, for the following reason. Let γ_n be the *label* associated to the class prototype of $\mathbf{v}_n$. Then, the image position of a feature vector $\mathbf{f}_k$, inherits the class label γ_n if $d(\mathbf{f}_k, \mathbf{v}_n)$ is smallest compared to all N_c classes. However, because the clustering process does not "know" from which image location the feature vectors come, the spatial (i.e., in the *label image*, which is Γ) continuity is not evident, although in practice most pixels will cluster together even spatially (in the image). At the coarsest resolution

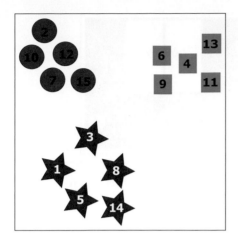
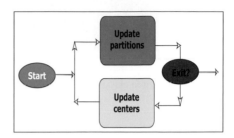

Fig. 16.3. The graph on the *left* represents some feature vectors to be clustered. On the *right* the processing flow to achieve a partitioning is shown

level L, we can define a label image $\Gamma(\mathbf{r}_k, L)$ in which each pixel receives the label of the corresponding prototype vector in the feature space. Note that even the label image Γ is also defined as a pyramid, albeit the high-resolution levels are yet to be determined. At the lowest resolution level, rough boundaries can be observed between the different classes in Γ. Let $\mathcal{N}_8(\mathbf{r}_k)$ be the 8-*connected neighborhood* of a pixel at location $\mathbf{r}_k$ composed of the 8-closest neighbors on a square grid.[3] A pixel $\mathbf{r}_k$ is considered as spatially misclassified if $\Gamma(\mathbf{r}_k, L)$ is different from all the labels in its $\mathcal{N}_8(\mathbf{r}_k)$. In that case, it is reassigned to the most common class in $\mathcal{N}_8(\mathbf{r}_k)$.

Next, we consider *insignificant classes*, which are small and scattered classes that need to be reassigned to their neighboring classes. This is because a class in the feature space is distributed in one or more subregions in $\Gamma(\mathbf{r}_k, L)$, even the largest of which may not be more than a few pixels. A class is considered as "insignificant" if its largest subregion contains no more than a threshold (nine in the discussions here). Thus, the idea is to give a preference to classes that are spatially distributed into large and compact subregions rather than the inverse. It is, however, clear that the meaning of "insignificant classes" is closely related to the height of the pyramid, i.e., for a low value of L, the actual size of an insignificant class is smaller than for higher values of L. We discuss a possible implementation of this by the illustration in Fig. 16.4. Representing the largest regions of their respective classes, the hatched regions are assumed to be insignificant classes on the grounds of their size (because they are fewer than 9 pixels each) and have to be reassigned to either one of the surrounding classes or, in the cases of (b) and (c), possibly even between themselves. Thus there are two types of insignificant classes: *isolated insignificant class*, as exemplified by (a), and *touching insignificant class*, as exemplified by (b) and (c). The isolated class

---

[3] That is, $\mathcal{N}_8(\mathbf{r}_k)$ represents the set of points in a 3×3 neighborhood around the current point $\mathbf{r}_k$, excluding the latter.

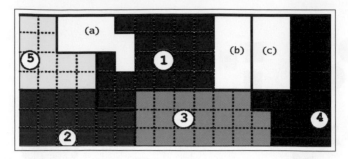

Fig. 16.4. Three classes (*white*) that have to be reassigned to a neighbor class (*colored*)

can be reassigned in one pass, whereas multipasses are needed if the insignificant classes are touching each other. First, one can reassign the isolated classes, like those (a) in Fig. 16.4, by generating a set of candidate classes obtained from the neighboring classes. This is determined for each pixel of the region. If a label in the $\mathcal{N}_8(\mathbf{r}_k)$ neighborhood of a point of such a region is different from $\Gamma(\mathbf{r}_k, L)$, then a new candidate class is obtained. Otherwise, a different point (outside of $\mathcal{N}_8(\mathbf{r}_k)$, but in the same direction) is examined until their corresponding class labels differ. These operations are repeated for the eight possible directions of the $\mathcal{N}_8(\mathbf{r}_k)$. For each pixel $\mathbf{r}_k$, the Euclidean distances between the feature vector prototype of the region, e.g., that of (a), and the prototypes of the candidate classes are computed. Each pixel is then assigned to the closest candidate class. To extend the search outside of the mask is necessary because not all points share boundaries with a point of a different class. This can be illustrated by region (a) in Fig. 16.4, where the central pixel near the border of the image lacks neighbors belonging to significant classes. For these pixels, an extended search beyond the $\mathcal{N}_8(\mathbf{r}_k)$ neighborhood is necessary to obtain relevant candidate classes.

Next, the remaining insignificant classes consisting of touching regions need to be reassigned. However, the reassignment order leads to different results, and multipasses are needed to avoid this problem. The process can be illustrated by considering the classes (b) and (c) in the figure. In the first pass, these two classes are temporarily reassigned as if they were isolated insignificant classes. The same initial state of $\Gamma(\mathbf{r}_k, L)$ is used for both classes. The maximum number of pixels K_{n_max} reassigned to the same candidate class is determined for each class label γ_n equaling to the class of (b), and ditto for class (c). Then a quotient q_n between K_{n_max} and the respective population K_n of the classes is computed, i.e., $q_n = K_{n\_max}/K_n$. A value of $q_n = 1$ means that the class n tends to be reassigned to a single candidate class. The reassignment of the region with the largest quotient q_n is then taken first. Suppose that (b) is reassigned first. Before reassigning (c) in a second pass, one needs to check if (c) is still considered as an insignificant class. It might occur that pixels of (b) were reassigned to (c), making it a significant class. If this is not the case, (c) is reassigned using the procedure described for isolated classes. A similar procedure can be applied to reassign more than two touching insignificant classes,

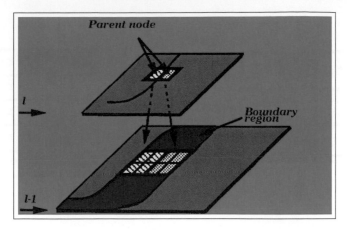

Fig. 16.5. The parent node at a boundary point of level l in the pyramid and its corresponding children nodes at level $l - 1$. Such children nodes define the boundary region

although it will take as many passes as the number of touching classes to terminate the reassignment procedure.

16.6 Boundary Refinement by Oriented Butterfly Filters

The final step of the pyramidal unsupervised segmentation is a boundary refinement procedure that gradually improves the spatial resolution of the label image $\Gamma(\mathbf{r}_k, L)$. As stated by the spatial principle of uncertainty [227], accurate prototypes can be obtained only at the expense of the spatial resolution. Therefore, at the coarsest level, the class uncertainty at the boundaries is higher than other points. One can reduce it by means of orientation-adaptive filters [195]. Other methods for boundary enhancements include [88, 149, 150]. The spatial resolution is gradually restored by projecting down the class labels, smoothing around the boundaries, and reassigning the boundary pixels to their closest neighboring class. We also assume that the prototypes have constant values across the different levels of the pyramid. First, at the coarsest level L the boundary pixels are determined. Each pixel $\mathbf{r}_k$ is considered as a boundary pixel if at least one label in $\mathcal{N}_8(\mathbf{r}_k)$ is different from $\Gamma(\mathbf{r}_k, L)$. Next, $\Gamma(\mathbf{r}_k, L - 1)$ is obtained by projecting down the label of each nonboundary parent node to its four respective children nodes, i.e., $\Gamma(n, j, l) = \Gamma(n/2, j/2, l+1)$, where / is integer division.[4] The children of the boundary nodes define a boundary region β (see Fig. 16.5) in which the oriented smoothing will be performed. For smoothing we need an estimate of local boundary directions. We use the linear symmetry algorithm, discussed in Sect. 10.11, which defines the dominant orientation in the TLS sense. The direction is determined by the complex number

---

[4] Integer division is the integer part of ordinary division, e.g. 5/3 would yield 1.

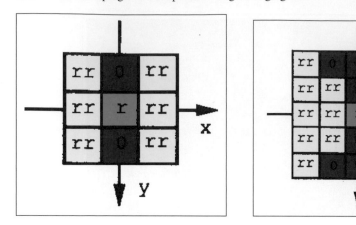

Fig. 16.6. The coefficients of a butterfly filter with orientation $\theta = 0$ for a 3×3 (*left*) and 5×5 (*right*) neighborhood

$$z = \urcorner(\Gamma) * g \qquad (16.10)$$

where Γ is the image of labels, $\urcorner(\Gamma)$ is the squared complex image $(D_x\Gamma + iD_y\Gamma)^2$, and $*g$ the convolution with an averaging filter. The argument of z obtained at every pixel location represents

$$\arg(z) = two\arg(k_{\min}) \qquad (16.11)$$

where $k_{\min}$ is the complex number whose argument is the dominant boundary gradient direction. In this case, the magnitude of the gradient of Γ is 1 at the transition between two classes, and 0 within a class. The smoothing filter g is of size $s \times s$ and is given by a Gaussian, which in this presentation is $s = 7$ induced by use of $\sigma = 1.8$. The direction is computed for the boundary pixels at the parent node level and is propagated to the children level. For each dominant local direction, a butterfly-like filter is defined. The butterflylike shape, Fig. 16.2, reduces the influence of vectors along the boundaries that have high uncertainty. The shape and the weights of a 3×3 and 5×5 filter for the horizontal direction ($\theta = 0$) are given in Fig. 16.6, where r is a function of the dissimilarity d, described below, between the two classes that define the boundary and $rr = (1 - r)/N_\nu$ with N_ν being the number of weights different from 0 or r. The dissimilarity d is given by

$$d = \frac{|\mu_m^A - \mu_m^B|}{\sqrt{(\sigma_m^A)^2 + (\sigma_m^B)^2}} \qquad (16.12)$$

where μ_m^A, μ_m^B and $(\sigma_m^A)^2, (\sigma_m^B)^2$ are the means and the variances of the two classes on both sides of the boundary in the mth component image of the feature vector $\mathbf{f}(\mathbf{r}_k, l)$. Note that the latter can be viewed as consisting of M layers of (scalar) images. Then $r = r(d)$ is defined empirically as an increasing function, here as in Fig. 16.7 [229]. If the dissimilarity d is large then $r = 1$ and no smoothing is applied, whereas a stronger smoothing is performed ($r = rr$) for low values of d.

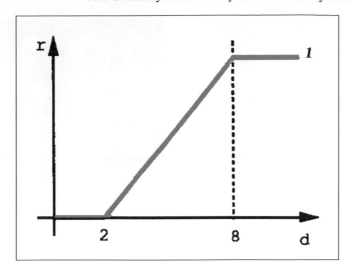

Fig. 16.7. The function relating the dissimialrity to the central coefficient of the butterfly filter

We still need to define the filters for different orientations θ. This is done by rotating the mask defined for the horizontal direction $\theta = 0$ and redistributing the weights for matching the grid of an image. The filters are computed for a fixed number of directions (eight here) and are stored in the memory, i.e., a lookup table.Starting from the coarsest level $l = L$, the boundary refinement procedure can be enumerated as follows:

1. Project down the labels, i.e., $\Gamma(\mathbf{r}_k, l - 1) = \Gamma(\mathbf{r}_k/2, l)$ and define the boundary region β at level $l - l$. Compute the mean μ^n and variance $(\sigma^n)^2$ for all classes.
2. For each boundary pixel at level l compute the dominant direction θ using Eqs. (16.10) and (16.11). Determine the two classes A, and B on both sides of the boundary. Choose the corresponding butterfly filter, and propagate its identity to the corresponding pixels at level $l - 1$.
3. For each pixel $\mathbf{r}_k \in \beta$, apply the filter corresponding to the current position to (all components of) the feature vectors $\mathbf{f}(\mathbf{r}_k, l - 1)$. If a feature-vector participating to averaging is outside of β, take the vector of its corresponding prototype. The left and right halves of the filters are applied separately, reducing the risk of smoothing across the boundaries. Two responses are obtained, $\mathbf{h}^A(\mathbf{r}_k, l - 1)$ and $\mathbf{h}^B(\mathbf{r}_k, l - 1)$. This smoothing is repeated a certain number of times found empirically (four in the examples) using a small filter size (3×3 here). This is equivalent to use a large filter size in one iteration, but it is computationally faster.
4. For each pixel $\mathbf{r}_k$, compute the four distances between the two filter responses $\mathbf{h}^A$, $\mathbf{h}^B$ and the prototypes μ^A, μ^B corresponding to the classes A, B on each side of the boundary region, i.e., $\|\mu^A - \mathbf{h}^A\|$, $\|\mu^A - \mathbf{h}^B\|$, $\|\mu^B - \mathbf{h}^A\|$, $\|\mu^B -$

 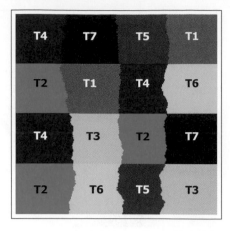

Fig. 16.8. A texture patch (P3) and the ground truth class identities (T1, T2,...,T7) represented by color at each point

$\mathbf{h}^{B}\|$. Each boundary pixel receives the label of the class that gives the minimum distance.

5. Decrease the value of l by one and repeat from step 1, until the bottom of the pyramid is reached.

16.7 Texture Grouping and Boundary Estimation Integration

Here we integrate the unsupervised clustering and the boundary estimation process discussed in the previous sections, by illustrating its effect on a patch image of textures (Fig. 16.8). It consists of regions corresponding to real texture images, as for those in [37], put in a 256×256 image. Each texture patch was photometrically normalized with respect to the mean and the variance, i.e., if the texture patch is represented by f, the normalized texture is obtained as $f' = af + b$ for some constants a and b to force the mean and the standard deviation f' to equal specific values (128 and 30 here, for all patches). This was done to illustrate the multidimensional grouping because, otherwise, the textures would be too easy to separate by gray values, being 1D features. There are, in total, seven different textures tiled as a 4×4 matrix, and the patches are repeated to obtain all texture combinations astride the boundaries for a maximal diversity. Accordingly, every texture is a neighbor of any other at least once.

Exercise 16.1. *Is there an alternative way to construct Fig. 16.8, i.e., to obtain 4×4 patches of T-textures with $T = 7, 8 \cdots 16$, so that all texture crossovers are present? A rotation of the result or a change of texture identities is not counted.*

The symmetry derivatives applied to a log–polar decomposition of the spatial frequencies discussed in Chap. 14 have been used as features. To be precise, there

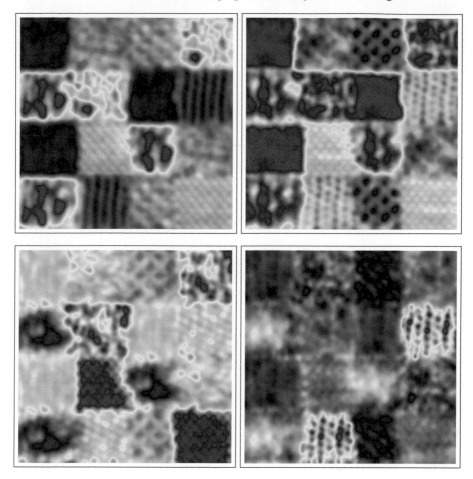

Fig. 16.9. (*Reading direction*) In pseudocolor, the images represent the four most significant texture features of P3 after dimension reduction

were three frequencies in the decomposition, at each of which I_{11} (real), I_{20} (complex), I_{40} (complex) were computed. Taking the real and the imaginary parts of the complex features, one thus obtains 15 real features. Furthermore, the dimension of this space was reduced to five by projecting these features to the five most significant eigenvectors from a KL transform, see Chap. 15. Consequently, the clustering and the boundary refinement procedure here were fed with 5 dimensional features, of which are 4 are shown in Fig. 16.9. We required $N_c = 7--9$ classes from the clustering with $L = 3$, i.e., the clustering had to be performed on a 16×16 image, and the correct patch identities (up to a texture name permutation) as well as reasonably accurate boundaries could be obtained in an unsupervised manner, i.e., there was no training. The boundaries are nondeterministic i.e., they show small variations between the runs because the prototypes display small variations. This depends on the

 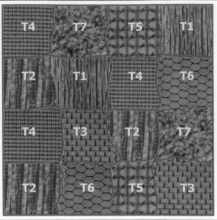

Fig. 16.10. On the *left* the found texture groups are shown in color, i.e., points sharing the same color belong to the same group. On the *right*, the boundaries and class labels implied by this grouping are superimposed the original image

initialization of the clustering, which was random, as well as the required classes, which was varied from 7 to 9. Figure 16.9 illustrates such a result, the boundaries and the texture classes of which agree fairly well with the ground truth, shown in Fig. 16.8.

16.8 Further Reading

A word of caution is in place at this point. A disadvantage of the simple structure of a resolution pyramid, is its incapacity to keep track of small regions while building up the pyramid. In other words, this scheme assumes that the regions to be segmented do not have extreme size variations because very small regions will be considered noise if one attempts to reduce the noise within large regions by using a too high L. On the other hand, a more adequate solution would need general automatic scale selection, an issue which has proven to be a significant challenge [151].

In the vast clustering literature available for general feature vectors, it can be noted that different criteria might lead to different results depending if the shape of the metrics is not reasonable as compared to cluster shapes. It is clear that no clustering criterion or measure of dissimilarity is universally applicable. For instance, the single link method is more suitable for elongated clusters while WGSS (within-group sum of squares) criterion gives good results for ball-shaped structures [87].However, it can be argued that clustering used to achieve perceptual grouping or image segmentation should not critically depend on the choice of a clustering technique if features are sufficiently strong to characterize groups. If they are not, it is often more worthwhile to devote efforts to redesign the features by using human experts than to the clustering scheme, because humans are formidable vision experts on their own. The feature vectors discussed here are easy to interpret by a human expert, as they encode the geometric properties. If human expertise is involved, features that are easy to interpret by humans is preferable. In that case, it will likely be easier to use or to devise novel features fitting to the clustering schemes that are available, than to design novel clustering schemes for the available features. An overview of classical clustering techniques can be found in [87, 135].

To restore the spatial continuity in Sect. 16.5, alternative approaches to the scheme discussed above could be used. These include the use of stochastic relaxation and Markov random fields MRF [81].

17

Region and Boundary Descriptors

In this chapter we discuss images consisting of two colors (or gray values), so called *binary images*. Such images are typically the result of a computer processing, e.g., applying a threshold, but they can also be semiautomatically or manually produced by humans in artistic activities, e.g., graphics with regions having the same color, such as logos or cartoon images. First, we introduce some general tools from mathematical morphology that can be useful in binary image processing. Then we elaborate on how to label the regions with the purpose to identify them for further processing. This will be done within the framework of morphological filtering. Subsequently, we describe elementary shape features of a region, assuming that the regions have been individually labelled. In the final two sections we present a more systematic approach to describe the shape of a region with possibility to increase the description power systematically, i.e., complex moments for regions and Fourier descriptors for boundaries.

17.1 Morphological Filtering of Regions

A binary image is an image which has only two colors in it. Without loss of generality we call these as 0 and 1 respectively.

We start with continuous morphological operations and then focus on binary versions of them, as this has some pedagogical benefits. The functions $\max(f)$ and $\min(f)$ correspond to the largest and smallest value of the image f within the region, where the image is defined as the basic operators in this theory. In applications, typically f is a local image, which means that max and min will be applied to all points of a local image, producing a (nonlinearly) filtered image. In consequence, there is a *neighborhood function*, $g(\mathbf{r})$, also called a *morphological filter*, or *structure element*, associated with such uses of max and min, delivering the result after a pointwise product with the original function. To mark the filtering aspect that necessitates the filter g in their use, these operators are called *dilation* and *erosion*, respectively, and are defined as:

Fig. 17.1. (*Left*) A binary image and (*right*) a morphological filter. *Red* and *blue* represent 1 and 0 respectively

$$(f \oplus g)(\mathbf{r}) = \max_{\mathbf{r}'} f(\mathbf{r} - \mathbf{r}')g(\mathbf{r}') \qquad \text{Dilation} \qquad (17.1)$$

$$(f \ominus g)(\mathbf{r}) = \min_{\mathbf{r}'} f(\mathbf{r} - \mathbf{r}')g(\mathbf{r}') \qquad \text{Erosion} \qquad (17.2)$$

where $\mathbf{r}'$ is varied through all admissible points of the image f. With this definition, dilation and erosion differ from convolution only in that the integration/summation is replaced by max and min functions. Each of these operations qualifies to be called a filtering. These filterings are nonlinear and can be applied several times, one after the other and in various orders, to achieve a particular purpose, e.g., to remove a specific type of noise or to smooth boundaries of regions. Collectively, these operations are called *morphological filtering*. Since max and min functions are nonlinear, with scant knowledge on them compared to linear functions, more user experience on morphological operators in discrete image representation is both necessary and crucial for their effective application, compared to linear operators. Accordingly, we first assume that $\mathbf{r}$ obtains discrete values, i.e., we have a grid by $\mathbf{r}_k$, and we discuss only images and filters that are binary, assuming that they take the values 1 or 0. Examples of filters yield

$$\mathbf{G}_1 = \begin{pmatrix} 1\ 1\ 1 \\ 1\ 1\ 1 \\ 1\ 1\ 1 \end{pmatrix} \quad \mathbf{G}_2 = \begin{pmatrix} 1\ 1\ 1 \\ 1\ 0\ 1 \\ 1\ 1\ 1 \end{pmatrix} \quad \mathbf{G}_3 = \begin{pmatrix} 1\ 1\ 1 \end{pmatrix} \quad \mathbf{G}_4 = \begin{pmatrix} 1 \\ 0 \\ 1 \end{pmatrix} \qquad (17.3)$$

The definition of $\oplus$ and $\ominus$ for discrete, multidimensional functions follows from Eqs. (17.1) and (17.2) except that now f and g are discrete binary images that we choose to represent by $\mathbf{F}$ and $\mathbf{G}$ for convenience. In 2D, these functions are defined on a 2D grid, a matrix, whereas in 3D and higher dimensions the functions are defined on discrete arrays that are multidimensional grids. Given a discrete *grid* and the binary image defined on it, the *morphological operations* are then defined as:

$$(\mathbf{F} \oplus \mathbf{G})(\mathbf{r}_k) = \max_{\mathbf{r}_l} \mathbf{F}(\mathbf{r}_k - \mathbf{r}_l)\mathbf{G}(\mathbf{r}_l) \qquad (17.4)$$

$$(\mathbf{F} \ominus \mathbf{G})(\mathbf{r}_k) = \min_{\mathbf{r}_l} \mathbf{F}(\mathbf{r}_k - \mathbf{r}_l)\mathbf{G}(\mathbf{r}_l) \qquad (17.5)$$

where r_l is the index vector of the grid (row and column indices for a matrix) of an element, G, with $G(r_l)$ representing the value of the element. When $F(r_k - r_l)$ is undefined for a given $r_k - r_l$, e.g., because it is a point outside of an image, usually an assumption such as that it is zero is made. The definition above has provisions that morphological operations can be applied to the discrete gray image F. This is because one can measure the local maximum and minimum after their elements are modified due to a pointwise multiplication by G. However, we will refrain from discussing gray images here to limit the scope only to images where both F as well as G are binary. In all cases, the purpose of the neighborhood is twofold:

- to mark the points that participate into max and min operations
- to weight them within the allowed region

Some care should be taken when marking and computing the points of local images because weighting with zero is not the same as excluding the point from the local image, although this is normally what the user wishes.

Whether or not a zero of the filter is to be interpreted as a weight or a region marker matters in morphological operations. For example, a weight of zero forces the min operation to *always* produce zero on binary images, which consist of ones and zeros, regardless of the other coefficients of the filter and even regardless the input image, whereas interpreting the same zeros as region excluders would mean that the corresponding points do not participate in min calculations. The zero coefficients of filters used under max, min thus behave differently in comparison to filters used in summation operations of convolutions. This is because a sum does not change, regardless of whether the zero coefficients are markers, or that the respective image points participate to summation with the weight zero. To avoid the interpretation ambiguities, it is practical to assume that the morphological filter coefficients as markers rather than weights, i.e., a coefficient zero is a region-excluding marker, whereas a coefficient one is a region-including marker.

What $\oplus$ and $\ominus$ really do to images becomes more intuitive when they are applied to an image. Figure 17.1 shows a binary image F and a filter G, which is the filter shown as G_3 in Eq. (17.3). The color code is such that blue represents 0 and red represents 1.

For convenience we used light red and light blue colors in illustrations. They represent 1 and 0, respectively, too, but they are chosen so to mark points that have changed their values as compared to the original. As seen from these results, the dilation operator dilates the image in a way that is consistent with the direction of the filter. There is an enlargement of the objects, but it is systematically in the direction of the filter. There is no dilation of horizontal edges. The erosion filtering acts analogously, but it reduces the objects. We can draw several conclusions from this:

1. Objects that are too close to each other are merged after a dilation, e.g., the letter Ä and the dot "." in the example.
2. If there is a dominant filter direction, then dilation has a directional preference too, e.g., a horizontal filter dilates edges with horizontal direction.
3. Objects smaller than the filter disappear under erosion, e.g., the dot "." in the example.

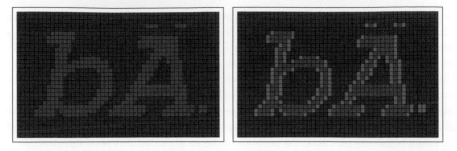

Fig. 17.2. (*Left*) The result of applying the dilation, $\mathbf{F} \oplus \mathbf{G}$ and (*right*) the erosion $\mathbf{F} \ominus \mathbf{G}$ operators when using the image $\mathbf{F}$ and the filter $\mathbf{G}$ shown in Fig. 17.1. *Red* and *blue* represent 1 and 0, respectively. The *light red* and the *light blue* represent 1 and 0, respectively, but they are chosen so to mark points that have changed their values as compared to the original

4. If there is a dominant filter direction, then erosion has a directional preference too, e.g., a horizontal filter erodes edges with horizontal direction.
5. If the background pixels are called 1 and the object pixels 0, then the roles of dilation and erosion are reversed.

Hit–Miss Transform

To detect binary objects of a specific shape and size, the *hit–miss transform* can be utilized. It exploits the fact that erosion removes only objects "smaller" than the "target". To fix the ideas we illustrate it by:

$$\mathbf{G} = (1, \mathbf{1}, 1)^T \tag{17.6}$$

so that our goal is to erase every object that does not equal $\mathbf{G}$. The center pixel is marked as boldface for convenience to mark that it is the point that represents the position of the object. The hit–miss transform is implemented as follows.

1. Delete all objects smaller than the target. This is achieved by

$$\mathbf{H} = \mathbf{F} \ominus \mathbf{G} \tag{17.7}$$

where $\mathbf{G}$ is the target. This step is illustrated by Fig. 17.2 (right).

2. Delete all "objects" larger than the target. This is achieved by

$$\mathbf{R} = \overline{\mathbf{F}} \ominus \overline{\mathbf{G}} \tag{17.8}$$

where $^-$ represents the conjugate operation $\overline{\mathbf{F}} = 1 - \mathbf{F}$, having the effect that all ones become zeros and zeros become ones. For the filter this is true too, but it would result in trivial filters if certain rules were not observed. To avoid this, the filter is assumed to be limited by a thin boundary of zeros that has no effect on

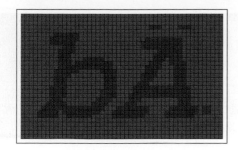

Fig. 17.3. (*Left*) The complement of the original image, $\overline{\mathbf{F}}$ and (*right*) the complement of the target to be used as (an erosion) filter $\overline{\mathbf{G}}$

the operation of $\mathbf{G}$ it is important when computing the complement $\overline{\mathbf{G}} = 1 - \mathbf{G}$. Accordingly, our example $\overline{\mathbf{G}}$ yields

$$\overline{\mathbf{G}} = 1 - \mathbf{G} = \begin{pmatrix} 1\,1\,1\,1\,1 \\ 1\,0\,0\,0\,1 \\ 1\,1\,1\,1\,1 \end{pmatrix} \tag{17.9}$$

The binary image $\overline{\mathbf{F}}$ and the morphological filter $\overline{\mathbf{G}}$ are illustrated in Fig. 17.3. This image, eroded with the shown filter, results in the binary image $\mathbf{R}$, which is shown in Fig. 17.4 (*left*).

3. Multiply pointwise the images $\mathbf{H}$ and $\mathbf{R}$ obtained in the steps above:

$$\mathbf{S} = \mathbf{H} \odot \mathbf{R} \tag{17.10}$$

where $\odot$ represents a pointwise multiplication. Note that pointwise multiplication of two matrices (which must have the same size) is different than multiplication of two matrices, in that the former is achieved by multiplying two entries from the same row and column from the input matrices and assigning it to the corresponding entry of the resulting matrix. Because only ordinary multiplication between ones and zeros are involved, this step is equivalent to a logical AND operation[1] between the pixel values of $\mathbf{H}$ and $\mathbf{R}$. The step is illustrated by Fig. 17.4 (*right*), which is the result of applying $\odot$ between Fig. 17.2 (*right*) and Fig. 17.4 (*left*). It shows the result of the hit–miss transform, which marks the found target object (as defined by the filter) red.

For a detailed discussion of morphological operators and skeletonization we refer to [32, 33, 35, 183–185, 197].

---

[1] In some studies AND, alternatively $\odot$, is represented by $\cap$.

Fig. 17.4. (*Left*) The image **R** that is obtained by erosion between the image and the filter shown in Fig. 17.3. The *light red*, representing 1, and the *light blue* representing 0, show the changes introduced by the erosion. (*Right*) The image **S** that is obtained by a pointwise multiplication between the *left* image and the *right* image of Fig. 17.2

17.2 Connected Component Labelling

For convenience, let 0 represent the color of the background. A connected component is the collection (region) of all points having the color 1 such that they are attached, i.e., if two points of the region are not already neighbors, there is a path through neighboring points belonging to the region that can join any two points of the region. We assume 8-*connected neighbors*, $\mathcal{N}_8(\mathbf{r}_k)$, for convenience but a 4-connected[2] neighbors assumption is also possible. Evidently, in 3D binary images, the neighborhood assumptions are more complex, because a point in a discrete 3D image, called a voxel, can have face, edge, and vertex neighbors. We refer to [34, 35] for a detailed discussion of binary image processing in higher dimensions. To fix the ideas, a connected component is typically a region, also called an object, which has one closed boundary with pixels having the same color. However, more than one closed boundary is possible, e.g., the letter "e" in Fig. 17.5. Accordingly, it is possible to assign a unique label to each component. For simplicity one can use colors to represent the labels e.g., as in Fig. 17.5. A method achieving this task is called *connected component labelling*. This is not just an image-understanding problem but also a computational physics problem, e.g., [7].

There are several techniques achieving connected component labelling [213] in 2D, differing in computational efficiency and computer architectures for which they are intended. The differences are at the implementation level because there is, given the neighborhood type, noroom for ambiguity on whether or not the labelling has been achieved correctly. We outline the conventional method of doing such a labelling in 2D below because of its simplicity although there are other ways of achieving connected components labelling that may be more efficient [213].

The algorithm in [95] scans through a binary image $f(\mathbf{r}_i)$ in the forward and the backward raster directions alternately. Assume that $f(\mathbf{r}_i)$ is a 2D (discrete) image

---

[2] In 4-*connected neighbors*, $\mathcal{N}_4(\mathbf{r}_k)$, only points that share an edge with a pixel as neighbors, i.e., top, down, left, and right pixels, are neighbors. An 8-connected neighborhood assumption admits, additionally, the adjacent 4-diagonal pixels.

that consists of pixel values 1, indicating connected components (object regions), and 0, indicating the background. The first two labels, $m = 0$ and $m = 1$, are assigned to the background and to the first object as soon as the first 0-valued pixel and the first 1-valued pixel, respectively, are encountered in the forward raster direction (reading direction). These pixels will obtain the labels 0 and 1, which is the same as their pixel values. Then the following operations in the forward raster scan order are performed using the neighborhood filter, as is shown in Eq. (17.11) for 8-connected neighborhoods. The filter is to be interpreted as a mask, i.e., the 0 elements mark the "don't care" pixels, which do not participate into the min operation below:

$$\mathbf{L}(\mathbf{r}_i) = \begin{cases} 0, & \text{if } \mathbf{F}(\mathbf{r}_i) = 0, \\ m; \quad m = m + 1, & \text{if } \{\text{for all } k : \mathbf{L}(\mathbf{r}_{i-k})\mathbf{G}(\mathbf{r}_k) = 0\}, \\ \min_k \mathbf{L}(\mathbf{r}_{i-k})\mathbf{G}(\mathbf{r}_k), & \text{otherwise.} \end{cases} \quad (17.11)$$

where the mask G for the forward scan (right–down direction) yields:

$$\mathbf{G} = \begin{pmatrix} 1 & 1 & 1 \\ 1 & 0 & 0 \end{pmatrix} \quad (17.12)$$

Here the 0 elements of the mask mark the "don't care" pixels, that is, they are not meant to be multiplied with the (label) values of $\mathbf{L}$ in Eq. (17.11). Conversely, the ones of the mask mark the points to be formally multiplied by one, although this may be replaced by other instructions for efficiency. The boldfaced zero element represents the current point, $\mathbf{r}_i$ when moving the mask over the label image $\mathbf{L}$. Notice that the mask is only moved over the label image being constructed. Consequently, the current point as well as the future point in the scan direction are "don't care" values since they are marked with zero.

The top condition of Eq. (17.11) is usually the one that is most frequently entered. This is because it can be activated by the current point $\mathbf{r}_i$ being a background point, without any regard to the values of the neighboring pixels within the $G(\mathbf{r}_{i-k})$. This branch aims to keep a background pixel as a background pixel in the scan direction. A necessity for the second condition to be active is that the current pixel value in the binary image, $\mathbf{F}(\mathbf{r}_i)$, be nonzero, even if this is not tested in this branch. This is because if the current pixel is zero then it is caught by the condition of the first branch. Accordingly, the second condition is active if and only if the current pixel value is an object pixel, and all of the neighboring pixels in the mask marked with 1, are already assigned to the background. When this happens the current point gets the label m, which is immediately incremented so that every time this branch is entered, the output label is a new label. The third condition is active when we have that the current pixel is an object pixel and one of the neighboring pixels, marked with 1 by the mask, has an (old) *object label*, i.e., $m > 1$. In this case no new labels are generated for the current point, but the minimum label of the neighboring pixels is assigned to the current point. Accordingly, the smaller labels invade the connected component.

After the forward pass, the following operations in the backward raster scan order are applied:

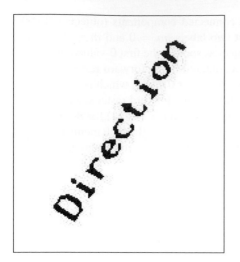 

Fig. 17.5. (*Left*) A binary image with several regions. (*Right*) A connected component labelling where colors represent labels

$$\mathbf{L}(\mathbf{r}_i) = \begin{cases} 0, & \text{if } \mathbf{L}(\mathbf{r}_i) = 0, \\ \min_k(\mathbf{L}(\mathbf{r}_{i-k})\mathbf{G}(\mathbf{r}_k)), & \text{otherwise,} \end{cases} \qquad (17.13)$$

where the mask G for the backward scan (right–down) is given by

$$\mathbf{G} = \begin{pmatrix} 0 & \mathbf{0} & 1 \\ 1 & 1 & 1 \end{pmatrix} \qquad (17.14)$$

The forward and the backward scans are alternately applied until no labels change in the label image $\mathbf{L}$.

Figure 17.5 shows a binary image consisting of the word "Direction". Each letter consists of one connected component except the letter "i" which consists of two. Accordingly, the connected component labelling comes out with 11 regions, each having its own unique label, shown as color of pixels. Using connected components, one can apply further processing to individual regions, e.g., measuring their shape properties, as we will discuss in what follows.

17.3 Elementary Shape Features

In practice, there exist a number of adhoc measures that may be sufficient to resolve shape-related recognition issues. We list some of these descriptors in the following. We assume that the image is f and that it contains one connected component $\mathcal{D}$, which is to be described w.r.t. its shape primitives. If there is more than one region in the image, it is split up to several images so that there is no more than one region in each by connected component labelling. The points inside the region $\mathcal{D}$ are collectively called the object whereas those outside are called the background. For convenience, we also assume that $f = 1$ for object points and $f = 0$ for the background.

Area

Formally, this is given by

$$A = \int_{\mathcal{D}} f(x, y) dx dy \qquad (17.15)$$

which in practice, translates to the count of points with $f = 1$.

Perimeter

The formal definition of the perimeter is

$$P = \int_{\partial \mathcal{D}} f(x(s), y(s)) ds \qquad (17.16)$$

where $\partial \mathcal{D}$ is the boundary of the region $\mathcal{D}$ and $(x(s), y(s))$ is a parametrization of the boundary curve. In practice, one decides on a neighborhood connectivity type, e.g., 8- or 4- connectivity in 2D, then counts neighborhoods containing *both* points with $f = 1$ and $f = 0$.

Bounding Box

This is the tightest "cuboid" that contains an image volume. For a 2D region, this is given by the rectangle represented by a four-tuple:

$$\left(\min_{\mathbf{r}_k} X(\mathbf{r}_k), \min_{\mathbf{r}_k} Y(\mathbf{r}_k), \max_{\mathbf{r}_k} X(\mathbf{r}_k), \max_{\mathbf{r}_k} Y(\mathbf{r}_k) \right) \qquad (17.17)$$

where $X(\mathbf{r}_k)$ and $Y(\mathbf{r}_k)$ represent the coordinates of image points $\mathbf{r}_k$ in the region. Bounding boxes can be used for having a simple idea on the shape of a region, i.e., elongated versus compact shape. It is also used for avoiding collisions, i.e. if two regions that move have nonoverlapping bounding boxes then they are not in collision.

Circularity/Compactness

The dimensionless ratio defined via the perimeter and the area is called circularity or compactness, C:

$$C = \frac{P^2}{A} \qquad (17.18)$$

It has a lower bound. In 2D it satisfies $4\pi \leq C$ and reaches its minimum when the region is a circle.

17.4 Moment-Based Description of Shape

The region to be described here may contain varying gray shades as opposed to many simple shape descriptors. In that case, the entire function f is assumed to represent the region. Otherwise, the function takes the value $f = 1$ inside the region to be described and $f = 0$ outside. A function f can be translated via

$$f\left(x - \frac{m_{10}}{m_{00}}, y - \frac{m_{01}}{m_{00}}\right) \tag{17.19}$$

where m_{10}, m_{01}, m_{00} are the real moments of f given by Eq. (10.9). The centroid of the resulting image coincides with the new origin because $\left(\frac{m_{10}}{m_{00}}, \frac{m_{01}}{m_{00}}\right)^T$ represents the position of the centroid in the above equation. Accordingly, we assume that f's centroid is already brought to the origin, making f translation-invariant. Because of this, the complex moments of f,

$$I_{pq}(f) = \int\!\!\int (x + iy)^p (x - iy)^q f(x, y) dx dy$$

are translation-invariant, too. We investigate now how

$$I_{pq}(f) = \int\!\!\int r^p \exp(ip\theta) r^q \exp(-iq\theta) f(x(r, \theta), y(r, \theta)) r dr d\theta$$
$$= \int\!\!\int r^{p+q} \exp(i(p - q)\theta) f(x(r, \theta), y(r, \theta)) r dr d\theta \tag{17.20}$$

where $x(r, \theta) = r\cos(\theta)$ and $y(r, \theta) = r\sin(\theta)$, will be affected by a rotation and scaling. In consequence of Eq. (17.20), the complex moment of a function defined in polar coordinates, $f(r, \theta)$, will be given by

$$I_{pq}(f) = \int\!\!\int r^{p+q} \exp(i(p - q)\theta) f(r, \theta) r dr d\theta \tag{17.21}$$

For convenience, we assume below that f is given in polar representation as $f(r, \theta)$ with its centroid at the origin. To discuss how complex moments transform under a rotation and scaling transformation, we use primed variables for the variables after the CT, i.e., $r' = \alpha r$ where $\alpha > 0$ and $\theta' = \theta + \varphi$. The transformed function f', is thus obtained from f via $f'(r', \theta') = f(\frac{r'}{\alpha}, \theta' - \varphi)$. Using Eq. (17.21), a complex moment of the transformed image yields

$$I_{pq}(f') = \int\!\!\int r'^{p+q} \exp(i(p - q)\theta') f'(r', \theta') r' dr' d\theta'$$
$$= \int\!\!\int r'^{p+q} \exp(i(p - q)\theta') f(\frac{r'}{\alpha}, \theta' - \varphi) r' dr' d\theta'$$
$$= \int\!\!\int (\alpha r)^{p+q} \exp(i(p - q)\theta) \exp(i(p - q)\varphi) f(r, \theta) \alpha r \alpha dr d\theta$$
$$= \alpha^{p+q+2} \exp(i(p - q)\varphi) \int\!\!\int r^{p+q} \exp(i(p - q)\theta) f(r, \theta) r dr d\theta$$
$$= \alpha^{p+q+2} \exp(i(p - q)\varphi) I_{pq}(f) \tag{17.22}$$

We summarize this result as a theorem.

Theorem 17.1. *When $f(r,\theta)$, having its centroid at the origin, is transformed to $f'(r',\theta')$ through a CT consisting of $\theta' = \varphi + \theta$, and $r' = \alpha r$, the complex moments are invariant up to a multiplication with a constant (complex) scalar,*

$$I_{pq}(f') = \alpha^{p+q+2} \exp(i(p-q)\theta) I_{pq}(f) \tag{17.23}$$

◆

This result, albeit with different notation, is due to Reddi [188] and simplifies the derivation of scale and rotation-invariant forms of moments [42, 89, 112, 190, 234] discussed below. Most important, however, the theorem shows that the complex moments are nearly invariant to rotation and scaling because the effects of these are no more than a multiplicative scalar on the complex moments. Furthermore, one can always increase the descriptive power of the invariants systematically, by using high-order complex moments.

Hu's moment invariants

The following quantities

$$
\begin{aligned}
H_1 &= m_{20} + m_{02} \\
H_2 &= (m_{20} - m_{02})^2 + 4m_{11}^2 \\
H_3 &= (m_{30} - 3m_{12})^2 + (3m_{21} - m_{03})^2 \\
H_4 &= (m_{30} + m_{12})^2 + (m_{21} + m_{03})^2 \\
H_5 &= (m_{30} - 3m_{12})(m_{30} + m_{12}) \cdot [(m_{30} + m_{12})^2 - 3(m_{21} + m_{03})^2] + \cdots \\
&\quad + (3m_{21} - m_{03})(m_{21} + m_{03}) \cdot [3(m_{30} + m_{12})^2 - (m_{21} + m_{03})^2] \\
H_6 &= (m_{20} - m_{02}) \cdot [(m_{30} + m_{12})^2 - (m_{21} + m_{03})^2] + \cdots \\
&\quad + 4m_{11}(m_{30} + m_{12})(m_{21} + m_{03}) \\
H_7 &= (3m_{21} - m_{03})(m_{30} + m_{12}) \cdot [(m_{30} + m_{12})^2 - 3(m_{21} + m_{03})^2] + \cdots \\
&\quad - (m_{30} - 3m_{12})(m_{21} + m_{03}) \cdot [3(m_{30} + m_{12})^2 - (m_{21} + m_{03})^2]
\end{aligned}
$$

where m_{pq} represent the real (central) moments, were suggested by Hu [112] as rotation-invariant measures. We will shortly see that this is indeed the case. We write down the complex moments so as to express $H_1 \cdots H_7$ via the complex moments:

$$
\begin{aligned}
I_{11} &= m_{20} + m_{02} = d \\
I_{20} &= m_{20} - m_{02} + i2m_{11} = c + ic' \\
I_{12} &= m_{30} + m_{12} - i(m_{21} + m_{03}) = b - ib' \\
I_{30} &= m_{30} - 3m_{12} + i(3m_{21} - m_{03}) = a + ia'
\end{aligned}
$$

yielding

$$I_{12}^2 = (b^2 - b'^2) - i2bb'$$
$$I_{20}I_{12}^2 = (c + ic')I_{12}^2 = c(b^2 - b'^2) + 2c'bb' + i[c'(b^2 - b'^2) - i2cbb']$$
$$I_{12}^3 = (b^3 - 3bb'^2) + i(b'^3 - 3b^2b')$$
$$I_{30}I_{12}^3 = (a + ia')I_{12}^3 = ab(b^2 - 3b') - a'b'(b'^2 - 3b^2) +$$
$$+i[a'b(b^2 - 3b') - ab'(3b^2 - b'^2)]$$

We identify then $H_1 \cdots H_7$

$$H_1 = d = I_{11}$$
$$H_2 = c^2 + c'^2 = |I_{20}|^2$$
$$H_3 = a^2 + a'^2 = |I_{30}|^2$$
$$H_4 = b^2 + b'^2 = |I_{12}|^2$$
$$H_5 = ab(b^2 - 3b') - a'b'(b'^2 - 3b^2) = \Re(I_{30}I_{12}^3)$$
$$H_6 = c(b^2 - b'^2) + 2c'bb' = \Re(I_{20}I_{12}^2)$$
$$H_7 = a'b(b^2 - 3b') - ab'(3b^2 - b'^2) = \Im(I_{30}I_{12}^3)$$

so that

$$H_1 = I_{11} \tag{17.24}$$
$$H_2 = |I_{20}|^2 \tag{17.25}$$
$$H_3 = |I_{30}|^2 \tag{17.26}$$
$$H_4 = |I_{12}|^2 \tag{17.27}$$
$$H_5 = \Re(I_{30} \cdot I_{12}^3) \tag{17.28}$$
$$H_6 = \Re(I_{20} \cdot I_{12}^2) \tag{17.29}$$
$$H_7 = \Im(I_{30} \cdot I_{12}^3) \tag{17.30}$$

By inspection and using the above theorem, it then follows that these scalars are rotation-invariant. Using the theorem again, the normalized quantities

$$H_2' = \frac{H_2}{H_1^2} \tag{17.31}$$

$$H_3' = \frac{H_3}{H_1^{2.5}} \tag{17.32}$$

$$H_4' = \frac{H_4}{H_1^{2.5}} \tag{17.33}$$

$$H_5' = \frac{H_5}{H_1^5} \tag{17.34}$$

$$H_6' = \frac{H_6}{H_1^{3.5}} \tag{17.35}$$

$$H_7' = \frac{H_7}{H_1^5} \tag{17.36}$$

will make these invariants also scale-invariant.

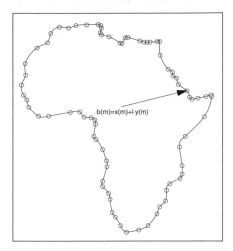

Fig. 17.6. The *circles* mark the data points represented by $b(m)$. They sample the boundary of a region, a closed curve

17.5 Fourier Descriptors and Shape of a Region

Let

$$b(m) = x(m) + iy(m), \quad \text{with} \quad m \in 0, \cdots N-1, \qquad (17.37)$$

be a sequence of complex numbers representing coordinates of boundary points of a region see Fig. 17.6. Regions have closed curves as boundaries and as such they are periodic,

$$b(m) = b(m + kN), \quad \text{with} \quad k \in 0, \pm 1, \pm 2, \cdots. \qquad (17.38)$$

They can accordingly be expanded in a suitable Fourier basis. Notice that $b(m)$ does not necessarily sample the boundary in an equidistant manner. Since it is a discrete sequence, the expansion coefficients are easily found by DFT. Accordingly, we have the DFT pair:

$$B(n) = \frac{1}{N} \sum_{m=0}^{N-1} b(m) \exp\left(-imn\frac{2\pi}{N}\right) \qquad (17.39)$$

with the inverse transform that can synthesize $b(m)$ from $B(n)$

$$b(m) = \sum_{n=0}^{N-1} B(n) \exp\left(imn\frac{2\pi}{N}\right) \qquad (17.40)$$

without loss. The coefficients $B(n)$ are called the *Fourier descriptors*[3] (FD) of the boundary $b(m)$. Because, b is periodic with N, m follows a circular topology, allowing us to write the synthesis formula symmetrically:

---

[3] Because FDs are DFT coefficients, the index n in $B(n)$ is cyclical, i.e., it follows the rules of modulo arithmetic when it is an integer other than $0, \cdots, (N-1)$.

$$b(m) = \sum_{n=-\frac{N-1}{2}}^{\frac{N-1}{2}} B(n) \exp\left(imn\frac{2\pi}{N}\right), \quad \text{with} \quad K \in 1, 2, \cdot \leq \frac{N-1}{2}, \quad (17.41)$$

where we assumed that N is odd, for simplicity. The synthesis can be truncated at $K \ll N/2$, in which case we obtain a smoother approximation of the boundary curve

$$\tilde{b}(m) = \sum_{n=-K}^{K} B(n) \exp\left(imn\frac{2\pi}{N}\right), \quad \text{with} \quad K \in 1, 2, \cdot \ll \frac{N-1}{2}, \quad (17.42)$$

because $\tilde{b}$ lacks high-frequency terms. Such a truncation is often utilized in image analysis, e.g., to reduce the number of FDs, which are in turn utilized in shape-based recognition or discrimination. Smoothing via truncations is illustrated by Fig. 17.7 where we show the boundary approximation of the continents by FDs. The total available FDs is equal to the number of boundary samples. Note that intricate curves need more samples, for a faithful reconstruction.

The FDs have a number of desirable properties that we list below.

- The centroid of the boundary is given by $B(0)$

$$B(0) = \frac{1}{N} \sum_{m=0}^{N-1} b(m) \exp\left(-im0\frac{2\pi}{N}\right) = \frac{1}{N} \sum_{m=0}^{N-1} b(m) \quad (17.43)$$

- All FDs except $B(0)$ are translation-invariant. To see this, we apply a translation $\Delta b = \Delta x + i\Delta y$ to the boundary and obtain the new coordinates as $b'(m) = b(m) + \Delta b$ having the FDs:

$$
\begin{aligned}
B'(n) &= \frac{1}{N} \sum_{m=0}^{N-1} b'(m) \exp(-imn\frac{2\pi}{N}) \\
&= \frac{1}{N} \sum_{m=0}^{N-1} (b(m) + \Delta b) \exp(-imn\frac{2\pi}{N}) \\
&= \frac{1}{N} \sum_{m=0}^{N-1} b(m) \exp(-imn\frac{2\pi}{N}) + \Delta b \cdot \frac{1}{N} \sum_{m=0}^{N-1} \exp(-imn\frac{2\pi}{N}) \\
&= \frac{1}{N} \sum_{m=0}^{N-1} b(m) \exp(-imn\frac{2\pi}{N}) + \Delta b\delta(n) \\
&= B(n) + \Delta b \cdot \delta(n) \quad (17.44)
\end{aligned}
$$

Accordingly, to obtain shift-invariant FDs one can ignore $B(0)$, the centroid.
- Assume that b' is a scaled version of the boundary, so that $b' = \alpha b$ with α being a positive real scalar. Then

$$B'(n) = \frac{1}{N} \sum_{m=0}^{N-1} b'(m) \exp\left(-imn\frac{2\pi}{N}\right)$$

$$= \frac{1}{N} \sum_{m=0}^{N-1} \alpha b(m) \exp\left(-imn\frac{2\pi}{N}\right)$$

$$= \alpha B(n) \tag{17.45}$$

Provided that $|B'(1)| > 0$, one can compute

$$\frac{B'(n)}{|B'(1)|} = \frac{\alpha B(n)}{\alpha|B(1)|} \tag{17.46}$$

which is scale-invariant. If $|B'(1)| \approx 0$ then another FD can be used in the normalization, yielding more stable scale-invariant features.

- Assume that b' is a rotated version of the boundary, so that $b' = \exp(i\theta)b$ with θ being an arbitrary angle. Then

$$B'(n) = \frac{1}{N} \sum_{m=0}^{N-1} b'(m) \exp\left(-imn\frac{2\pi}{N}\right)$$

$$= \frac{1}{N} \sum_{m=0}^{N-1} \exp(i\theta)b(m) \exp\left(-imn\frac{2\pi}{N}\right)$$

$$= \exp(i\theta)B(n) \tag{17.47}$$

From this it follows that $|B'(1)| = |B(1)|$, so that if $|B'(1)| > 0$, one can, by division of both sides of Eq. (17.47), obtain for $n = 1$:

$$\frac{B'(1)}{|B'(1)|} = \exp(i\theta)\frac{B(1)}{|B(1)|} \tag{17.48}$$

Accordingly, we can use this ratio to normalize $B'(n)$ for $n > 1$:

$$\frac{B'(n)}{\frac{B'(1)}{|B'(1)|}} = \exp(-i\theta)\frac{B^*(1)}{|B(1)|}\exp(i\theta)B(n) = \frac{B^*(1)}{|B(1)|}B(n) \tag{17.49}$$

to obtain rotation-invariant FD features. As before, if $|B'(1)| \approx 0$ then another FD can be used in the normalization, yielding more stable rotation-invariant features. One can also use $|B'(n)|$ to achieve rotation-invariance, but these features annihilate $\approx \frac{N}{2}$ freedoms, far too much description power of FDs than necessary for many applications.

- Because, $B(1)$ contains both scale and rotation parameters, to achieve both *scale and rotation-invariance* in FDs one can normalize the other FDs by this complex scalar

$$\frac{B'(n)}{B'(1)}, \quad \text{with} \quad n = 2, 3, \cdots N - 1, \tag{17.50}$$

provided that it is nonzero (else one can use another FD that has a large magnitude).

To summarize, to achieve translation, rotation, and scale-invariance jointly, one can ignore $B(0)$ and normalize all other FDs by another $B(n)$ that has a large magnitude, usually $B(1)$. In that one ignores $B(0)$ and $B(1)$, containing four freedoms. This is economical because it corresponds precisely to translation- (two freedoms), rotation- (1 freedom), and scale-invariance (one freedom). Note that the index $n \in 0 \cdots N - 1$ runs over the full range of the coefficients so that the negative frequencies have the indices $N - n$ according to the cyclical translation rules of DFT. Whereas $|B(n)| \neq |B(-n)|$ because $b(k)$ are not real, in practical applications $|B(n)|$ still decays as the index n increases through $0, \pm 1, \pm 2, \pm 3, \cdots$. Accordingly, when truncating $B(n)$, the coefficients are kept symmetrical w.r.t. $n = 0$, as is done in Fig. 17.7.

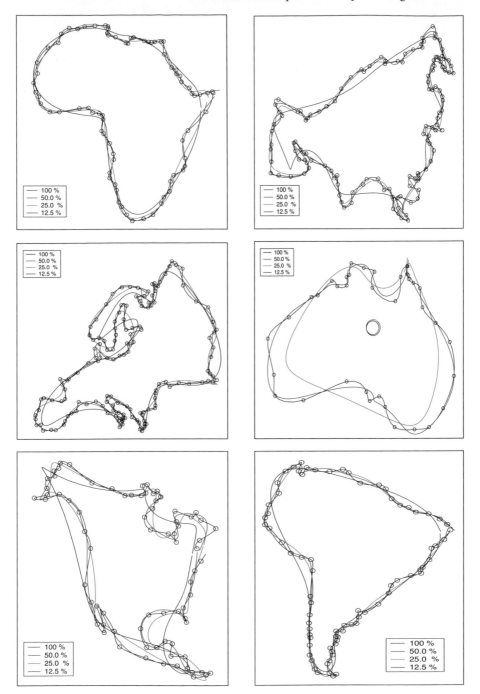

Fig. 17.7. The images illustrate some region boundary curves synthesized by FDs truncated at 100%, 50%, 25%, and 12.5% of the available FDs

18

Concluding Remarks

The term *vision* encompasses much more than registering the distribution of light by our eyes or by a camera. It refers to a process that analyzes what has been captured, so as to facilitate actions and inferences. In this book, we anatomized the observed light distribution by using multidimensional signal analysis tools and provided some illustrations of the latter in terms of applications, e.g., tracking, feature extraction, segmentation, texture description, 3D geometry reconstruction, face recognition, and fingerprint recognition. In many of these applications humans are experts, almost by birth, and yet we do not have detailed knowledge on how this amazing performance is achieved. Biased by this performance, the uninitiated can therefore not be blamed for underestimating the challenges of image analysis, whether the issues are to be addressed by her/his own visual system or by a computer.

When a computing system has to solve the same vision problem for multiple applications in changing environments, the versatility and robustness demands quickly become challenging. In this book, we discussed certain image analysis concepts and tools in an attempt to make them useful in many scenarios. Intelligent vision systems in changing environments are increasingly in demand not only because humans are increasingly online while moving, but also because we want mobile machines to be online and autonomous. Just as a hunting man must carefully consider what tools to carry, so must a vision system that must function in changing environments be selective in its choice of processing tools because of the extreme resource limitations the mobility demands.

In the face of undisputable evidence being accumulated, there is hardly a need to argue about the existence of sophisticated signal processing in human vision, solely devoted to *directional processing*. Such a mechanism, which is the sine qua non of the most advanced mobile organisms, including the human, hunting or not, is at first counterintuitive due to its computational "heaviness". This is because directional processing amounts to increasing the dimension of the visual signal in a hefty manner. It is almost an explosion of data that takes place—and that routinely, before any simplification, e.g., decision, takes place. Motivated by its imposing presence in biological vision, this book attempts to simplify and unify the modern directional signal processing. It attempted to reconcile some of the most utilized principles in machine vision, e.g., the generalized Hough transform, with directional processing. In the text, the belief that most vision problems can be solved by systems mimicking the known processing of human vision in their frontend is not concealed. In this euphoria, it would nevertheless be a serious mistake to pretend that all methods and principles to achieve human visual intelligence are covered, not least because there are too many of these yet to be discovered.

The dimension explosion in the visual pathways is not an unknown phenomenon in the mathematical treatment of signals. It goes under the general notion of *feature extraction*. It is nearly routinely demanded in all advanced decision support systems because each dimension brings a simplification, not available in the original signal. In comparison, it is worth noting that an image filtered through a particular directional filter is more predictable than the original, and therefore less complex. The filtered image, in its essence 1D, e.g., see [163], is more likely constant in the tune-on direction than the orthogonal direction because of the specific filtering. Several chapters of the book treat therefore principles that are also found in statistics literature. However, the visual signals have unique properties which can allow a vision system to limit its feature extraction to directional processing as a resource management strategy to cope with multiple environments.

In English and in numerous other languages, the notion of *vision* is also synonymous with qualified imagination about the future. However, predicting the future of visual signal analysis as a science is a feat even larger than solving the problems of vision. Accordingly, what I expressed in this book cannot be anything but my vision of the matter, which may not be perceived as neutral—it is necessarily directional.

References

1. Y.I. Abdel-Aziz and H.M. Karara. Direct linear transformation into object space coordinates in close-range photogrammetry. In *Proc. Symposium on Close-Range Photogrammetry, Urbana, IL, Jan.*, pages 1–18, 1971.
2. Y.S. Abu-Mostafa and D. Psaltis. Recognitive aspects of moment invariants. *IEEE-PAMI*, 6:698–706, 1984.
3. E.H. Adelson and J.R. Bergen. Spatiotemporal energy models for the perception of motion. *J. of the Optical Society of America A*, 2(2):284–299, 1985.
4. D.G. Albrecht and R.L. de Valois. Striate cortex responses to periodic patterns with and without the fundamental harmonics. *J. Physiol., (London)*, 319:497–514, 1979.
5. A. Aldroubi and H. Feichtinger. Exact iterative reconstruction algorithm for multivariate irregularly sampled functions in spline-like spaces: The Lp-theory. *Proceedings of the Amer. Math. Soc.*, 126(9):2677–2686, 1998.
6. B.W. Andrews and D.A. Pollen. Relationship between spatial frequency selectivity and receptive field profile of simple cells. *J. Physiol., (London)*, 287:163–176, 1979.
7. J. Apostolakis, P. Coddington, and E. Marinari. New SIMD algorithms for cluster labeling on parallel computers. *Int. J. Mod. Phys.*, C4:749, 1993.
8. K. Åström, R. Cipolla, and P.J. Giblin. Generalised epipolar constraints. In J. O. Eklundh, editor, *Computer Vision—ECCV 1996*, LNCS 1065, pages 97–108, Springer, Heidelberg, 1996.
9. S. Ayer, P. Schroeter, and J. Bigun. Segmentation of moving objects by robust motion parameter estimation over multiple frames. In J. O. Eklundh, editor, *Computer Vision—ECCV 1994*, LNCS 800, pages 316–327, Springer, Heidelberg, 1994.
10. R. Balian. Un principe d'incertitude fort en théorie du signal ou en mécanique quantique. *C. R. Acad. Sci. Paris*, 292(2):1357–1363, 1981.
11. G. Ball and D. Hall. ISODATA, an iterative method of multivariate analysis and pattern classification. In *IFIPS Congress*, 1965.
12. B.H. Ballard. Generalizing the Hough transform to detect arbitrary shapes. *Pattern Recognition*, 13(2):111–112, 1981.
13. G.C. Baylis, E.T. Rolls, and C.M. Leonard. Functional divisions of the temporal lobe neocortex. *J. Neuroscience*, 7:330–342, 1987.
14. P.A. Beardsley, A. Zisserman, and D.W. Murray. Sequential updating of projective and affine structure from motion. *Int. Journal of Computer Vision*, 23(3):235–297, 1997.
15. M. Bertero, T. Poggio, and V Torre. Illposed problems in early vision. *Proceedings of the IEEE*, 76(8):869–889, 1988.

16. J.C. Bezdek. *Pattern recognition with fuzzy objective function algorithm*. Plenum, New York, 1981.
17. J. Bigun. Pattern recognition by detection of local symmetries. In E.S. Gelsema and L.N. Kanal, editors, *Pattern recognition and artificial intelligence*, pages 75–90. North-Holland, 1988.
18. J. Bigun. Recognition of local symmetries in gray value images by harmonic functions. In *Ninth International Conference on Pattern Recognition, Rome, Nov. 14–17*, pages 345–347. IEEE Computer Society, 1988.
19. J. Bigun. A structure feature for some image processing applications based on spiral functions. *Computer Vision, Graphics, and Image Processing*, 51(2):166–194, 1990.
20. J. Bigun. Gabor phase in boundary tracking and region segregation. In *Proc. DSP & CAES Conf. Nicosia, Cyprus, July 14-16*, pages 229–237. Univ. of Nicosia, 1993.
21. J. Bigun. Speed, frequency, and orientation tuned 3-D Gabor filter banks and their design. In *Proc. International Conference on Pattern Recognition, ICPR, Jerusalem*, pages C–184–187. IEEE Computer Society, 1994.
22. J. Bigun. Pattern recognition in images by symmetries and coordinate transformations. *Computer Vision and Image Understanding*, 68(3):290–307, 1997.
23. J. Bigun, T. Bigun, and K. Nilsson. Recognition by symmetry derivatives and the generalized structure tensor. *IEEE-PAMI*, 26:1590–1605, 2004.
24. J. Bigun, K. Choy, and H. Olsson. Evidence on skill differences of women and men concerning face recognition. In J. Bigun and F. Smeraldi, editors, *Audio and Video Based Biometric Person Authentication—AVBPA 2001*, LNCS 2091, pages 44–51, Springer, Heidelberg, 2001.
25. J. Bigun and J.M.H. du Buf. N-folded symmetries by complex moments in Gabor space. *IEEE-PAMI*, 16(1):80–87, 1994.
26. J. Bigun and J.M.H. du Buf. Symmetry interpretation of complex moments and the local power spectrum. *Visual Communication and Image Representation*, 6(2):154–163, 1995.
27. J. Bigun, H. Fronthaler, and K. Kollreider. Assuring liveness in biometric identity authentication by real-time face tracking. In *International Conference on Computational Intelligence for Homeland Security and Personal Safety, CIHSPS, Venice, July 21–22*, pages 104–112. IEEE, 2004.
28. J. Bigun and G.H. Granlund. Optimal orientation detection of linear symmetry. In *First International Conference on Computer Vision, ICCV, London, June 8–11*, pages 433–438. IEEE Computer Society, 1987.
29. C.M. Bishop. *Neural Networks for Pattern Recognition*. Oxford University Press, Oxford, 1995.
30. M.J. Black and P. Anandan. A framework for the robust estimation of optical flow. In *ICCV-93, Berlin, May 11-14*, pages 231–236. IEEE Computer Society, 1993.
31. G.W. Bluman and S. Kumei. *Symmetries and differential equations*. Springer, Heidelberg, 1989.
32. G. Borgefors. Distance transformations in arbitrary dimensions. *Computer Vision, Graphics, and Image Processing*, 27(3):321–345, 1984.
33. G. Borgefors. Hierarchical chamfer matching: (A) parametric edge matching algorithm. *IEEE-PAMI*, 10(6):849–865, 1988.
34. G. Borgefors. Weighted digital distance transforms in four dimensions. *Discrete Applied Mathematics*, 125(1):161–176, 2003.
35. G. Borgefors, I. Nyström, and G. Sanniti di Baja. Computing skeletons in three dimensions. *Pattern Recognition*, 32(7):1225–1236, 1999.

36. R.K. Bothwell, J.C. Brigham, and R.S. Malpass. Cross-racial identification of faces. *Personality and Social Psychology Bulletin*, 15:19–25, 1989.

37. P. Brodatz. *Textures*. Dover, New York, 1966.

38. A.J. Bruce and K.W. Beard. African Americans, and Caucasian Americans recognition and likability responses to african american and caucasian american faces. *Journal of General Psychology*, 124:143–156, 1997.

39. V. Bruce and A. Young. Understanding face recognition. *British Journal of Psychology*, 77:305–327, 1986.

40. V. Bruce and A. Young. *In the eye of the beholder*. Oxford University Press, Oxford, 1998.

41. A.H. Bunt, A.E. Hendrickson, J.S. Lund R.D. Lund, and A.F. Fuchs. Monkey retinal ganglion cells: Morphometric analysis and tracing of axonal projections, with a consideration of the peroxidase technique. *J. Comp. Neurol.*, 164:265–286, 1975.

42. H. Burkhardt. *Transformationen zur lageinvarianten Merkmalgewinnung*. Habilitationsschrift, Universität Karlsruhe, VDI-Verlag, 1979.

43. P. Burt. Fast filter transforms for image processing. *Computer graphics and image processing*, 16:20–51, 1981.

44. S.R.Y. Cajal. *Histologie du systeme nerveux de l'homme et des vertebres*. Maloine, Paris, 1911.

45. F.W. Campbell, G.F. Cooper, and C. Enroth-Cugell. The spatial selectivity of the visual cells of the cat. *J. Physiol. (London)*, 203:223–235, 1969.

46. F.W. Campbell and J.G. Robson. Application of Fourier analysis to the visibility of gratings. *J. Physiol. (London)*, 197:551–566, 1968.

47. V.A. Casagrande and J.H. Kaas. The afferent, intrinsic, and efferent connections of primary visual cortex in primates. In A. Peters and K. Rockland, editors, *Cerebral Cortex, Vol. 10*, pages 201–259, Plenum, New York, 1994.

48. J. W. Cooley and J. W. Tukey. An algorithm for the machine calculation of complex Fourier series. *Mathematics of Computation*, 19:297–301, 1965.

49. C. Cortes and V. Vapnik. Support-vector networks. *Machine Learning*, 20:273–297, 1995.

50. J. Crowley and H. Christensen. *Hardware for active vision systems*. Kluwer, 1995.

51. P.E. Danielsson. Rotation invariant linear operators with directional response. In *Fifth International Conference on Pattern Recognition, ICPR-80, Miami Beach, Florida, Dec.*, pages 1171–1176. IEEE Computer Society, 1980.

52. P.E. Danielsson, Q. Lin, and Q.Z. Ye. Efficient detection of second-degree variations in 2D and 3D images. *Visual Communication and Image Representation*, 12:255–305, 2001.

53. I. Daubechies. The wavelet transform, time-frequency localization and signal analysis. *IEEE Trans. on Inf. Theory*, 36(5):961–1005, 1990.

54. E.R. Davies. Finding ellipses using the generalised Hough transform. *Pattern Recognition Letters*, 9:87–96, 1989.

55. A. Dick, P. Torr, S. Ruffle, and R. Cipolla. Combining single view recognition and multiple view stereo for architectural scenes. In *Eighth International Conference On Computer Vision ICCV–01, July, 9–12*, pages 268–274. IEEE Computer Society, 2001.

56. F. Doyle. The historical development of analytical photogrammetry. *Photogrammetric Engineering*, 2:259–265, 1964.

57. J.M.H. du Buf. Responses of simple cells: Events, interferences, and ambiguities. *Biol. Cybernet.*, 68:321–333, 1993.

58. B. Duc. Motion estimation using invariance under group transformations. In *12th International Conference on Pattern Recognition, ICPR–94, Jerusalem, Oct.*, pages 159–163. IEEE Computer Society, 1994.

59. B. Duc. *Feature design: Applications to motion analysis and identity verification.* PhD thesis, Ecole Polytechnique Fédérale de Lausanne, 1997.

60. B. Duc, P. Schroeter, and J. Bigun. Motion estimation and segmentation by fuzzy clustering. In *International Conference on Image Processing, ICIP-95, Washington D.C., Oct. 23–26*, pages III–472–475. IEEE Computer Society, 1995.

61. B. Duc, P. Schroeter, and J. Bigun. Spatio-temporal robust motion estimation and segmentation. In Hlavac and Sara, editors, *Computer Analysis of Images and Patterns–CAIP 1995*, LNCS 970, pages 238–245, Springer, Heidelberg, 1995.

62. R.O. Duda, P.E. Hart, and D.G. Stork. *Pattern Classification.* Wiley, New York, 2000.

63. D.E. Dudgeon and R.M. Mersereau. *Multi dimensional digital signal processing.* Prentice Hall, Englewood Cliffs, N.J., 1981.

64. R. Eagleson. Measurement of the 2D affine Lie group parameters for visual motion analysis. *Spatial Vision*, 6(3):183–198, 1992.

65. E.S. Elliott, E.J. Wills, and A.G. Goldstein. The effects of discrimination training on the recognition of white and oriental faces. *Bulletin of the Psychonomic Soceity*, 2:71–73, 1973.

66. H.D. Ellis, D.M. Ellis, and J.A. Hosie. Priming effects in children's face recognition. *British Journal of Psychology*, 84:101–110, 1993.

67. B.S. Everitt. *Cluster analysis.* Heinemann, London, 1974.

68. M.J. Farah. Is face recognition special? evidence from neuropsychology. *Behavioral Brain Research*, 76:181–189, 1996.

69. M.J. Farah. *The cognitive neuroscience of vision.* Blackwell, Maden, 2000.

70. O. Faugeras. *Three-Dimensional Computer Vision: A Geometric Viewpoint.* MIT Press, Cambridge, MA, 1993.

71. M. Felsberg and G. Sommer. Image features based on a new approach to 2D rotation invariant quadrature filters. In A. Heyden et. al., editor, *Computer Vision–ECCV 2002*, LNCS 2350, pages 369–383, Springer, Heidelberg, 2002.

72. M. Ferraro and T.M. Caelli. Relationship between integral transform invariances and Lie group theory. *J. Opt. Soc. Am. A*, 5(5):738–742, 1988.

73. D.J. Fleet and A.D. Jepson. Computation of component image velocity from local phase information. *International Journal of Computer Vision*, 5(1):77–104, 1990.

74. W. Forstner and E. Gulch. A fast operator for detection and precise location of distinct points, corners and centres of circular features. In *Proc. Intercommission Conference on Fast Processing of Photogrammetric Data, Interlaken*, pages 281–305, 1987.

75. W.T. Freeman and E.H. Adelson. The design and use of steerable filters. *IEEE-PAMI*, 13(9):891–906, 1991.

76. H. Fronthaler, K. Kollreider, and J. Bigun. Local feature extraction in fingerprints by complex filtering. In S. Z. Li et. al., editor, *International Workshop on Biometric Recognition Systems–IWBRS 2005, Beijing, Oct. 22–23*, LNCS 3781, pages 77–84, Springer, Heidelberg, 2005.

77. D. Gabor. Theory of communication. *Journal of the IEE*, 93:429–457, 1946.

78. A. Gagalowicz. Texture modelling applications. *The Visual Computer*, 3(4):186–200, 1987.

79. I. Gauthier and M.J. Tarr. Becoming a "greeble" expert: Exploring mechanisms for face recognition. *Vision Research*, 37:1673–1682, 1997.

80. K.R. Gegenfurtner. Cortical mechanisms of colour vision. *Nature Reviews Neuroscience*, 4(7):563–572, 2003.

81. D. Geman and S. Geman. Stochastic relaxation, Gibbs distribution and bayesian restoration of images. *IEEE-PAMI*, 6(6):721–741, 1984.
82. P. Gerard and A. Gagalowicz. Three dimensional model-based tracking using texture learning and matching. *Pattern Recognition Letters*, 21(13-14):1095–1103, 2000.
83. H. Goldstein. *Classical Mechanics*. Addison-Wesley, Reading, MA, 1986.
84. G.H. Golub and C.F. Van Loan. *Matrix Computations*. Johns Hopkins University Press, Baltimore, 1989.
85. R.C. Gonzalez and R.E. Woods. *Digital image processing*. Addison-Wesley, Reading, MA, 1992.
86. L. Van Gool, T. Tuytelaars, V. Ferrari, C. Strecha, J. Vanden Wyngaerd, and M. Vergauwen. 3D modeling and registration under wide baseline conditions. In *Proceedings ISPRS Symposium on Photogrammetric Computer Vision*, pages 3–14, 2002.
87. A. Gordon. *Classification*. Chapman and Hall, New York, 1981.
88. A. Grace and M. Spann. Edge enhancment and fine feature restoration of segmented objects using pyramid based adaptive filters. In *British Machine Vision Conf., Guildford*, 1993.
89. G.H. Granlund. Fourier preprocessing for hand print character recognition. *IEEE Trans. Computers*, 21:195–201, 1972.
90. G.H. Granlund. In search of a general picture processing operator. *Computer Graphics and Image Processing*, 8(2):155–173, 1978.
91. W.C. Graustein. *Introduction to Higher Geometry*. MacMillan, New York, 1930.
92. W. Grosky and R. Jain. Optimal quadtrees for image segments. *IEEE-PAMI*, 5:77–83, 1983.
93. H. Gruner. Photogrammetry: 1776–1976. *Photogrammetric Engineering and Remote Sensing*, 43(5):569–574, 1977.
94. O. Hansen and J. Bigun. Local symmetry modeling in multi-dimensional images. *Pattern Recognition Letters*, 13:253–262, 1992.
95. R.M. Haralick. *Some neighborhood operations*. Real Time/Parallel Computing Image Analysis. Plenum, New York, 1981.
96. R.M. Haralick and I. Dinstein. Textural features for image classification. *IEEE Trans. Syst. Man and Cybern.*, 3:610–621, 1973.
97. C. Harris and M. Stephens. A combined corner and edge detector. In *Proceedings of the fourth Alvey Vision Conference*, pages 147–151, 1988.
98. R. Hartley and A. Zisserman. *Multiple View Geometry in Computer Vision*. Cambridge University Press, Cambridge, 2004.
99. M.E. Hasselmo, E.T. Rolls, G.C. Baylis, and V. Nalwa. Object-centered encoding by face-selective neurons in the cortex in the superior temporal sulcus of the monkey. *Experimental Brain Research*, 75:417–429, 1989.
100. M. Hauta-Kasari, W. Wang, S. Toyooka, J. Parkkinen, and R. Lenz. Unsupervised filtering of munsell spectra. In R.T. Chin and T.-C. Pong, editors, *Computer Vision—ACCV 1998*, LNCS 1351, pages 248–255, Springer, Heidelberg, 1998.
101. D. J. Heeger. Optical flow from spatio-temporal filters. In *First International Conference on Computer Vision, ICCV-1, London, June*, pages 181–190. IEEE Computer Society, 1987.
102. J. Heikkilä and O. Silven. Calibration procedure for short focal length off-the-shelf ccd cameras. In *International Conf. on Pattern Recognition, ICPR-96*, pages 166–170. IEEE Computer Society, 1996.
103. W. Heisenberg. Über den anschaulichen Inhalt der quantentheoretischen Kinematik und Mechanik. *Z. für Phys.*, 43:172–198, 1927.

104. S.H. Hendry and R.C. Reid. The koniocellular pathway in primate vision. *Annual Review of Neuroscience*, 23:127–153, 2000.

105. E. Hering. *Outlines of a theory of the light sense*. Harvard University press, Cambridge, MA, 1964. (Translation).

106. A. Heyden, G. Sparr, and K. Åström. Perception and action using multilinear forms. In G. Sommer and J. Koenderink, editors, *Algebraic Frames for the Perception–action Cycle—AFPAC 1997*, LNCS 1315, pages 54–65, Springer, Heidelberg, 1997.

107. W.C. Hoffman. The Lie algebra of visual perception. *J. Math. Psychol.*, 3:65–98, 1966.

108. L. Hong, Y. Wand, and A.K. Jain. Fingerprint image enhancement: Algorithm and performance evaluation. *IEEE-PAMI*, 20(8):777–789, 1998.

109. S.D. van Hooser and S.B. Nelson. Visual system. *Encyclopedia of life sciences (online)*, 2005. Wiley.

110. B.K.P. Horn and B.G. Schunck. Determining optical flow. *Artificial intelligence*, 17:185–203, 1981.

111. P.V.C. Hough. *Method and means for recognizing complex patterns*. U.S. patent 3,069,654, 1962.

112. M.K. Hu. Visual pattern recognition by moment invariants. *IRE Trans. on Information Theory*, pages 179–187, 1962.

113. D.H. Hubel. *Eye, brain and vision*. Scientific American Library, 1988.

114. D.H. Hubel and T.N. Wiesel. Receptive fields of single neurons in the cat's striate cortex. *J. physiol. (London)*, 148, 1959.

115. S.van Huffel and J. Wandewalle. The total least squares problem: Computational aspects and analysis. *Frontiers in applied mathematics*, 1991.

116. R.A. Hummel. Feature detection using basis functions. *Computer Graphics and Image Processing*, 9:40–55, 1979.

117. A. Hyvärinen and E. Oja. Independent component analysis: Algorithms and applications. *Neural Networks*, 13(4–5):411–430, 2000.

118. J. Illingworth and J. Kittler. The adaptive hough transform. *IEEE-PAMI*, 9(5):690–698, 1987.

119. M. Irani, B. Rousso, and S. Peleg. Detection and tracking multiple moving objects using temporal integration. In G. Sandini, editor, *Computer Vision—ECCV 1992*, LNCS 588, pages 282–287, Springer, Heidelberg, 1992.

120. ISO-IEC. *International standard 13818-2, Information technology generic coding of moving pictures and associated audio information: Video. Ref. No.: ISO/IEC:13818-2*. ISO, 1996.

121. Special Issue. Brain. *Scientific American*, 241(3), 1979.

122. B. Jähne. *Spatio-Temporal Image Processing*. LNCS 751. Springer, Heidelberg, 1993.

123. B. Jähne. *Digital Image Processing*. Springer, Heidelberg, 1997.

124. A.K. Jain and F. Farrokhnia. Unsupervised texture segmentation using Gabor filters. *Pattern Recognition*, 24(12):1167–1186, 1991.

125. M. James. *Classification algorithms*. Collins, London, 1985.

126. S.C. Jeng and W.H. Tsai. Scale-invariant and orientation-invariant generalized Hough transform - a new approach. *Pattern Recognition*, 24(11):1037–1051, 1991.

127. D.J. Jobson, Z. Rahman, and G.A. Woodell. Properties and performance of a center/surround retinex. *IEEE Trans. on Image Processing*, 6(3):451–462., 1997.

128. P. Johansen, M. Nielsen, and O. F. Olsen. Branch points in one-dimensional Gaussian scale space. *Journal of Mathematical Imaging and Vision*, 13(3):193–203, December 2000.

129. B. Johansson. *Multiscale curvature detection in computer vision*. Lic. thesis no. 877; LIU-TEK-LIC-2001:14, Linköping University, 2001.

130. M.H. Johnson. *Developmental cognitive neuroscience*. Blackwell, 1997.
131. B. Julesz. Visual pattern discrimination. *IRE Trans. on Information Theory*, 8:84–92, 1962.
132. K. Kanatani. *Group-theoretical methods in image understanding*. Springer Series in Information Sciences. Springer, Heidelberg, 1990.
133. K. Karhunen. Zur Spectraltheorie stochasticher Prozesse. *Ann. Acad. Sci. Fennicae*, 37, 1947.
134. S. Karlsson and J. Bigun. Texture analysis of multiscal complex moments of the local power spectrum. Technical Report IDE0574, Halmstad University, November 2005.
135. L. Kaufman and P.J. Rousseeuw. *Finding Groups in Data: An Introduction to Cluster Analysis*. Wiley, New York, 1990.
136. J.P. Keener. *Principles of applied mathematics*. Addison-Wesley, New York, 1988.
137. H. Knutsson. *Filtering and reconstruction in image processing*. PhD Thesis no:88, Linköping University, 1982.
138. H. Knutsson. Representing local structure using tensors. In *Proceedings 6th Scandinavian Conf. on Image Analysis, Oulu, June*, pages 244–251, 1989.
139. H. Knutsson and G. H. Granlund. Texture analysis using two-dimensional quadrature filters. In *IEEE Computer Society Workshop on Computer Architecture for Pattern Analysis and Image Database Management - CAPAIDM, Pasedena, Oct.*, pages 206–213, 1983.
140. H. Knutsson, M. Hedlund, and G.H. Granlund. *Apparatus for Determining the Degree of Consistency of a Feature in a Region of an Image that is Divided into Discrete Picture Elements*. U.S. patent, 4.747.152, 1988.
141. J.J. Koenderink and A.J. van Doorn. The structure of images. *Biological Cybernetics*, 50:363–370, 1984.
142. T. Kohonen, E. Oja, O. Simula, A. Visa, and J. Kangas. Engineering applications of the self-organizing map. *Proceedings of the IEEE*, 84(10):1358–84, 1996.
143. S.W. Kuffler. Discharge patterns and functional organization of mammalian retina. *Journal of Neurophysiology*, 16:37–68, 1953.
144. M. Lades, J.C. Vorbruggen, J. Buhmann, J. Lange, C. von der Malsburg, R. P. Hurtz, and W. Konen. Distortion invariant object recognition in the dynamic link architectures. *IEEE Trans. on Computers*, 42(3):300–311, 1993.
145. E.H. Land. The retinex theory of colour vision. *Sci. Am.*, 237:108–129, 1977.
146. E.H. Land. An alternative technique for the computation of the designator in the retinex theory of color vision. *Proc. Nat. Acad. Sci.*, 83:3078–3086, 1986.
147. D.C. Lay. *Linear Algebra and Its Applications*. Addison–Wesley, Reading, MA, 1994.
148. F. Leymarie and M.D. Levine. Tracking deformable objects in the plane using an active contour model. *IEEE-PAMI*, 15(6):617–634, 1993.
149. Q. Liang and T. Gustavsson. Isointensity directional smoothing for edge-preserving noise reduction. In *ICIP*, pages 867–871, 1998.
150. Q. Liang, I. Wendelhag, J. Wikstrand, and T. Gustavsson. A multiscale dynamic programming procedure for boundary detection in ultrasonic artery images. *IEEE Trans. Medical Imaging*, 19(2):127–142, February 2000.
151. T. Lindeberg. On automatic selection of temporal scales in time-causal scale-space. In G. Sommer and J. Koenderink, editors, *Algebraic Frames for the Perception-Action Cycle–AFPAC 1997*, LNCS 1315, pages 94–113, Springer, Heidelberg, 1997.
152. T. Lindeberg and B. ter H. Romeny. *Linear scale-space: I. Basic theory: II. Early visual operations*. Kluwer Academic, Dordrecht, 1994.
153. D. V. Lindley. *Making decisions*. Wiley, London, 1990.

154. C. Liu and H. Wechsler. Evolutionary pursuit and its application to face recognition. *IEEE-PAMI*, 22:570–582, 2000.

155. M.S. Livingstone and D.H. Hubel. Anatomy and physiology of a color system in the primate visual cortex. *Journal of Neuroscience*, 4:309–356, 1984.

156. H.C. Longuet-Higgins. A computer algorithm for reconstructing a scene from two projections. *Nature*, 293(133-135), Sep 1981.

157. B.D. Lucas and T. Kanade. An iterative image registration technique with an application to stereo vision. In *Proc. of the seventh Int. Joint Conf. on Artificial Intelligence, Vancouver*, pages 674–679, 1981.

158. T.S. Luce. The role of experience in inter-racial recognition. *Personality and Social Psychology Bulletin*, 1:39–41, 1974.

159. L. Maffei and A. Fiorentini. The visual cortex as a spatial frequency analyser. *Vision Res.*, 13:1255–1267, 1973.

160. D. Marr and E. Hildreth. Theory of edge detection. *Proc. Royal Society of London Bulletin*, 204:301–328, 1979.

161. G. Medioni, M.S. Lee, and C.K. Tang. *A computational framework for segmentation and grouping*. Elsevier, 2000.

162. K. Messer, J. Matas, J. Kittler, J. Luettin, and G. Maitre. XM2VTSDB: The extended m2vts database. In *Audio and Video based Person Authentication - AVBPA99*, pages 72–77. University of Maryland, 1999.

163. S. Michel, B. Karoubi, J. Bigun, and S. Corsini. Orientation radiograms for indexing and identification in image databases. In *Eusipco-96, European Conference on Signal Processing*, pages 1693–1696, 1996.

164. F.M. de Monasterio, P. Gouras, and J. Tolhurst. Spatial summation, response pattern and conduction velocity of ganglion cells of the rhesus monkey retina. *Vision Research*, 16(6):674–678, 1976.

165. J.A. Movshon, I.D. Thompson, D.J., and Tolhurst. Spatial and temporal contrast sensitivity of neurons in areas 17 and 18 of the cat's visual cortex. *J. Physiol. (London)*, 283:101–120, 1978.

166. G. Murch. Physiological principles for the effective use of color. *IEEE CG&A*, 4(11):49–54, 1984.

167. H.H. Nagel. On the estimation of optical flow: relations between different and some new results. *Artificial intelligence*, 33:299–324, 1987.

168. K. Nassau. *Physics and Chemistry of Color*. Wiley, New York, 1983.

169. K. Nilsson and J. Bigun. Localization of corresponding points in fingerprints by complex filtering. *Pattern Recognition Letters*, 24:2135–2144, 2003.

170. D. Nister. *Automatic dense reconstruction from uncalibrated video sequences*. PhD thesis, Royal Inst. of Technology, KTH, 2001.

171. T.E. Ogden. The receptor mosaic of aotus trivirgatus: Distribution of rods and cones. *J. Comp. Neurol.*, 163:193–202, 1975.

172. E. Oja, H. Ogawa, and J. Wangviwattana. Principal component analysis by homogeneous neural networks, part I: The weighted subspace criterion. *IEICE Transactions on Information and Systems*, E75-D(3):366–375, 1992.

173. G.A. Orban. *Neuronal operations in the visual cortex*. Springer, Heidelberg, 1984.

174. G.A. Orban and M. Callens. Influence of movement parameters on area 18 neurones in the cat. *Exp. Brain Res.*, 30:125–140, 1977.

175. J. Ortega-Garcia, J. Bigun, D. Reynolds, and J. Gonzalez-Rodriguez. Authentication gets personal with biometrics. *IEEE Signal Processing Magazine*, pages 50–62, March 2004.

176. G. Osterberg. Topography of the layer of rods and cones in the human retina. *Acta Ophtalmologica (Supplementum)*, 6(1):11–97, 1935.

177. P.L. Palmer, M. Petrou, and J. Kittler. A Hough transform algorithm with a 2D hypothesis testing kernel. *Computer Vision, Graphics, and Image Processing. Image Understanding*, 58(2):221–234, 1993.

178. T.V. Papathomas and B. Julesz. Lie differential operators in animal and machine vision. In J.C. Simon, editor, *From pixels to features*, pages 115–126. Elsevier, 1989.

179. F. Pedersen, M. Bergstrom, E. Bengtsson, and B. Langstrom. Principal component analysis of dynamic positron emission tomography images. *European Journal of Nuclear Medicine*, 21(12):1285–1292, 1994.

180. P. Perona. Steerable–scalable kernels for edge detection and junction analysis. In G. Sandini, editor, *Computer Vision—ECCV 1992*, LNCS 588, pages 3–18, Springer, Heidelberg, 1992.

181. D.I. Perrett, P.A. Smith, D.D. Potter, A. Mistlin, A.S. Head, A.D. Milner, and M.A. Jeeves. Visual cells in the temporal cortex sensitive to face view and gaze direction. *Proc. of the Royal Society of London, Series B*, 223:293–317, 1985.

182. M. Persson and J. Bigun. Detection of spots in 2-d electrophoresis gels by symmetry features. In S. Singh et al., editor, *Pattern Recognition and Data Mining—ICAPR 2005*, LNCS 3686, pages 436–445, Springer, Heidelberg, 2005.

183. J. Piper and E. Granum. Computing distance transformations in convex and non-convex domains. *Pattern Recognition*, 20(6):599–615, 1987.

184. I. Pitas and A.N. Venetsanopoulos. Shape decomposition by mathematical morphology. In *First International Conference on Computer Vision, ICCV, London, June 8–11*, pages 621–625. IEEE Computer Society, Washington, DC,, 1987.

185. I. Pitas and A.N. Venetsanopoulos. *Nonlinear Digital Filters: Principles and Applications*. Kluwer Academic, 1990.

186. H.L. Premaratne and J. Bigun. A segmentation-free approach to recognise printed sinhala script using linear symmetry. *Pattern Recognition*, 37:2081–2089, 2004.

187. R. P. N. Rao, G. J. Zelinsky, M. M. Hayhoe, and D. H. Ballard. Eye movements in visual cognition: a computational study. Technical Report 97.1, National Resource Laboratory for the Study of Brain and Behaviour, 1997.

188. S.S. Reddi. Radial and angular invariants for image identification. *IEEE-PAMI*, 3(2):240–242, 1981.

189. R.C. Reid and J.M. Alonso. Specificity of monosynaptic connections from thalamus to visual cortex. *Nature*, 378:281–284, 1995.

190. T.H. Reiss. The revised fundamental theorem of moment invariants. *IEEE-PAMI*, 13(8):830–834, 1991.

191. H. Romesburg. *Cluster Analysis for Researchers*. Lifetime Learning, Belmont, 1984.

192. A. Rosenfeld and A.C. Kak. *Digital Picture Processing*. Academic Press, Inc., Orlando, USA, 1982.

193. W. Rudin. *Real and complex analysis*. Mc Graw-Hill, New York, 1987.

194. G. Sandini and M. Tistarelli. Recovery of depth information: Camera motion as an integration to stereo. In *Workshop on Motion: Representation and Analysis Charleston, SC, May 7–9*, pages 39–43. IEEE, 1986.

195. P. Schroeter and J. Bigun. Hierarchical image segmentation by multi-dimensional clustering and orientation adaptive boundary refinement. *Pattern Recognition*, 28(5).695–709, 1995.

196. E.L. Schwartz. Computational anatomy and functional architecture of striate cortex: A spatial mapping approach to perceptual coding. *Visual Research*, 20, 1980.

197. J. Serra. *Image Analysis and Mathematical Morphology.* Academic, 1982.
198. S.K. Sethi and R. Jain. Finding trajectories of feature points in a monocular image sequence. *IEEE-PAMI*, 9(1):56–73, 1987.
199. E.P. Simoncelli. A rotation invariant pattern signature. In *International Conference on Image Processing, ICIP-96*, pages III–185–188, 1996.
200. A. Singh. An estimation-theoretic framework for image-flow computation. In *Third International Conference on Computer Vision, ICCV*, pages 168–177. IEEE Computer Society, 1990.
201. P. Sinha and T. Poggio. Role of learning in three-dimensional form perception. *Nature*, 384(6608):460–463, 5 Dec 1996.
202. L. Sirovich and M. Kirby. Application of the Karhunen-Loève procedure for the characterization of human faces. *IEEE-PAMI*, 12(1):103–108, 1990.
203. D. Slepian. Prolate spheroidal wave functions, Fourier analysis and uncertainty–I. *The Bell System Technical J.*, pages 43–63, 1961.
204. D. Slepian. Prolate spheroidal wave functions, Fourier analysis and uncertainty–IV: Extensions to many dimensions; generalized prolate spheroidal functions. *The Bell System Technical J.*, 43:3009–3057, 1964.
205. F. Smeraldi and J. Bigun. Retinal vision applied to facial features detection and face authentication. *Pattern Recognition Letters*, 23:463–475, 2002.
206. H. Sompolinsky and R. Shapley. New perspectives on mechanisms for orientation selectivity. *Current Opinion in Neurobiology*, 7:514–522, 1997.
207. M. Spann and R. Wilson. A quadtree approach to image segmentation which combines statistical and spatial information. *Pattern Recognition*, 18:257–269, 1989.
208. E. Stein and G. Weiss. *Fourier Analysis on Euclidean Spaces.* Princeton University Press, Princeton, NJ, 1971.
209. M.H. Stone. Applications of the theory of boolean rings to general topology. *Transactions of the American Mathematical Society*, 41(3):375481, 1937.
210. G. Strang. *Linear algebra and its applications.* Harcourt Brace Jovanovich, San Diego, 1980.
211. C. Strecha and L. Van Gool. Dense matching of multiple wide-baseline views. *ICCV*, 2:1194–1201, 2003.
212. R. Strichartz. *A guide to distribution theory and Fourier transforms.* CRC, 1994.
213. K. Suzuki, I. Horiba, and N. Sugie. Fast connected-component labeling based on sequential local operations in the course of forward raster scan followed by backward raster scan. In *Proc. Int. Conf. on Patt. Rec. ICPR-15*, pages II–434–437. IEEE Computer Society, 2000.
214. P. Tchamitchian. Biorthogonalité et théorie des opérateurs. *Rev. Math. Iberoamer.*, 3:163–189, 1987.
215. D. Tell and S. Carlsson. Combining appearance and topology for wide baseline matching. In A. Heyden et. al., editor, *Computer Vision–ECCV 2002*, LNCS 2350, pages 68–81, Springer, Heidelberg, 2002.
216. M. Tistarelli and G. Sandini. On the advantages of log-polar mapping for direct estimation of time-to-impact from optical flow. *IEEE-PAMI*, 15(4):401–410, 1993.
217. R. B. H. Tootell, M. S. Silverman, E. Switkes, and R. L. de Valois. Deoxyglucose analysis of retinotopic organization in primate striate cortex. *Science*, 218:902–904, 1982.
218. L. Tran and R. Lenz. PCA based representation of color distributions for color based image retrieval. In *International Conference on Image Processing*, pages II: 697–700, 2001.

219. E. Trucco and A. Verri. *Introductory techniques for 3D computer vision.* Prentice-Hall, Upper Saddle River, NJ, 1998.
220. M. Turk and A. Pentland. Eigenfaces for recognition. *J. Cognitive Neuroscience (Winter),* 3(1):71–86, 1991.
221. M. Unser, A. Aldroubi, and M. Eden. Fast B–Spline transforms of continuous image representation and interpolation. *IEEE-PAMI,* 13(3):277–285, 1991.
222. S. Le Vay, D. H. Hubel, and T. N. Wiesel. The pattern of ocular dominance columns in macaque visual cortex revealed by a reduced silver stain. *J. Comp. Neurol.,* 159:559–576, 1975.
223. H. Wassle and R.B. Illing. The retinal projection to the superior colliculus in the cat: a quantitative study with HRP. *J. Comp. Neurol.,* 190:333–356, 1980.
224. J. Weickert. Coherence-enhancing shock filters. In *DAGM-Symposium,* pages 1–8, 2003.
225. C.F. Westin. *A tensor framework for multidimensional signal processing.* PhD thesis, Linkoping University, 1994.
226. R.L. Wheeden and A. Zygmund. *Measure and integral.* Marcel Dekker, Basel, 1977.
227. R. Wilson and G. Granlund. The uncertainty principle in image processing. *IEEE-PAMI,* 6:758–767, 1984.
228. R. Wilson and M. Spann. Finite prolate spheroidal sequences and their applications II: Image feature description and segmentation. *IEEE-PAMI,* 10(2):193–203, 1988.
229. R. Wilson and M. Spann. *Image Segmentation and Uncertainty.* Wiley, New York, 1988.
230. A.P. Witkin. Scale-space filtering. In *8th Int. Joint Conf. on Artificial Intelligence, Karlsruhe, Aug. 8–12,* pages 1019–1022, 1983.
231. A. Wouk. *A course of applied functional analysis.* Wiley, New York, 1979.
232. S. Yamane, S. Kaji, and K. Kawano. What facial features activate face neurons in the inferotemporal cortex of the monkey. *Experimental Brain Research,* 73:209–214, 1988.
233. K. Yogesan, H. Schulerud, F. Albregtsen, and H.E. Danielsen. Ultrastructural texture analysis as a diagnostic tool in mouse liver carcinogenesis. *Ultrastructural Pathology,* 22(1):27–37, 1998.
234. C.T. Zahn and R.Z. Roskies. Fourier descriptors for plane closed curves. *IEEE-T on computers,* C-21(3):269–281, 1972.
235. S. Zeki. *A vision of the brain.* Blackwell, London, 1993.

Index